T0145342

Quantum Science and Technology

Series editors

Nicolas Gisin, Geneva, Switzerland
Raymond Laflamme, Waterloo, Canada
Gaby Lenhart, Sophia Antipolis, France
Daniel Lidar, Los Angeles, USA
Gerard J. Milburn, Brisbane, Australia
Masanori Ohya, Noda, Japan
Arno Rauschenbeutel, Vienna, Austria
Renato Renner, Zürich, Switzerland
Maximilian Schlosshauer, Portland, USA
H.M. Wiseman, Brisbane, Australia

Aims and Scope

The book series Quantum Science and Technology is dedicated to one of today's most active and rapidly expanding fields of research and development. In particular, the series will be a showcase for the growing number of experimental implementations and practical applications of quantum systems. These will include, but are not restricted to: quantum information processing, quantum computing, and quantum simulation; quantum communication and quantum cryptography; entanglement and other quantum resources; quantum interfaces and hybrid quantum systems; quantum memories and quantum repeaters; measurement-based quantum control and quantum feedback; quantum nanomechanics, quantum optomechanics and quantum transducers; quantum sensing and quantum metrology; as well as quantum effects in biology. Last but not least, the series will include books on the theoretical and mathematical questions relevant to designing and understanding these systems and devices, as well as foundational issues concerning the quantum phenomena themselves. Written and edited by leading experts, the treatments will be designed for graduate students and other researchers already working in, or intending to enter the field of quantum science and technology.

More information about this series at http://www.springer.com/series/10039

Robert H. Hadfield · Göran Johansson
Editors

Superconducting Devices in Quantum Optics

Editors
Robert H. Hadfield
School of Engineering
University of Glasgow
Glasgow
UK

Göran Johansson
Microtechnology and Nanoscience
Chalmers University of Technology
Gothenburg
Sweden

ISSN 2364-9054 ISSN 2364-9062 (electronic)
Quantum Science and Technology
ISBN 978-3-319-79578-2 ISBN 978-3-319-24091-6 (eBook)
DOI 10.1007/978-3-319-24091-6

Springer Cham Heidelberg New York Dordrecht London
© Springer International Publishing Switzerland 2016
Softcover reprint of the hardcover 1st edition 2016
This work is subject to copyright. All rights are reserved by the Publisher, whether the whole or part of the material is concerned, specifically the rights of translation, reprinting, reuse of illustrations, recitation, broadcasting, reproduction on microfilms or in any other physical way, and transmission or information storage and retrieval, electronic adaptation, computer software, or by similar or dissimilar methodology now known or hereafter developed.
The use of general descriptive names, registered names, trademarks, service marks, etc. in this publication does not imply, even in the absence of a specific statement, that such names are exempt from the relevant protective laws and regulations and therefore free for general use.
The publisher, the authors and the editors are safe to assume that the advice and information in this book are believed to be true and accurate at the date of publication. Neither the publisher nor the authors or the editors give a warranty, express or implied, with respect to the material contained herein or for any errors or omissions that may have been made.

Printed on acid-free paper

Springer International Publishing AG Switzerland is part of Springer Science+Business Media
(www.springer.com)

Preface

Over the past decade, superconducting devices have risen to prominence in the arena of quantum optics and quantum information processing. Superconducting detectors provide unparalleled performance for the detection of infrared photons. These devices enable fundamental advances in quantum optics, the realization of quantum secure communication networks, and open a direct route to on-chip optical quantum information processing. Superconducting circuits based on Josephson junctions provide a blueprint for scalable quantum information processing as well as opening up a new regime for quantum optics at microwave wavelengths. We have endeavored to provide a timely compilation of contributions from top groups worldwide across this dynamic field. This volume provides both an introduction to this area of growing scientific and technological interest, and a snapshot of the global state-of-the-art. Future advances in this domain are anticipated.

Part I of this volume focuses on the technology and applications of superconducting single-photon detectors for near infrared wavelengths.

Chapter 1 provides an authoritative introduction to superconducting nanowire single photon detectors by leading researchers from the Massachusetts Institute of Technology, the NASA Jet Propulsion Laboratory, and the National Institute of Standards and Technology (NIST), USA. The superconducting nanowire device principle is discussed in detail. Key concepts are introduced such as detection efficiency, dark count rate, and timing jitter. The use of amorphous superconducting materials to achieve high device yield and extended mid-infrared sensitivity is highlighted. Parallel wire architectures are presented and an outlook is given on the scale-up to large area arrays.

Chapter 2 introduces another key superconducting detector technology, the superconducting transition edge sensor. The chapter is contributed by the team at NIST, USA, who has been in the vanguard of developments in this technology. The device operation principle is introduced, including key considerations for maximizing sensitivity for the detection of infrared photons and photon-number resolving capability. A range of important quantum optics experiments exploiting the unique capabilities of transition edge sensors are reviewed. Finally, the integration of transition edge sensors with optical waveguide circuits is discussed.

Chapter 3 focuses on superconducting nanowire detectors integrated with GaAs photonic circuits. The chapter is contributed by a team of authors from the Technical University of Eindhoven, the Netherlands, Consiglio Nazionale delle Richerche (CNR) Rome, Italy and the University of Bristol, UK. This approach is compatible with quantum dot single photon emitters, opening the pathway to fully integrated quantum photonic circuits.

Chapter 4 reviews progress on superconducting nanowire detectors on silicon-based photon circuits. This work is contributed by researchers at Yale University, USA, and the University of Muenster, Germany. The marriage of superconducting detectors with mature planar lightwave circuits is a compelling alternative for on-chip quantum information processing. A range of technological applications are discussed, including single photon characterization of on-chip resonators and optical time domain reflectometry for long haul fiber links.

Chapter 5 gives an overview of applications of superconducting nanowire single photon detectors in the realm of quantum communications. The chapter is authored by a team of researchers from the National Institute of Information and Communication Technology (NICT) in Japan. The chapter describes how low noise superconducting nanowire single photon detectors enabled the world's most ambitious quantum cryptography network to be realized in Tokyo, Japan. This chapter also describes developments in heralded single-photon source and quantum interface technology, enabled by high-performance superconducting detectors.

Part II switches the emphasis to quantum optics in the microwave regime using superconducting circuits at millikelvin temperatures.

Chapter 6 gives an authoritative introduction to the emerging field of microwave quantum photonics. It describes the basic building blocks including transmission lines, cavities, artificial atoms (qubits), and measurement setups. This chapter is authored by researchers from the University of Queensland, Australia, and Jiangxi Normal University, China. This chapter highlights the potential of superconducting circuits to access and control quantum phenomena in the realm of microwave photons.

Chapter 7 explores the role of continuous weak measurements as a probe of quantum dynamics in superconducting circuits. In particular, it discusses the possibility to experimentally characterize the systems evolution in terms of quantum trajectories and also how to use the feedback from weak measurements to stabilize Rabi oscillations. This chapter is contributed by experts from Washington University, St. Louis, USA, the University of California, Berkeley, USA, and the Tata Institute of Fundamental Research, Mumbai, India.

Chapter 8 focuses on digital feedback control methods for superconducting qubits. In particular, it describes the use of projective measurements and feedback for fast qubit reset and deterministic entanglement generation. This chapter is contributed by leading researchers at the Delft University of Technology, the Netherlands, and Raytheon BBN Technologies, USA. This high fidelity projective measurement-based technique now allows fast initialization of superconducting qubits and brings deterministic generation of entangled states by parity measurement within reach.

Chapter 9 highlights the use of surface acoustic waves (SAWs) in connection with superconducting quantum circuits. In particular, it describes the coupling of a superconducting artificial atom to propagating SAWs and also how to form SAW cavities in the relevant parameter regime of high frequency and low temperatures. This chapter is authored by a team from Chalmers University, Sweden, Columbia University, USA, RIKEN, Japan, and the University of Oxford, UK. This new technique enables exploitation of single phonons as carriers of quantum information between superconducting qubits as well as providing a method of storage of quantum information in high quality phononic cavities.

This book arose out of a special symposium on 'Superconducting Optics' at the joint Conference on Lasers and Electro-Optics—International Quantum Electronics Conference (CLEO ®/Europe—IQEC conference) which took place in Munich, Germany, in May 2013. One of us (R.H.H.) was a chair of the symposium and the other (G.J.) was an invited speaker. We thank Prof. Dr. Jürgen Eschner of Universität des Saarlandes, Germany, and the CLEO ®/Europe—IQEC programme committee for proposing the special symposium. We undertook the task of editing this book with encouragement from Dr. Claus Ascheron at Springer and Prof. Gerard J. Milburn of the University of Queensland, Australia, Springer Quantum Science and Technology Series Editor. We thank Praveen Kumar and his team at the Springer production office in Chennai, India, for their diligent handling of the proofs. We thank the authors for their high-quality contributions and their dedication in meeting challenging deadlines. We also acknowledge colleagues who proofread parts of the volume, including Dr. Chandra Mouli Natarajan, Dr. Jian Li and Dr. Robert Heath of the University of Glasgow, and Dr. Matti Silveri of Yale University, USA. Finally, we thank our respective families of their patience and forbearance as we guided this volume towards completion.

Glasgow, UK
Gothenburg, Sweden
December 2015

Robert H. Hadfield
Göran Johansson

Contents

Part I Superconducting Single Photon Detectors: Technology and Applications

1 Superconducting Nanowire Architectures for Single Photon Detection 3
Faraz Najafi, Francesco Marsili, Varun B. Verma, Qingyuan Zhao, Matthew D. Shaw, Karl K. Berggren and Sae Woo Nam

2 Superconducting Transition Edge Sensors for Quantum Optics 31
Thomas Gerrits, Adriana Lita, Brice Calkins and Sae Woo Nam

3 Waveguide Superconducting Single- and Few-Photon Detectors on GaAs for Integrated Quantum Photonics 61
Döndü Sahin, Alessandro Gaggero, Roberto Leoni and Andrea Fiore

4 Waveguide Integrated Superconducting Nanowire Single Photon Detectors on Silicon 85
Wolfram H.P. Pernice, Carsten Schuck and Hong X. Tang

5 Quantum Information Networks with Superconducting Nanowire Single-Photon Detectors 107
Shigehito Miki, Mikio Fujiwara, Rui-Bo Jin, Takashi Yamamoto and Masahide Sasaki

Part II Superconducting Quantum Circuits: Microwave Photon Detection, Feedback and Quantum Acoustics

6 Microwave Quantum Photonics 139
Bixuan Fan, Gerard J. Milburn and Thomas M. Stace

7 Weak Measurement and Feedback in Superconducting Quantum Circuits 163
Kater W. Murch, Rajamani Vijay and Irfan Siddiqi

8 Digital Feedback Control 187
Diego Ristè and Leonardo DiCarlo

9 Quantum Acoustics with Surface Acoustic Waves 217
Thomas Aref, Per Delsing, Maria K. Ekström, Anton Frisk Kockum,
Martin V. Gustafsson, Göran Johansson, Peter J. Leek,
Einar Magnusson and Riccardo Manenti

Index 245

Contributors

Thomas Aref Microtechnology and Nanoscience, Chalmers University of Technology, Göteborg, Sweden

Karl K. Berggren Department of Electrical Engineering and Computer Science, Massachusetts Institute of Technology, Cambridge, MA, USA

Brice Calkins National Institute of Standards and Technology, Boulder, CO, USA

Per Delsing Microtechnology and Nanoscience, Chalmers University of Technology, Göteborg, Sweden

Leonardo DiCarlo QuTech Advanced Research Center and Kavli Institute of Nanoscience, Delft University of Technology, Delft, The Netherlands

Maria K. Ekström Microtechnology and Nanoscience, Chalmers University of Technology, Göteborg, Sweden

Bixuan Fan College of Physics, Communication and Electronics, Jiangxi Normal University, Nanchang, China

Andrea Fiore COBRA Research Institute, Eindhoven University of Technology, Eindhoven, The Netherlands

Mikio Fujiwara National Institute of Information and Communication Technology, Koganei, Tokyo, Japan

Alessandro Gaggero Istituto di Fotonica e Nanotecnologie, Consiglio Nazionale delle Richerche (CNR), Roma, Italy

Thomas Gerrits National Institute of Standards and Technology, Boulder, CO, USA

Martin V. Gustafsson Microtechnology and Nanoscience, Chalmers University of Technology, Göteborg, Sweden; Department of Chemistry, Columbia University, New York, NY, USA

Rui-Bo Jin Advanced ICT Research Institute, National Institute of Information and Communication Technology, Koganei, Tokyo, Japan

Göran Johansson Microtechnology and Nanoscience, Chalmers University of Technology, Göteborg, Sweden

Anton Frisk Kockum Microtechnology and Nanoscience, Chalmers University of Technology, Göteborg, Sweden; Center for Emergent Matter Science, RIKEN, Wako, Japan

Peter J. Leek Clarendon Laboratory, Department of Physics, University of Oxford, Oxford, UK

Roberto Leoni Istituto di Fotonica e Nanotecnologie, Consiglio Nazionale delle Richerche (CNR), Roma, Italy

Adriana Lita National Institute of Standards and Technology, Boulder, CO, USA

Einar Magnusson Clarendon Laboratory, Department of Physics, University of Oxford, Oxford, UK

Riccardo Manenti Clarendon Laboratory, Department of Physics, University of Oxford, Oxford, UK

Francesco Marsili Jet Propulsion Laboratory, California Institute of Technology, Pasadena, CA, USA

Shigehito Miki Advanced ICT Research Institute, National Institute of Information and Communications Technology, Nishi-ku, Kobe, Hyogo, Japan

Gerard J. Milburn ARC Centre for Engineered Quantum Systems, University of Queensland, Brisbane, Australia

Kater W. Murch Department of Physics, Washington University, St. Louis, MO, USA

Faraz Najafi Department of Electrical Engineering and Computer Science, Massachusetts Institute of Technology, Cambridge, MA, USA

Sae Woo Nam National Institute of Standards and Technology, Boulder, CO, USA

Wolfram H.P. Pernice Institute of Physics, University of Muenster, Muenster, Germany

Diego Ristè Raytheon BBN Technologies, Cambridge, MA, USA

Döndü Sahin H. H. Wills Physics Laboratory, Centre for Quantum Photonics, University of Bristol, Bristol, UK

Masahide Sasaki National Institute of Information and Communication Technology, Koganei, Tokyo, Japan

Carsten Schuck Department of Electrical Engineering, Yale University, New Haven, CT, USA

Matthew D. Shaw Jet Propulsion Laboratory, California Institute of Technology, Pasadena, CA, USA

Irfan Siddiqi Quantum Nanoelectronics Laboratory, Department of Physics, University of California, Berkeley, CA, USA

Thomas M. Stace ARC Centre for Engineered Quantum Systems, University of Queensland, Brisbane, Australia

Hong X. Tang Department of Electrical Engineering, Yale University, New Haven, CT, USA

Varun B. Verma National Institute of Standards and Technology, Boulder, CO, USA

Rajamani Vijay Department of Condensed Matter Physics and Materials Science, Tata Institute of Fundamental Research, Mumbai, India

Takashi Yamamoto Graduate School of Engineering Science, Osaka University, Toyonaka, Osaka, Japan

Qingyuan Zhao Department of Electrical Engineering and Computer Science, Massachusetts Institute of Technology, Cambridge, MA, USA

Part I
Superconducting Single Photon Detectors: Technology and Applications

Chapter 1
Superconducting Nanowire Architectures for Single Photon Detection

Faraz Najafi, Francesco Marsili, Varun B. Verma, Qingyuan Zhao, Matthew D. Shaw, Karl K. Berggren and Sae Woo Nam

Abstract Over the past decade, superconducting nanowire single photon detectors (SNSPDs) have emerged as a key enabling technology for quantum optics and free-space optical communication. We review the operating principle and the latest advances in the performance of SNSPDs, such as extending sensitivity into the mid infrared, and the adoption of amorphous superconducting films. We discuss the limits and trade-offs of the SNSPD architecture and review novel device designs, such as parallel and series nanowire detectors (PNDs and SNDs), superconducting nanowire avalanche photodetector (SNAPs), and nanowire arrays with row-column readout, which have opened the pathway to larger active area, higher speed and photon-number resolution.

F. Najafi (✉) · Q. Zhao · K.K. Berggren
Department of Electrical Engineering and Computer Science, Massachusetts Institute of Technology, 77 Massachusetts Avenue, Cambridge, MA 02139, USA
e-mail: f_najafi@mit.edu

Q. Zhao
e-mail: qyzhao@mit.edu

K.K. Berggren
e-mail: berggren@mit.edu

F. Marsili · M.D. Shaw
Jet Propulsion Laboratory, California Institute of Technology, 4800 Oak Grove Drive, Pasadena, CA 91109, USA
e-mail: Francesco.Marsili.Dr@jpl.nasa.gov

M.D. Shaw
e-mail: matthew.d.shaw@jpl.nasa.gov

V.B. Verma · S.W. Nam
National Institute of Standards and Technology, 325 Broadway, Boulder, CO 80305, USA
e-mail: varun.verma@nist.gov

S.W. Nam
e-mail: nams@boulder.nist.gov

© Springer International Publishing Switzerland 2016

R.H. Hadfield and G. Johansson (eds.), *Superconducting Devices in Quantum Optics*, Quantum Science and Technology,
DOI 10.1007/978-3-319-24091-6_1

1.1 Introduction

Superconducting detectors can outperform other photon-counting technologies in a variety of performance metrics such as detection efficiency, dark count rate, timing jitter, reset time, and photon-number resolution [1]. As a result, superconducting detectors have found use at the cutting edge of basic research in astronomy, quantum optics, and free-space optical communication. In particular, single photon detectors based on superconducting nanowires [2] have been widely adopted due to their unrivaled performance in the infrared and due to advances in practical cryogenics.

In this chapter we discuss the design and performance of various types of detectors based on the superconducting nanowire concept, such as superconducting nanowire single photon detectors (SNSPDs or SSPDs) [3], superconducting nanowire avalanche photodetectors (SNAPs or Cascade-Switching SSPDs) [4, 5], multi wire photon-number-resolving (PNR) detectors [6, 7], and SNSPD arrays with scalable multi-pixel readout [8–10].

1.2 Performance Metrics for Photon Counting Detectors

In this section we review the definitions of the key metrics used to quantify the performance of photon-counting detectors, in order to assist with their direct comparison.

1.2.1 Detection Efficiency

The term *system detection efficiency* (*SDE*) describes the probability of registering an electrical signal produced by a photon once the photon has entered into the *input aperture* of the photon detection system. For example, for a fiber-coupled detector system, the input aperture would be defined as the fiber penetrating into the cryogenic vacuum chamber, which delivers photons to the detector itself. For a free-space coupled system, the input aperture would be the optical window into the cryogenic system. Conceptually, *SDE* can be thought of as the product of multiple efficiencies:

$$SDE = \eta_{coupling} \cdot \eta_{absorb} \cdot \eta_{internal} \cdot \eta_{trigger}, \tag{1.1}$$

where $\eta_{coupling}$ is the efficiency for the photons at the input aperture to be delivered to the active area of the device within the detector system, η_{absorb} is the probability that a photon incident on the active area of the detector is absorbed by the detector, $\eta_{internal}$ is the probability that an absorbed photon generates an observable electrical signal, and $\eta_{trigger}$ is the efficiency with which the counting electronics actually registers the electrical signal as a count. Note that for the photon-number-resolving (PNR) nanowire detectors discussed below, a separate $\eta_{trigger}$ must be defined for each photon number. While it is difficult to independently measure the four parameters in

Eq. (1.1) with high accuracy, the SDE is often obtained from measured quantities in the following way:

$$SDE = \frac{R_{light} - R_{dark}}{R_{incident}} \tag{1.2}$$

where R_{light} is the number of recorded counts per second when the detector is illuminated, R_{dark} is the number of recorded counts per second when the input aperture is blanked off, and $R_{incident}$ is the number of photons per second coupled into the input aperture of the system. Another important definition of the efficiency which occurs in literature is the *device detection efficiency*

$$DDE = \frac{SDE}{\eta_{coupling}}. \tag{1.3}$$

When quoting the *SDE*, care must be taken to describe the conditions of the optical input beam used to illuminate the detector. For example, the coupling, internal and absorption efficiencies depend on wavelength, while the absorption efficiency also depends on polarization. For free-space coupled detectors and detectors coupled through multimode fiber, $\eta_{coupling}$ and η_{absorb} may depend strongly on the mode structure and Fresnel number of the incoming beam, so it is important to consider the conditions of the experiment when evaluating the *SDE* of a detector in a particular application. Another important consideration when extracting *SDE* from Eq. 1.2 is that $\eta_{internal}$ and $\eta_{trigger}$ can appear artificially high when the detector exhibits afterpulsing (where one incident photon generates multiple electrical pulses at later times) and retriggering (where a single electrical pulse is counted multiple times due to electrical noise or unusual pulse shapes). It is therefore important to quantify these effects using time domain methods such as inter-arrival-time histograms [11] when evaluating the *SDE*.

1.2.2 Dark Count Rate

We define the dark count rate of a detector system as the response pulse rate that is not due to incident signal photons. However, when minimizing the dark count rate in a particular experiment, it is useful to draw a distinction between intrinsic dark count rate $R_{intrinsic}$, background count rate $R_{background}$, and dark count rate due to electrical noise R_{noise}. Intrinsic dark counts occur when the detector produces a response pulse when no photon was incident on the detector. Background counts are photodetection events due to stray photons coupled to the detector system. Since many superconducting nanowire detectors are single-photon-sensitive at mid-infrared wavelengths [12, 13], background counts are often dominated by black-body radiation emitted either from inside or outside the cryostat. Electrical noise can cause the counting electronics to record the presence of a response pulse when the detector did not produce one. In practice, the observed dark count rate that is measured in an experiment is

$$DCR = (R_{intrinsic} + R_{background} + R_{electronic}). \quad (1.4)$$

Depending on the superconducting material, the detector operation conditions, and the experimental setup, the *DCR* can be limited be any of these three mechanisms. When the detector is limited by background counts, it is important to engineer the cryogenic shielding and the optical system to minimize background counts while maintaining high coupling efficiency in the wavelength band of interest, which can involve using spectral and spatial filters at cryogenic temperatures.

1.2.3 Timing Jitter

Superconducting nanowire detectors are often used for time-correlated single-photon counting (TCSPC) applications, which are critically dependent on the timing accuracy of the detector. The time resolution of the detector is characterized by the timing jitter, which measures the fluctuation in the time delay (δt) between the instant a photon is absorbed in the detector and the instant a photodetection pulse is registered. The photodetection delay fluctuation can be quantified by acquiring many δt samples and measuring the full-width-at-half-maximum (FWHM) of the δt histogram. The δt histogram is called the Instrument response function (IRF), and the FWHM of the IRF is referred to as the timing jitter of the detector.[1] Timing jitter can arise from mechanisms intrinsic to the detector, from amplifier noise, and from jitter in the time-tagging electronics. The measured IRF is the convolution of the detector and system IRFs. For Gaussian IRFs, the jitter from all of these sources adds in root-sum-square (RSS) fashion.

The time delay corresponding to the maximum of the IRF is referred to as the *latency*. The measured latency includes delays due to the optical path length between the input aperture and the detector (free space or fiber), the internal latency of the photodetection process itself, the electrical path length (primarily RF cables), and the processing time of the readout electronics. For certain applications, such as laser time transfer, the long-term stability of the latency is an important parameter.

1.2.4 Recovery Time

Another important performance metric is the recovery time t_D. This is the time needed for the system to recover before it can efficiently detect another single photon following an initial photodetection event. While it is often sufficient to characterize this quantity by a single number, the recovery process can be thought of as a variable detection efficiency $\eta_{internal}(t)$ which depends on the time t since the last detection event, the form of which will depend on the physics of the recovery process

[1] Other definitions of timing jitter also exist in the literature, such as the 1/e width of the IRF.

and of the current dependence of the detection efficiency. The maximum count rate of a single detector is ultimately limited by $1/t_D$. However, in practice the maximum count rate of the detector system is limited by the readout architecture, with DC-coupled amplifiers resulting in higher maximum count rates than AC-coupled configurations [14].

1.3 Superconducting Nanowire Single Photon Detectors (SNSPDs)

Since their invention in 2001 [3], SNSPDs have emerged as the leading technology for time-resolved single-photon detection, with applications in free-space optical communication [15, 16], quantum optics [17–19], quantum key distribution [20], ranging [21], and life sciences [22, 23]. SNSPDs have recently been demonstrated with detection efficiencies up to 93 % [24] for 1550 nm-wavelength photons, background-limited dark count rates of 4 counts per second (at 70 % SDE) [25], timing jitter down to 18 ps [26] and wideband sensitivity between visible and mid-infrared wavelengths [13]. The typical structure of an SNSPD is shown in Fig. 1.1. The SNSPD consists of a narrow (<150 nm) and thin (<10 nm) superconducting nanowire. In order to achieve a large active area with significant overlap with the optical mode, the long superconducting nanowire is often folded into a meander shape.

Among the materials used for SNSPDs are niobium nitride (NbN) [3, 27–29], niobium titanium nitride (NbTiN) [30, 31], tungsten silicide (WSi) [32], tantalum nitride (TaN) [33], Molybdenum silicide (MoSi) superconductiong nanowire, SNSPD [34,

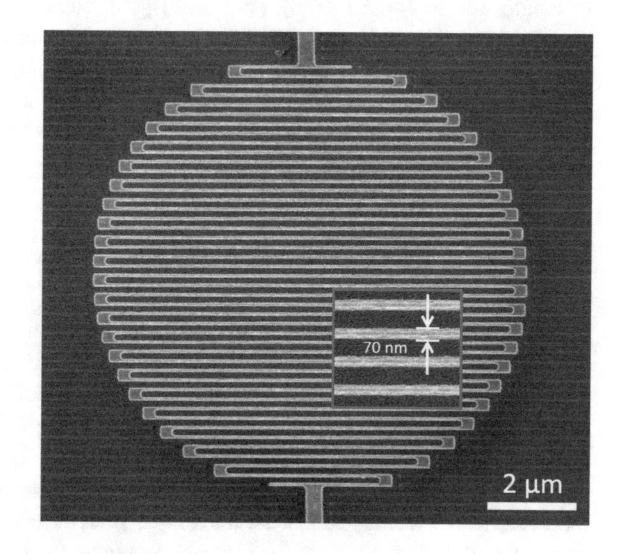

Fig. 1.1 Top-down scanning-electron micrograph (SEM) of an SNSPD based on 70-nm-wide nanowires. The *inset* shows a magnified SEM of the nanowires

35], molybdenum germanium (MoGe) [36], niobium silicide (NbSi) [37] and magnesium diboride (MgB_2) [38, 39].

The optimal operating temperature of most SNSPDs ranges from <1 K for WSi to <4 K for the nitrides. MgB_2 SNSPDs achieved the operating temperature of 10 K [38], which is the highest reported to date for this type of detectors. Since SNSPDs operate at higher temperature than transition edge sensors (TESs, discussed in Chap. 2) [40], they require less complicated cryogenics and are easier to use in practical applications.

1.3.1 Photodetection Mechanism

Figure 1.2 illustrates the current understanding of the physical process of single-photon detection in superconducting nanowires. When a photon is absorbed in the superconducting nanowire, it generates an energetic quasiparticle which quickly thermalizes through electron-electron and electron-phonon scattering into a population of quasiparticles at the gap edge, generating a spatially localized region where the superconducting gap is suppressed. This region of suppressed superconductivity, illustrated in Fig. 1.2a, is referred to as the *hotspot* [41]. Several models have been proposed to describe the evolution of the system once the hotspot is formed [41–44]. According the model that to date has most successfully explained the experimental data [44, 45], once the superconducting gap has been sufficiently suppressed, a vortex can enter the nanowire (Fig. 1.2b). A sufficiently high bias current through the nanowire can exert a force on the vortex, driving it across the nanowire and resulting in a resistive slab across the nanowire (Fig. 1.2b). The resistive region will then grow in length due to Joule heating (Fig. 1.2c) [46], causing the resistance of the wire to grow from zero to several kΩ. The sudden change in impedance causes the current to be diverted out of the nanowire into the readout electronics (see Sect. 1.2.2), allowing the resistive slab to cool down (Fig. 1.2d) and return to the superconducting state. Once the current flowing through the nanowire has recovered, the nanowire can detect another photon.

1.3.2 Detection Efficiency and Constrictions

In sub-50-nm-wide nanowire SNSPDs based on Niobium nitride (NbN) superconducting nanowire, SNSPD [5, 13] and in SNSPDs based on amorphous superconductors such as WSi [24, 32], Molybdenum germanium (MoGe) superconducting nanowire, SNSPD [36], and MoSi [35], the detection efficiency at 1550 nm wavelength generally exhibits a sigmoidal dependence on the bias current. At low bias currents (I_B), the *DDE* increases exponentially with I_B while at high bias currents the detection efficiency saturates, reaching a plateau where the detection efficiency shows weak dependence on I_B. Figure 1.3 shows: (a) the *DDE* versus bias current

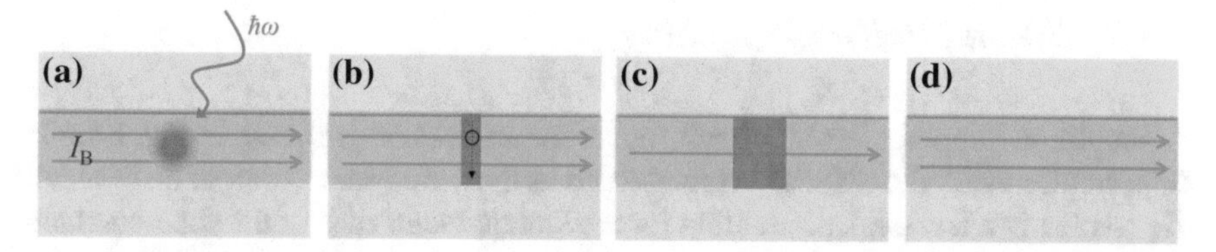

Fig. 1.2 Sketch of photodetection process inside a current-biased superconducting nanowire according to the model proposed in Ref. [44]. **a** A photon of energy $\hbar\omega$ is absorbed in the nanowire, suppressing the superconductivity locally (hotspot). **b** A vortex crosses the region with suppressed superconductivity, resulting in a resistive slab across the nanowire. **c** Joule heating results in a growth of the resistive region. In this state, the resistance of the wire rises to several kΩ, and the current is shunted into the readout amplifier. **d** The resistive region eventually cools down and returns to the superconducting state

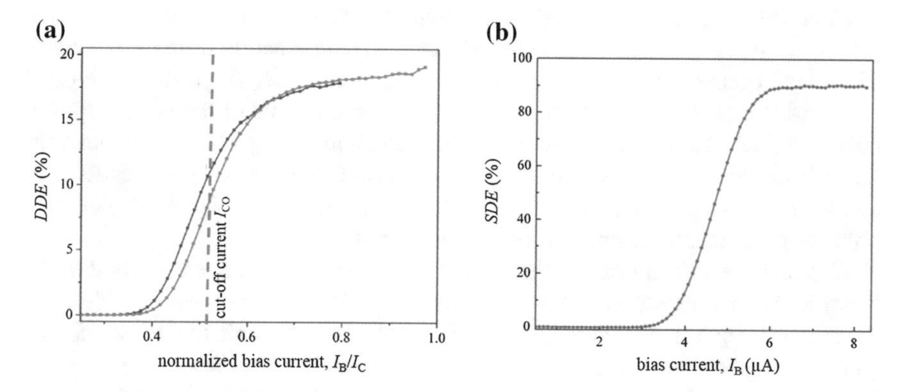

Fig. 1.3 **a** Device detection efficiency versus normalized bias current of SNSPDs based on 30-nm-wide nanowires illuminated with 1550-nm-wavelength light [5]. The bias current of both detectors is normalized by the critical current of the less constricted detector (*red curve*). **b** System detection efficiency versus bias current for a WSi SNSPD, reproduced from [47] with permission of SPIE

of NbN SNSPDs based on 30-nm-wide nanowires [5], and (b) the *SDE* versus bias current of WSi SNSPDs embedded in a vertical optical stack to enhance absorption (as discussed in detail in Sect. 2.2.2.1, in the context of TESs) [24, 47]. Saturated detection efficiency versus I_B curves are a sign of high internal efficiency [24] and is predicted by the model described in [41]. Non uniformities and defects along the nanowire introduced by lithography or film growth (referred to as *constrictions*) can limit the switching current of the detector, resulting in reduced detection efficiency [48]. Geometries (ultra-narrow nanowires) and materials (WSi, MoGe, MoSi) that yield saturated detectors are more robust towards constrictions. For comparison, the purple curve in Fig. 1.3a shows a constricted 30-nm-nanowire NbN SNSPD. In the presence of a large saturation plateau, the constricted detector can still reach an efficiency value comparable to a constriction-free detector (red curve in Fig. 1.3a). The inflection point of the *DDE* versus I_B curve is referred to as the cutoff current I_{CO}.

1.3.3 Speed Limit and Latching

A simplified lumped-element circuit model for an SNSPD is shown in Fig. 1.4a. In this circuit model, the SNSPD is represented as the series connection of an inductor L_K (typically a few μH), accounting for the kinetic inductance of the superconducting nanowire (which is orders of magnitude larger than the geometric inductance), and a resistor R_N (typically a few kΩ) in parallel with a switch. R_N represents the normal-state resistance of the SNSPD. The impedance of the readout electronics is represented by a load resistor R_L (typically 50 Ω) connected in parallel to the SNSPD. In steady state, represented by a closed switch, the detector is superconducting, biased at I_B and exhibits no significant resistance. After the detection event, the switch opens, and since $R_L \ll R_N$, I_B redistributes into R_L. However, the rate at which the current in the detector can change is determined by the inductance L_K [28] of the superconducting nanowire. The current redistribution from the nanowire to the load occurs within a time constant of $\tau_{rise} \sim L_K / R_N$ (typically <100ps in NbN and <1 ns in WSi), which we refer to as the rise time of the photoresponse pulse. The detector eventually returns to its superconducting state and the current redistributes back into the SNSPD. The current recovery in the detector occurs with a time constant $\tau_{fall} \sim L_K/R_L$ (~1–2 ns for NbN, ~20–40 ns for WSi), which is referred to as the fall time of the photoresponse pulse.

If the recovery is so fast that the current flows back into the nanowire before the nanowire has sufficiently cooled—i.e. while the gap is still suppressed, a stable, self-sustaining hotspot is created [46, 49]. In this state, the detector is said to be *latched*, and is unavailable for photon detection until the bias current is reset. The recovery time, also referred to as dead time or reset time, is illustrated in Fig. 1.4d, which shows the histogram of the inter-arrival time of two subsequent detector pulses [5, 11]. The dead time t_D is often defined as the time delay at which the inter-arrival-time histogram reaches 90 % of its peak value. While intuitive, the simplified circuit model is not sufficient to explain the current dynamics in SNSPDs and more complex multi-nanowire detectors. Furthermore, the lumped-element circuit cannot explain latching.

A more advanced approximation is the one-dimensional electrothermal (ET) model [46], a macroscopic model that quantitatively describes the formation of a measurable photodetection pulse in the readout circuit, the current dynamics in multi-wire configurations [50], and latching. The ET model is based on two coupled differential equations. The first equation is the time-dependent heat equation

$$J^2\rho + \kappa\frac{\partial^2 T}{\partial x^2} - \frac{\alpha}{d}(T - T_{sub}) = \frac{\partial}{\partial t}cT, \quad (1.5)$$

where the three terms on the left hand side respectively describe the thermal dynamics of the resistive region governed by Joule heating ($J^2\rho$, where J is the current density and ρ is the resistivity of the superconducting material), cooling through diffusion inside the superconductor (κ and c are the thermal conductivity and specific heat of the superconducting material, respectively), and cooling through the substrate

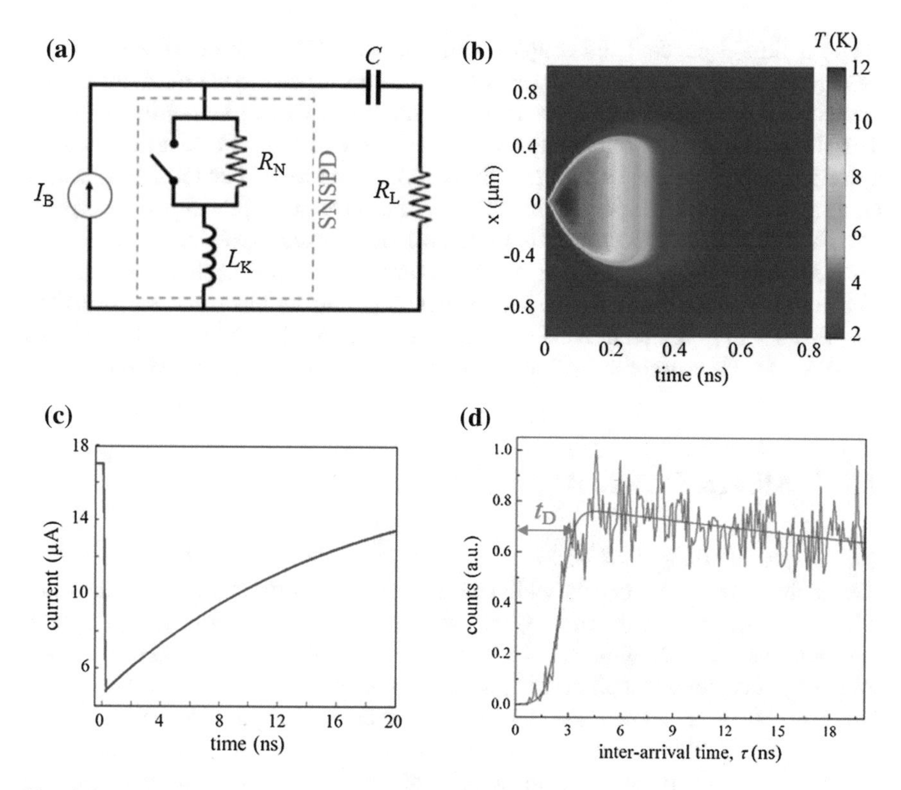

Fig. 1.4 **a** Simple circuit model for an SNSPD after photon absorption (open switch). The impedance of the readout electronics is modeled with the load resistor $R_L = 50$. The inductor in series models the kinetic inductance of the SNSPD. **b** Calculated time-dependent temperature distribution along an NbN nanowire. After the creation of a resistive slab at $t = 0$ s, the nanowire temperature increases due to Joule heating and reaches a maximum value of 12 K. After ~100–200 ps the resistive region cools back down to the substrate temperature. **c** Simulated time-dependent detector current (t) after the creation of a resistive slab at $t = 0$ s. **d** Measured pulse-to-pulse inter-arrival time histogram of a 4-SNAP (shown in *green*) and a fit (shown in *red*) [5]. The dead time $t_D \sim 3.3\,ns$ was extracted from the fit as the inter-arrival time at which the count rate reaches 90 % of its maximum value

(α is the thermal conductivity between the superconductor at temperature T and the substrate at temperature T_{sub}). The second equation,

$$C\left(\frac{d^2}{dt^2}L_K I + \frac{d}{dt}IR_N + R_L\frac{dI}{dt}\right) = I_B - I \quad (1.6)$$

is the differential equation for the circuit shown in Fig. 1.4a. In this equation, C is the AC coupling capacitance of the readout amplifier, L_K is the kinetic inductance of the nanowire, R_L is the load impedance of the amplifier (typically 50 Ω) and I_B is the bias current provided by the source. Note that in this equation the detector resistance

$R_N(t)$ is time-dependent and couples both equations. The time-dependent normal-state resistance is a more accurate approximation of the SNSPD behavior during photodetection than the constant R_N in the simplified Circuit model (of SNSPD). The ET model is now frequently used in the field to design detectors based on superconducting nanowires. Figure 1.4b shows the time-dependent local temperature along an NbN nanowire after the formation of a resistive slab following the absorption of a photon. Joule heating results in the growth of the resistive region, which in turn results in the diversion of I_B out of the SNSPD and into the load, as shown in Fig. 1.4c. The resistive region returns to the superconducting state within a characteristic time on the order of ~100 ps in NbN—however, due to the kinetic inductance, it takes ~1–10 ns for the current through the nanowire to return to its initial value I_B.

1.3.4 Mid-IR Detection

One of the advantages of SNSPDs over other single-photon detector technologies is their sensitivity over a wide spectral range. Figure 1.5 shows the *DDE* of NbN nanowires as a function of width, wavelength and bias current [13]. At high bias currents ($I_B > I_{CO}$) the *DDE* curve shows saturation—a sign of high internal efficiency. The detection efficiency plateau shrinks with decreasing photon energy. However, narrower nanowires are sensitive to longer wavelengths. As an example, 30-nm-wide NbN SNSPDs could be operated above I_{CO} for wavelengths beyond 2 μm.

Although ultranarrow (<50-nm-wide) NbN nanowires are sensitive in the mid infrared (mid-IR), they are challenging to fabricate over large areas with high yield. Amorphous WSi has a number of properties that make it a desirable superconducting material for fabrication of large-area SNSPDs [9, 24, 51]. The reduced carrier density and larger hotspot size in WSi allows the nanowires to be wider than NbN- or NbTiN-based nanowires, which considerably improves device yield due to a lower probability of constriction, making WSi a more promising choice for mid-IR single-photon detection. Figure 1.6 shows the photoresponse count rate (*PCR*) of WSi SNSPDs based on nanowires of different widths (w =180, 140 and 100 nm) in the wavelength range $\lambda = 2.1-5.5$ μm [52]. As shown in Fig. 1.6a, 100 nm wide nanowire SNSPDs showed saturated detection efficiency at wavelengths as long as 5.5 μm. To date, SNSPDs have not been optimized for high efficiency in the mid-IR, but given the intrinsic sensitivity of the material we expect there to be no fundamental limitation to engineering near-unity-detection efficiency SNSPDs in the 2–5 μm wavelength range.

1.3.5 Performance Trade-Offs

The performance metrics discussed in the previous sections are difficult to independently optimize because they are subject to several trade offs. In this section we discuss the limitations that must be considered when designing SNSPDs.

Active area and dead time: In order to achieve high *SDE*, it is desirable to maximize the active area of the detector to simplify optical coupling. However, as the kinetic inductance of a nanowire is proportional to its length, the overall dead time of an SNSPD increases with increasing active area. As we will discuss in the next section, multi-nanowire architectures can result in a speed-up compared to conventional SNSPDs in the limit of large active areas.

Active area and detector yield: While a larger detector area can enable larger optical coupling efficiency, it also increases the probability of constrictions along the nanowire. Detectors based on amorphous superconducting films (WSi [24], MoGe [36], MoSi [34, 35]) are more robust towards constrictions than NbN, as discussed in the previous section.

Sensitivity and timing jitter: SNSPDs based on narrow NbN nanowires and amorphous superconductors offer large saturation plateaus and sensitivity to longer wavelengths. However, narrower nanowires have a smaller cross-section, resulting in smaller switching currents (I_{SW}) compared to wider NbN wires. In a similar fashion, WSi SNSPDs have switching currents comparable to 30-nm-wide NbN SNSPDs. Since the detector voltage signal amplitude is on the order of $R_L \cdot I_{SW}$, increased

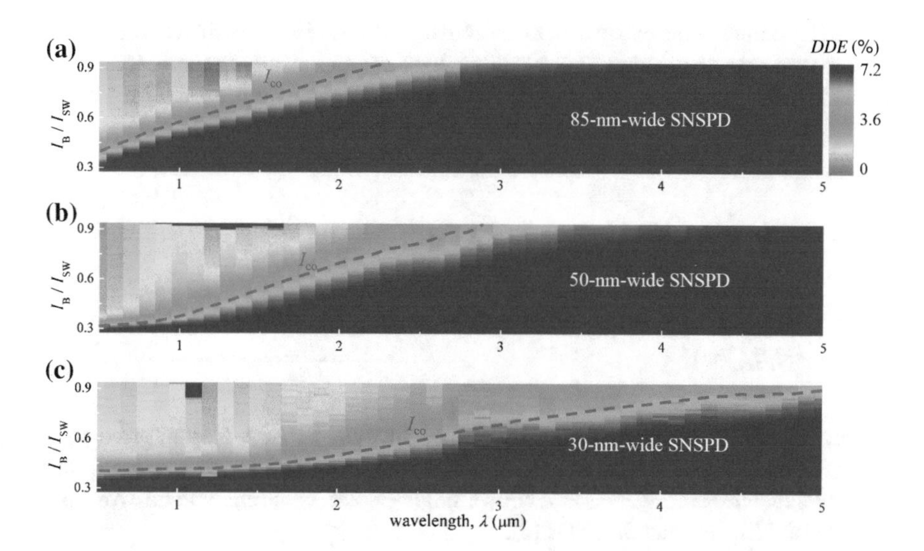

Fig. 1.5 Device detection efficiency (*DDE*, in *color scale*) versus wavelength λ and normalized bias current (I_B/I_{SW}) for SNSPDs based on **a** 85-nm-wide ($I_{SW} = 20.6\,\mu$A), **b** 50-nm-wide ($I_{SW} = 9.3\,\mu$A), and **c** 30-nm-wide NbN ($I_{SW} = 7.4\,\mu$A) nanowires. The λ-dependent cutoff current I_{CO} for each detector is marked with a *red dashed line*. Adapted from Ref. [13]

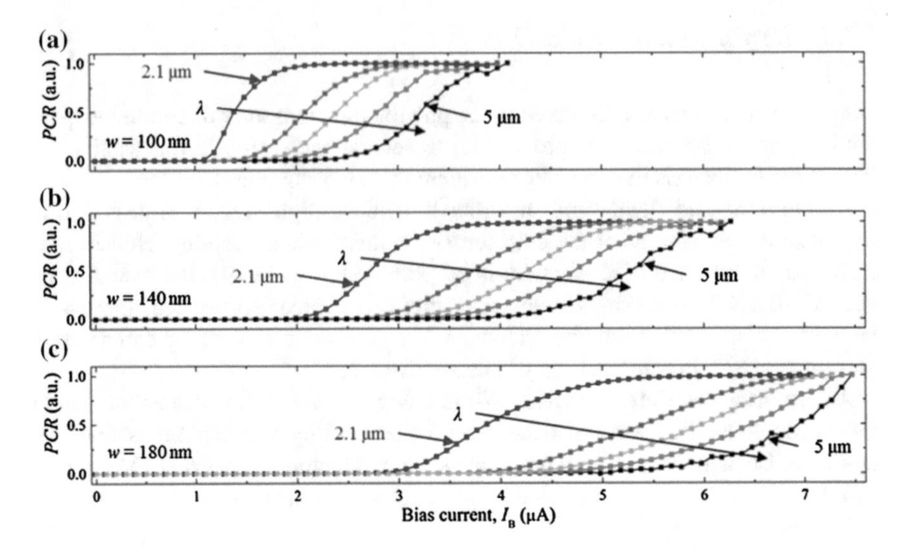

Fig. 1.6 Photoresponse count rate (*PCR*) as a function of the normalized bias current (I_B/I_{SW}) at $\lambda = 2.1, 3, 3.8, 4.2,$ and $5.5\ \mu$m for WSi SNSPD basedon **a** 100 nm wide nanowires; **b** 140 nm wide nanowires; and **c** 180 nm wide nanowires. The nanowires were 4 nm thick. The switching currents of the SNSPDs were: $I_{SW} = 4.1\ \mu$A for the 100 nm-wide-nanowire SNSPD; $I_{SW} = 6.4\ \mu$A for the 40-nm-wide-nanowire SNSPD; and $I_{SW} = 7.5\ \mu$A for the 180 nm-wide-nanowire SNSPD. Reproduced from Ref. [52] with permission of the Optical Society of America (OSA)

sensitivity comes at the expense of decreased signal-to-noise ratio (*SNR*). The limited *SNR* makes detector readout more challenging [24] and results in increased timing jitter [53]. By optimizing the nanowire width and employing multi-nanowire detector architectures (see next section), this trade off can be relaxed.

1.4 Multi-Nanowire Detector Architectures

1.4.1 Superconducting Nanowire Avalanche Photodetectors (SNAPs)

Superconducting nanowire avalanche photodetectors (SNAPs, also referred to as cascade switching superconducting single photon detectors) [4, 5] are based on a parallel nanowire architecture that allows single-photon counting with an *SNR* up to a factor of 4 higher than SNSPDs [5].

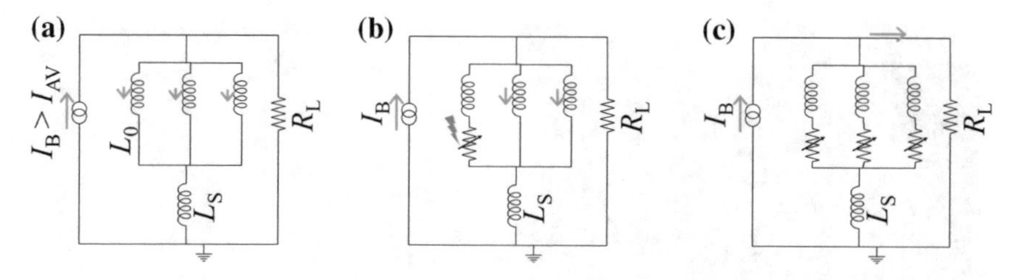

Fig. 1.7 Circuit model of a 3-SNAP biased above avalanche current I_{AV}. **a** All three sections are biased at $I_B/3$. **b** The absorption of a photon drives one of the SNAP sections (initiating section) into the normal state. **c** The current redistribution drives the remaining sections (secondary sections) into the normal state (avalanche), resulting in a current redistribution into the load and a measurable voltage pulse across R_L

1.4.1.1 The Avalanche Regime

SNAPs are the parallel connection of N superconducting nanowires (referred to as an N-SNAP) as shown in Fig. 1.7 for $N = 3$. For the correct operation of the device, the parallel nanowires are connected in series with a choke inductor L_S. In the steady state (Fig. 1.7a) all N sections are biased with a current I_B/N. The detection event begins with the absorption of a photon triggering a hotspot nucleation (HSN) event in one of the SNAP sections, as shown in Fig. 1.7b. This section, referred to as the *initiating* section, becomes resistive, diverting its current to the nanowires that are electrically connected in parallel to it. These sections are called *secondary* sections. If the initial bias current (I_B) of an N-SNAP is higher than the avalanche threshold current (I_{AV}), the current diverted to the $(N - 1)$ secondary sections is sufficient to switch these sections to the normal state (Fig. 1.7c). As one wire firing causes all the remaining wires to fire, this process is called an avalanche. As a result, a current N times higher than the current through an individual section is diverted to the readout [4, 5, 50], resulting to an N-fold increase in the *SNR*.

In the avalanche regime, SNAPs operate as single-photon detectors, because a single HSN event is sufficient to trigger an avalanche and therefore a measurable detector pulse. Figure 1.8a shows the SEM of a 3-SNAP resist mask based on 30-nm-wide nanowires, resulting in a threefold increase in *SNR*, as illustrated in the green trace in Fig. 1.8b [5].

Figure 1.8c shows the *DDE* of an SNSPD and a 2-, 3- and 4-SNAP as a function of normalized bias current I_B/I_{SW}. The last inflection point in the *DDE* versus I_B curve is the avalanche current I_{AV}. The normalized avalanche current I_{AV}/I_{SW} increases with increasing N. As will be discussed later, stable operation of SNAPs with $N > 4$ has not been demonstrated in the avalanche regime.

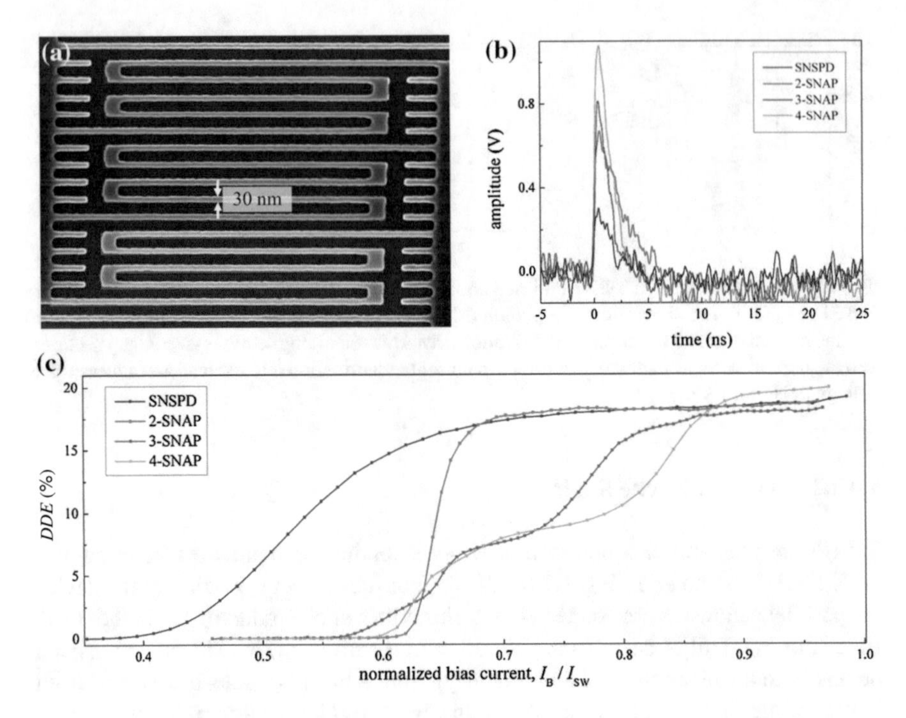

Fig. 1.8 **a** False-color SEM of a 3-SNAP resist mask on top of NbN with each section colored differently. **b** Voltage traces of detector pulses of an SNSPD, a 2-SNAP, a 3-SNAP, and a 4-SNAP based on 20-nm-wide nanowires (in *purple*, *red*, *green*, and *orange*, respectively), showing increasing signal-to-noise ratio as the number N of SNAP sections is increased. **c** *DDE* at 1550 nm wavelength versus normalized bias current (I_B/I_{SW}) for an SNSPD ($I_{SW} = 7.2\,\mu\text{A}$), a 2-SNAP ($I_{SW} = 13.4\,\mu\text{A}$), a 3-SNAP ($I_{SW} = 18.1\,\mu\text{A}$), and a 4-SNAP ($I_{SW} = 28.4\,\mu\text{A}$) based on 30-nm-wide nanowires (shown in *purple*, *red*, *green*, and *orange*, respectively). Adapted from Ref. [5]

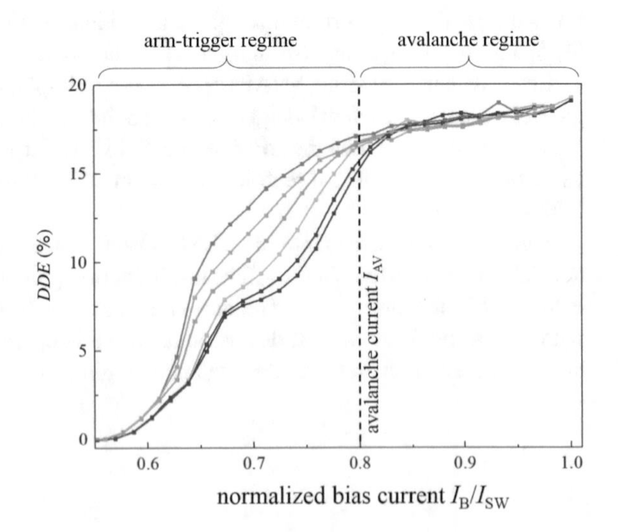

Fig. 1.9 Device detection efficiency versus bias current curves of a 3-SNAP measured at photon fluxes ranging from 0.6 (*red curve*) to 19 million photons per second (*purple curve*) Adapted from [5]. In the avalanche regime, the detection efficiency is independent of the incident photon flux

1.4.1.2 The Arm-Trigger Regime

When biased below I_{AV}, the current diverted from the initiating section is not sufficient to trigger an avalanche. For an avalanche to form, additional HSN events have to occur in the secondary sections. Figure 1.9 shows the detection efficiency versus bias current-curve for a 3-SNAP [5]. For $I_{\mathrm{B}} < I_{\mathrm{AV}}$, at least two HSN events are necessary to trigger an avalanche in the 3-SNAP. This operation condition is referred to as the *arm-trigger* regime, and is illustrated in Fig. 1.10. As shown in Fig. 1.10a, the first HSN event (which we called *arm* HSN event) diverts the current from the initiating section to the secondary sections, which do not switch to the normal state. After the arm HSN event, the still superconducting secondary sections operate as a 2-SNAP biased above the avalanche current, in that an avalanche will be created if an additional HSN event (which we called *trigger* HSN event) occurs in one of the secondary sections (see Fig. 1.10b, c). In this regime the SNAP operates as a multi-photon gate [54] rather than a single-photon detector. Figure 1.10d shows the simulated current dynamics for a 3-SNAP in the arm-trigger regime [5, 50]. The arm HSN event diverts the current from the initiating section to the secondary sections. The trigger HSN event in one of the secondary sections results in an avalanche, diverting the current from the detector into the load and resulting in a measurable voltage pulse across R_L.

1.4.1.3 The Unstable Regime and Afterpulsing

The N parallel nanowires of an N-SNAP have an N^2-times lower inductance than an SNSPD with the same area. It was initially believed that SNAPs had the potential to reach N^2-times reduced reset time compared to SNSPDs. However, part of the speed advantage is negated by the choke inductor L_{S} in series with the nanowires, resulting in a smaller effective speed-up. It has been shown [55] that a sufficiently large L_S is essential for stable operation in avalanche regime. When the initiating section fires, it diverts its current to the secondary sections and to the load. The current diverted to the load, referred to as leakage current I_{lk}, does not contribute to create an avalanche, so it should be minimized. We can approximate the ratio of I_{lk} and the current redistributed to all $(N-1)$ secondary sections after an HSN event as $r = L_0/(L_S(N-1))$ [55], where L_0 is the inductance of a single SNAP section. A decrease in L_{S}/L_0 results in an increase of I_{lk} (see I_{out} in Fig. 1.10d). A higher I_{lk} results in an increase of I_{AV}, so a larger bias current is necessary to trigger an avalanche. A large $I_{\mathrm{AV}}/I_{\mathrm{SW}}$ is undesirable because the detector needs to be biased closer to I_{SW} to operate in the avalanche regime, resulting in a higher dark count rate.

In addition to the undesirable effect of a higher avalanche current, a small L_S changes the behavior of the SNAP at low bias currents. Figure 1.11a shows the normalized photon count rate (*PCR*) versus bias current of a 3-SNAP with $r = 0.28$. The *PCR* is normalized by the incident photon flux, which corresponds to the device detection efficiency for $I_B > I_{AV}$. Between $I_B \sim 0.7 I_{SW}$ and I_{AV} the detector operates in arm-trigger regime. For $I_B < 0.7 I_{SW}$ the SNAP emitted many smaller-

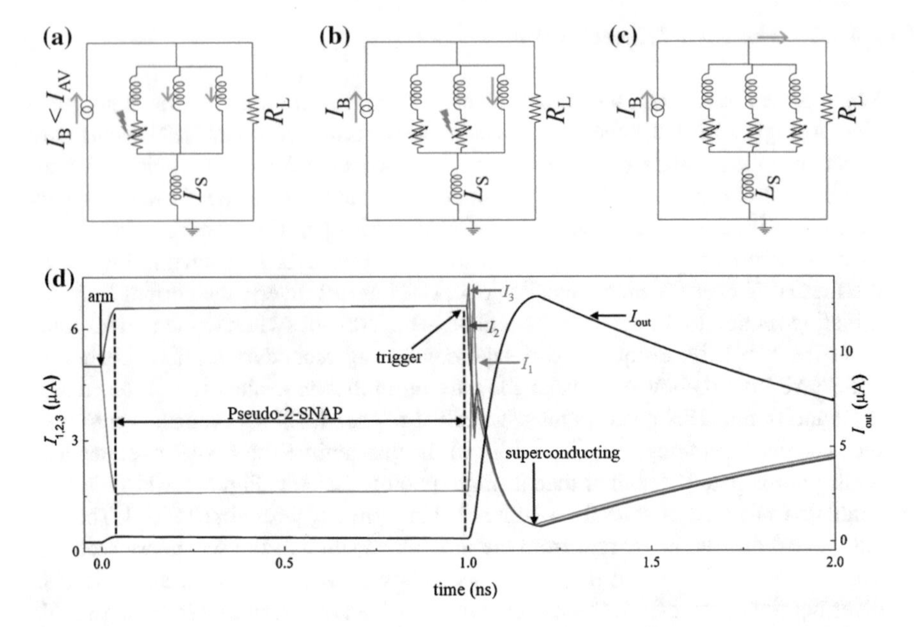

Fig. 1.10 **a** Circuit model of a 3-SNAP biased below I_{AV}. An initial HSN drives the initiating section into the normal state. However, the current redistribution is not sufficient to drive the secondary sections to the normal state. **b, c** A second HSN in a secondary section occurs, triggering an avalanche. **d** Simulated current dynamics through the initiating section (I_1), the secondary section where the second (trigger) HSN occurs (I_2) and the secondary section that switches to the resistive state (I_3) following current redistribution from the first two sections. The second HSN event results in an avalanche and current redistribution from the SNAP into the load (I_{out}, shown in *black*). The detector was biased below I_{AV}. Adapted from Ref. [5]

amplitude pulses in addition to the avalanche pulses, as shown in the lower inset (blue trace) of Fig. 1.11a. This is referred to as the *unstable regime*. The detector pulses in the unstable regime result in a spurious, photon-flux-dependent peak in the *DEE* versus I_B curve. We observed that SNAPs with $r > 0.1$ operated in the unstable regime at low bias currents. For some applications, a large r might be acceptable, and a question arises regarding the lowest achievable L_S value.

The reset time of SNAPs in NbN is ultimately limited to $\geq$1 ns [55]. SNAPs with shorter reset times showed afterpulsing, i.e. they generated a series of pulses for each detected photon, as shown in Fig. 1.11b, c. Therefore, the speed limit of SNAPs is similar to SNSPDs ($\sim$1 ns dead time for NbN [46]), although they can still be used to extend the active area of the device without increasing the dead time.

1.4.1.4 3D-SNAPs

The detection efficiency of conventional SNSPDs and SNAPs is dependent on polarization due to the polarization dependence of η_{absorb} [56], which is larger for light

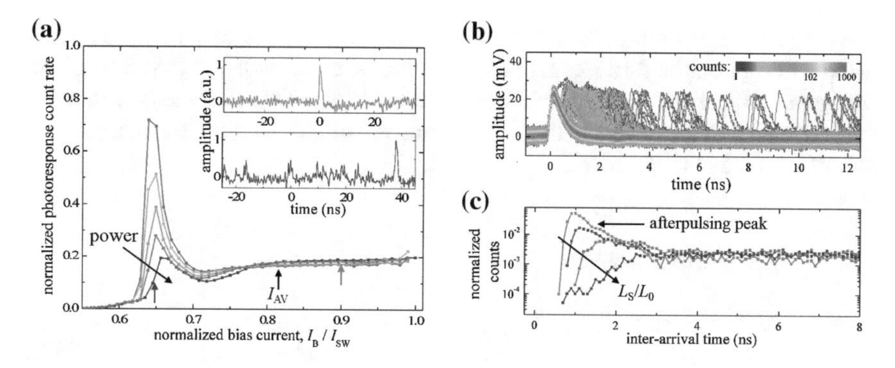

Fig. 1.11 **a** Normalized photon count rate (PCR) versus I_B/I_{SW} of a 30-nm-wide 3-SNAP at different incident photon fluxes. The PCR is normalized to the photon flux, corresponding to the device detection efficiency for $I_B > I_{AV}$. The *inset* shows traces of the detector output voltage in the unstable regime ($I_B = 0.65 I_{SW}$, *bottom panel, blue arrow*) and the avalanche regime ($I_B = 0.9 I_{SW}$, *top panel, red arrow*). **b** Persistence map of the traces of the detector output voltage for a 4-SNAP with leakage current parameter $r = 1$. The detector was biased close to the switching current. **c** Inter-arrival time delay histograms of the detector pulses for 4-SNAPs with r ranging from ~0.1 (*red*) to 1 (*crimson*). The detectors were biased close to the switching current. Adapted from Ref. [55]

polarized parallel to the nanowires compared to light polarized perpendicular to the nanowires. In order to construct a polarization-independent detector, two sections of a 2-SNAP were stacked vertically on top of each other forming a three dimensional SNAP (3D-SNAP) [51]. Figure 1.12a shows a sketch of the 3D-SNAP. The two nanowire meanders were separated by a $30 \times 30\,\mu$m, 75 nm-thick square pad of hydrogen silsesquioxane (HSQ), which served as the electrical insulation. To increase the detection efficiency, the 3D-SNAP was embedded in a stack of dielectric materials to optimize absorption at $\lambda = 1550$ nm (See also Sect. 2.2.2.1, where the same strategy is used to optimize TES performance) [24]. Since the two sections could be patterned independently, the nanowires in each section were oriented at orthogonal angles with respect to one another.

The vertical stacking of orthogonal nanowire meanders connected electrically in parallel enabled: (1) a factor of ~2 higher signal-to-noise ratio than with SNSPDs, (2) polarization independent system detection efficiency over a ~ 100 nm-wide wavelength range, as illustrated in Fig. 1.12b, and (3) system detection efficiencies greater than 85 %, comparable to the best results achieved to date with a planar SNSPD (see Fig. 1.12c) [24].

1.4.1.5 Series-SNAPs

A disadvantage of the traditional SNAP design is that a large series inductor $L_S > 3L_0$ is generally required for the correct operation of the device. For large-active-area SNAPs the requirement on L_S results in a detector with long reset time, long rise time and large timing jitter. The constraint on the series inductor can be relaxed using a design outlined in Fig. 1.13a, referred to as a series-SNAP [53, 57]. In this configuration, the series inductor is itself fashioned out of SNAP elements, extending the active area of the device without sacrificing the timing jitter nor the reset time of the detector (see Fig. 1.13c–e).

1.4.1.6 Multi-Stage SNAPs

As discussed in Sect. 1.4.1.1, in N-SNAPs the avalanche current increases with increasing N. Practically, the maximum number of SNAP sections in parallel with stable avalanche has been limited to 4 sections ($I_{AV}/I_{SW} \sim 0.9$ for a 4-SNAP) [5]. One approach to circumventing this limit is to implement a multi-stage SNAP structure that allows for a smaller avalanche current I_{AV} [58]. Figure 1.14a shows the structure of a three stage 2-SNAP. In this SNAP, a pulse is not generated through a single avalanche, but rather through three cascaded avalanches. An HSN event results in a first avalanche in a 2-SNAP, marked with a blue arrow. The resistive 2-SNAP

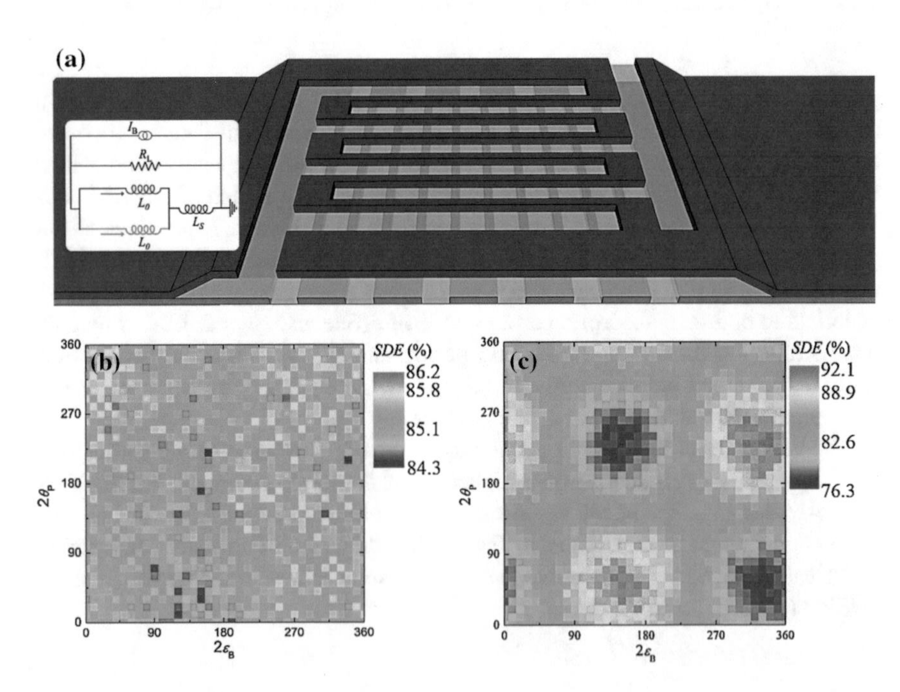

Fig. 1.12 **a** Three-dimensional sketch of a 3D-SNAP (not to scale) based on two stacked WSi-SNSPDs with orthogonal meander orientation. Both SNSPDs are connected electrically in parallel to each other and in series with an inductor L_S, comprising a 2-SNAP. The equivalent circuit is shown in the inset. L_0 represents the inductance of a single section. **b, c** SDE (in *color scale*) as a function of polarization (represented by orientation on the Poincaré sphere) for **b** a 3D-SNAP and **c** a standard single-layer WSi-SNSPD. Adapted from Ref. [51]

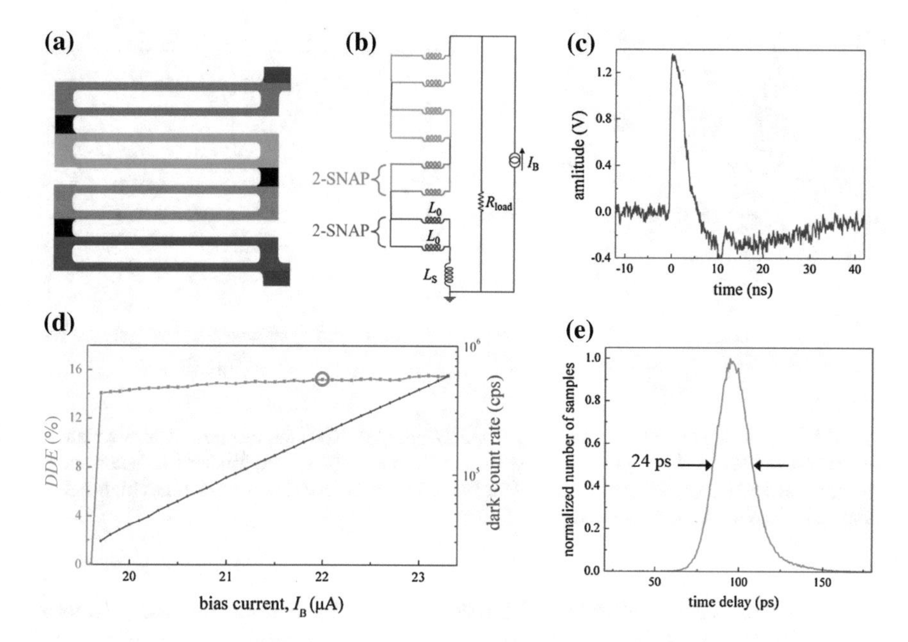

Fig. 1.13 **a** Nanowire arrangement sketch for a series-2-SNAP comprising four 2-SNAP units connected in series. **b** Electrical circuit equivalent of the series-2-SNAP shown in (**a**). **c** Detector output voltage trace of a series-2-SNAP with the design outlined in (**a, b**). **d** *DDE* versus I_B for a series-2-SNAP based on ~80-nm-wide nanowires. **e** IRF of the same device as (**d**). The IRF was measured at $I_B \sim 22\,\mu$A (see *red circle*). Adapted from Ref. [53, 57]

is connected in parallel to an identical 2-SNAP and the two parallel 2-SNAPs are connected in series to an inductor L_{S2}. The resistive 2-SNAP diverts its current to the still superconducting 2-SNAP connected in parallel to it (second avalanche, marked with a green arrow). In the avalanche regime, the current redistribution drives the second 2-SNAP into the resistive state as well, and subsequent avalanches follow the same scheme. The signal from the detector illustrated in Fig. 1.14a is shown as the red trace in Fig. 1.14b.

The SNR of the three-stage 2-SNAP, referred to as 2^3-SNAP, is comparable to a traditional 8-SNAP, with the advantage of a lower avalanche current. The avalanche current of the 2^3-SNAP is comparable to that of a 2-SNAP, while an 8-SNAP would be extremely challenging to realize due to its high expected avalanche current.

1.4.2 *Parallel- and Series-Nanowire Detectors (PNDs, SNDs)*

SNSPDs are not inherently photon-number-resolving (PNR) detectors. However, n photons in a short optical pulse can be resolved by spreading the incoming light onto $m \geq n$ SNSPDs, and adding up the number of detection events [6, 7, 59]. If

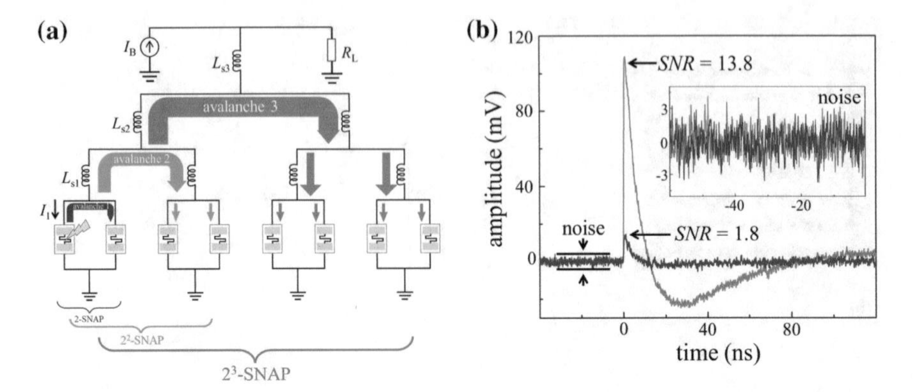

Fig. 1.14 **a** Equivalent circuit of a 3-stage 2-SNAP (2^3-SNAP). The first, second and third avalanche are illustrated with the *blue*, *green* and *red arrows*, respectively. **b** Amplitude of detector output voltage of an SNSPD (*blue*) and of a 2^3-SNAP (*red*). The 2^3-SNAP shows a ~8-times higher SNR compared to the SNSPD. Adapted from Ref. [58]

the pulses from each SNSPD can be recorded (e.g. by reading out each detector individually [16, 59]), this system can be used to measure multi-photon correlation functions [60].

For applications where counting the mean number of photons is of interest, PNR detectors based on multiple superconducting nanowires have been demonstrated. The first approach to be demonstrated was the parallel-nanowire detector (PND) [6, 61, 62], shown in Fig. 1.15a. The PND comprises the parallel connection of N superconducting nanowires in series with resistors (N-PND). At the steady state, each nanowire is biased at I_B/N. An HSN event in one of the sections (which is referred to as the *firing* section) results in the diversion of the current of that section into the load (I_{out}) and into the other still superconducting sections (which are referred to as *non-firing* sections). The current diverted to the non-firing section is referred to as leakage current δI. If δI is negligible, most of the current of the firing section is diverted to the load. Additional HSN events in other non-firing sections result in an increase of I_{out} proportional to the number of HSN events, as shown in Fig. 1.15b. When the firing sections relax, the series resistors ensure the current is equally redistributed amongst all the sections of the device.

The leakage current is significantly affects the correct operation of PNDs. For small N, δI limits the maximum bias current of the N-PND, resulting in lower detection efficiency when non-saturated nanowires are used. For large N, δI severely limits the signal to noise ratio of the detector, since most of the current through the firing section is diverted to the non firing sections, instead of the load. δI can be reduced by increasing the impedance of the sections, e.g. by increasing the resistance in series with each section. PND design is discussed in detail in Ref. [62].

A recent approach, the series-nanowire detector (SND) [7, 63, 64], aims at addressing the issue of the leakage current affecting PNDs. Figure 1.15c shows the equivalent circuit of an N-section SND (N-SND). Each section comprises a super-

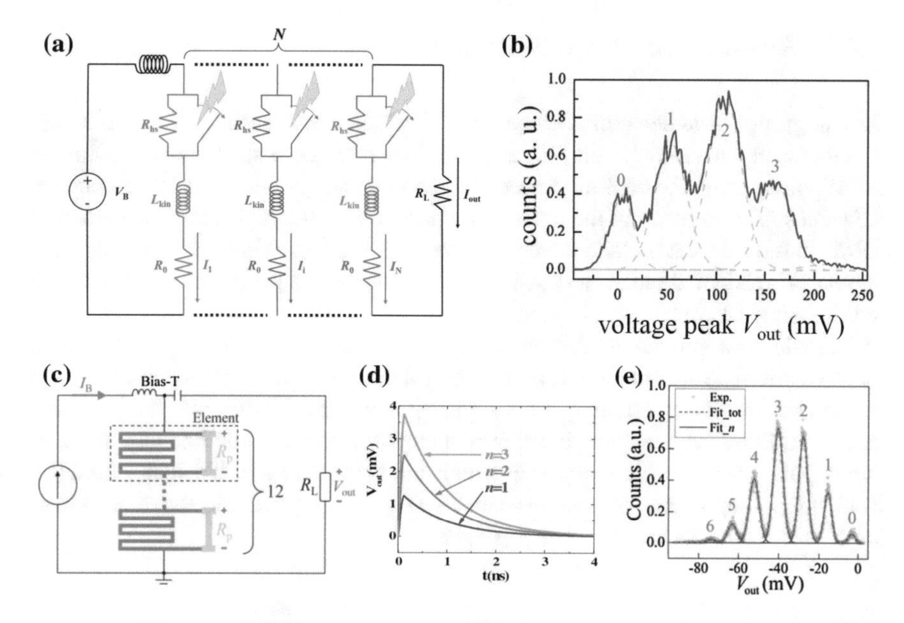

Fig. 1.15 **a** Circuit diagram of an N-PND, comprising N superconducting nanowires connected in parallel. Each nanowire is connected to a series resistor R_0 to minimize leakage current. **b** Detector output voltage histogram of a 5-PND. The influence of dark counts was minimized by using a pulsed laser source and gated detector readout [6]. The peaks correspond to 0, 1, 2 and 3 detected photons. **c** Sketch of the structure of an N-SND, comprising N superconducting nanowires connected in series. Each nanowire is connected to a parallel resistor R_p. **d** Calculated output voltage of an SND for 1, 2, and 3 detected photons. **e** Detector output voltage histogram of an SND. The peaks correspond to 0, 1, 2, 3, 4, 5, and 6 detected photons. Reproduced from Ref. [6] in accordance with Nature Publishing Group permissions, and from Refs. [63, 64] with permission of the Optical Society of America (OSA)

conducting nanowire in parallel with a resistor R_p. In the steady state, all of the nanowires are biased at the same bias current I_B. When an HSN event occurs in one of the sections, the current is diverted out of the nanowire and into the corresponding R_p, resulting in a voltage $V_1 \sim I_B \cdot R_p$ across the SND. Each independent HSN event increases the voltage across the SND by V_1, and thus if all HSNs are photodetection events, the total voltage across the SND is proportional to the number of detected photons, as shown in Fig. 1.15d, e.

The SND holds the potential for large dynamic range [7, 63] and high efficiency, since it is not affected by the leakage current. It should be noted that the proof-of-concept PND and SND data should be seen as pseudo photon number resolving detectors. True single-shot photon-number resolution (as opposed to the statistical data shown in Fig. 1.15b, e) requires a detector with intrinsic energy resolution, such as a transition-edge sensor [40] or microwave kinetic inductance detector [65].

1.4.3 Row-Column SNSPD Arrays

Several groups have developed arrays of SNSPDs in which each SNSPD has a dedicated bias and readout circuit [16, 66, 67]. However, these approaches are limited in scalability since the cooling power of the cryogenic system limits the number of RF cables used to read out the detectors. Furthermore, the high timing accuracy of SNSPDs requires each readout line to have >1 GHz bandwidth, which makes the number of readout channels a cost-intensive resource and precludes any frequency multiplexing scheme.

A scalable approach to this problem is the current-splitting concept, which includes the resistive *row-column* readout scheme, in which nanowires are used as switches to divert current in a two-dimensional array [9, 10], and the inductive current splitting scheme, where different nanowires produce qualitatively different output pulse shapes, which can be discriminated to identify which detector fired [8]. Both of these topologies are similar to the PND approach presented above. Schematics of these concepts are shown in Fig. 1.16a, b.

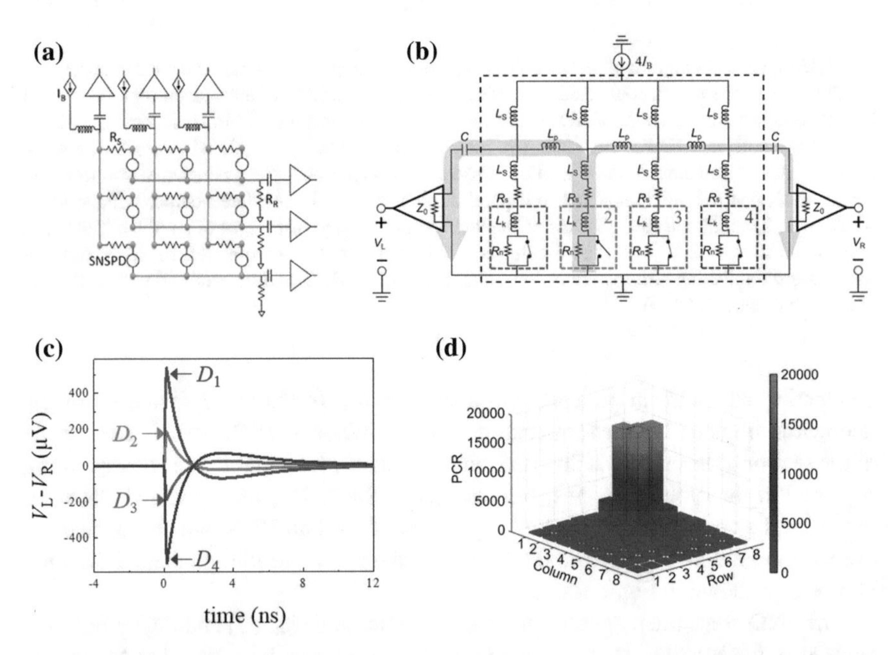

Fig. 1.16 **a** Circuit diagram illustrating the scalability of the detection scheme with a 3 × 3 array of SNSPDs. **b** Circuit diagram of a current-splitting array based on inductive splitting. This scheme was used to read out four SNSPDs $D_{1,2,3,4}$. The *red arrows* represent the current redistribution path after an HSN event in D_2. **c** Calculated time-dependent differential output voltage after an HSN at t = 0 in D_2 (*blue*), D_2 (*red*), D_3 (*green*) and D_4 (*purple*). The electrothermal simulation was performed using the following values: $R_S = 200$, $Z_0 = 50$, $L_S + L_K = 400$ nH, and $L_p = 50$ nH. **d** Normalized count rate for an 8 × 8 array read out using the resistive scheme shown in (**a**). The graph shows the photoresponse of the array to a centered laser spot. Reproduced from Refs. [8–10] with permission of the American Institute of Physics (AIP)

We will discuss the resistive current-splitting approach in further detail because it offers better prospects of scalability. The row-column readout scheme requires only 2 N high-speed readout channels for an N $\times$ N two dimensional array of SNSPDs, greatly reducing the cryogenic heat load requirements of the system, while maintaining the high timing accuracy of the detectors. The position of the detection event is not encoded in the pulse amplitude like in the inductive current splitting scheme, but rather in a combination of *row* and *column* pulses.

Figure 1.16a shows a circuit diagram for a 3 $\times$ 3 array. The SNSPDs are biased with a current I_B through a bias tee in each column. Each pixel consists of the SNSPD in series with a resistor R_S. In simplified terms, each SNSPD can be represented as an electrical switch in parallel with a resistor R_N (see Fig. 1.4). When the SNSPD is in the superconducting state, the electrical switch is closed, shorting the resistor R_N. In this steady-state situation, the circles in Fig. 1.16a are equivalent to short circuits, and I_B is equally distributed between all three pixels in the column. When an HSN event occurs in one of the SNSPDs, the electrical switch opens, and the current is diverted into R_N [28]. In this case, the circles in Fig. 1.16a can be replaced with resistors of magnitude $R_N \gg R_S = R_R$ in series with the total pixel inductance. Typically the value of R_N is on the order of 1–10 kΩ for SNSPDs fabricated from amorphous WSi. The high pixel resistance causes the current through the firing pixel to be diverted to the non firing pixels in the same column (this leakage current is analogous to that of PNDs), and to the column amplifier of the firing pixel (on top of the picture), creating a positive voltage pulse. The current through the row resistor R_R is simultaneously reduced, resulting in a negative voltage pulse in the row amplifier of the firing pixel. When the resistive SNSPD pixel relaxes to the superconducting state, the series resistance R_S ensures that the current is equally redistributed amongst all the pixels in the column, like in PNDs. By observing correlations between pulses in the row and column readout lines, it is possible to determine which detector fired, as long as multiple pixels are not firing simultaneously. As such, this technique is best suited to applications where the photon flux rate is low compared to the inverse recovery time of the detectors, or for applications where single photons arrive on a clock.

The pulse height, signal-to-noise ratio, and hence the jitter of the row pulse scales in proportion to R_R. Thus it is desirable to make R_R much larger than the input impedance of the row amplifier (typically 50 Ω), without causing the SNSPDs to latch into the normal state due to speedup of the current recovery into the pixel [46, 49]. Typical values of R_R for fabricated arrays range between 25 and 50. The pixel resistance R_S affects the leakage current like the section resistance of PNDs [62].

The first demonstration of the row-column readout was a four-pixel (2 $\times$ 2) array [10]. 8 $\times$ 8 SNSPD arrays have recently been fabricated using WSi, with >96 % pixel yield [9]. The latest generation consists of 30 $\times$ 30 μm pixels, with a pixel spacing of 60 μm. The 8 $\times$ 8 arrays are free-space-coupled and read out with only 16 coaxial cables. Figure 1.16d shows the pixel count rate with a laser spot centered on the active area of the array. Since all 64 pixels are simultaneously read out, real time imaging can be performed as the spot is moved across the array. Given the excellent yield of current 64-pixel arrays, 256-pixel and possibly even kilopixel-class arrays may be within reach in the near future.

1.5 Conclusions

In this chapter we have introduced the most widely used type of superconducting single photon detector, the superconducting nanowire single photon detector (SNSPD). Since its invention in 2001 [3] this device has undergone rapid development and has rapidly become the detector of choice for infrared quantum optics and many other photon-counting applications [2]. We have defined key performance metrics for benchmarking SNSPDs against off-the-shelf photon counting technologies. We have explored the mechanism of photodetection with a view to maximizing detection efficiency and minimizing timing jitter. We have discussed potential limits on device yield, spectral sensitivity and count rates. We highlight key recent developments which address these challenges. We introduce several promising multi-nanowire detector architectures which offer enhanced speed, photon-number resolution and imaging capability. SNSPDs are discussed further in this volume in the context of integration with photonic circuits (Chaps. 3 and 4) and as an enabling technology for quantum communication networks (Chap. 5).

Acknowledgments Part of the research was carried out at the Jet Propulsion Laboratory, California Institute of Technology, under a contract with the National Aeronautics and Space Administration.

References

1. M.D. Eisaman, J. Fan, A. Migdall, S.V. Polyakov, Invited Review Article: Single-photon sources and detectors. Rev. Sci. Instrum. **82**, 071101 (2011)
2. C.M. Natarajan, M.G. Tanner, R.H. Hadfield, Superconducting nanowire single-photon detectors: physics and applications. Supercond. Sci. Technol. **25**, 063001 (2012)
3. G.N. Gol'tsman, O. Okunev, G. Chulkova, A. Lipatov, A. Semenov, K. Smirnov, B. Voronov, A. Dzardanov, C. Williams, R. Sobolewski, Picosecond superconducting single-photon optical detector. Appl. Phys. Lett. **79**, 705 (2001)
4. M. Ejrnaes, R. Cristiano, O. Quaranta, S. Pagano, A. Gaggero, F. Mattioli, R. Leoni, B. Voronov, G. Gol'tsman, A cascade switching superconducting single photon detector. Appl. Phys. Lett. **91**, 262509 (2007)
5. F. Marsili, F. Najafi, E. Dauler, F. Bellei, X. Hu, M. Csete, R.J. Molnar, K.K. Berggren, Single-photon detectors based on ultra-narrow superconducting nanowires. Nano Lett. **11**, 2048 (2011)
6. A. Divochiy, F. Marsili, D. Bitauld, A. Gaggero, R. Leoni, F. Mattioli, A. Korneev, V. Seleznev, N. Kaurova, O. Minaeva, G. Gol'tsman, K. Lagoudakis, M. Benkhaoul, F. Lévy, A. Fiore, Superconducting nanowire photon-number-resolving detector at telecommunication wavelengths. Nat. Photonics **2**, 302 (2008)
7. S. Jahanmirinejad, G. Frucci, F. Mattioli, D. Sahin, A. Gaggero, R. Leoni, A. Fiore, Photon-number resolving detector based on a series array of superconducting nanowires. Appl. Phys. Lett. **101**, 072602 (2012)

8. Q. Zhao, A. McCaughan, F. Bellei, F. Najafi, D. De Fazio, A. Dane, Y. Ivry, K.K. Berggren, Superconducting-nanowire single-photon-detector linear array. Appl. Phys. Lett. **103**, 142602 (2013)
9. M.S. Allman, V.B. Verma, M. Stevens, T. Gerrits, R.D. Horansky, A.E. Lita, F. Marsili, A. Beyer, M.D. Shaw, D. Kumor, R. Mirin, S.W. Nam, A near-infrared 64-pixel superconducting nanowire single photon detector array with integrated multiplexed readout. Appl. Phys. Lett. **106**, 192601 (2015)
10. V.B. Verma, R. Horansky, F. Marsili, J.A. Stern, M.D. Shaw, A.E. Lita, R.P. Mirin, S.W. Nam, A four-pixel single-photon pulse-position array fabricated from WSi superconducting nanowire single-photon detectors. Appl. Phys. Lett. **104**, 051115 (2014)
11. J.A. Stern, W.H. Farr, Fabrication and characterization of superconducting NbN nanowire single photon detectors. IEEE Trans. Appl. Supercond. **17**, 306 (2007)
12. Y. Korneeva, I. Florya, A. Semenov, A. Korneev, G. Goltsman, New generation of nanowire NbN superconducting single-photon detector for mid-infrared. IEEE Trans. Appl. Supercond. **21**, 323 (2011)
13. F. Marsili, F. Bellei, F. Najafi, A. Dane, E.A. Dauler, R.J. Molnar, K.K. Berggren, Efficient single photon detection from 500 nanometer to 5 micron wavelength. Nano Lett. **12**, 4799 (2012)
14. A.J. Kerman, D. Rosenberg, R.J. Molnar, E.A. Dauler, Readout of superconducting nanowire single-photon detectors at high count rates. J. Appl. Phys. **113**, 144511 (2013)
15. D.M. Boroson, B.S. Robinson, D.V. Murphy, D.A. Burianek, F. Khatri, J.M. Kovalik, Z. Sodnik, D.M. Cornwell, Overview and results of the Lunar Laser Communication Demonstration. Proc. SPIE **8971**, 89710S (2014)
16. A. Biswas, J.M. Kovalik, M.W. Wright, W.T. Roberts, M.K. Cheng, K.J. Quirk, M. Srinivasan, M.D. Shaw, K.M. Birnbaum, LLCD operations using the Optical Communications Telescope Laboratory (OCTL). Proc. SPIE **8971**, 89710X (2014)
17. E. Saglamyurek, J. Jin, V.B. Verma, M.D. Shaw, F. Marsili, S.W. Nam, D. Oblak, W. Tittel, Quantum storage of entangled telecom-wavelength photons in an erbium-doped optical fibre. Nat. Photonics **9**, 83 (2015)
18. D.R. Hamel, L.K. Shalm, H. Hübel, A.J. Miller, F. Marsili, V.B. Verma, R.P. Mirin, S.W. Nam, K.J. Resch, T. Jennewein, Direct generation of three-photon polarization entanglement. Nat. Photonics **8**, 801 (2014)
19. F. Bussières, C. Clausen, A. Tiranov, B. Korzh, V.B. Verma, S.W. Nam, F. Marsili, A. Ferrier, P. Goldner, H. Herrmann, C. Silberhorn, W. Sohler, M. Afzelius, N. Gisin, Quantum teleportation from a telecom-wavelength photon to a solid-state quantum memory. Nat. Photonics **8**, 775 (2014)
20. H. Takesue, S.W. Nam, Q. Zhang, R.H. Hadfield, T. Honjo, K. Tamaki, Y. Yamamoto, Quantum key distribution over a 40-dB channel loss using superconducting single-photon detectors. Nat. Photonics **1**, 343 (2007)
21. A. McCarthy, N.J. Krichel, N.R. Gemmell, X. Ren, M.G. Tanner, S.N. Dorenbos, V. Zwiller, R.H. Hadfield, G.S. Buller, Kilometer-range, high resolution depth imaging via 1560 nm wavelength single-photon detection. Opt. Express **21**, 8904 (2013)
22. N.R. Gemmell, A. McCarthy, B. Liu, M.G. Tanner, S.D. Dorenbos, V. Zwiller, M.S. Patterson, G.S. Buller, B.C. Wilson, R.H. Hadfield, Singlet oxygen luminescence detection with a fiber-coupled superconducting nanowire single-photon detector. Opt. Express **21**, 5005 (2013)
23. T. Yamashita, D. Liu, S. Miki, J. Yamamoto, T. Haraguchi, M. Kinjo, Y. Hiraoka, Z. Wang, H. Terai, Fluorescence correlation spectroscopy with visible-wavelength superconducting nanowire single-photon detector. Optics Express **22**, 28783 (2014)
24. F. Marsili, V.B. Verma, J.A. Stern, S. Harrington, A.E. Lita, T. Gerrits, I. Vayshenker, B. Baek, M.D. Shaw, R.P. Mirin, S.W. Nam, Detecting single infrared photons with 93% system efficiency. Nat. Photonics **7**, 210 (2013)
25. J.D. Cohen, S.M. Meenehan, G.S. MacCabe, S. Groblacher, A.H. Safavi-Naeini, F. Marsili, M.D. Shaw, O. Painter, Phonon counting and intensity interferometry of a nanomechanical resonator. Nature **520**, 522 (2015)

26. W.H.P. Pernice, C. Schuck, O. Minaeva, M. Li, G.N. Goltsman, A.V. Sergienko, H.X. Tang, High-speed and high-efficiency travelling wave single-photon detectors embedded in nanophotonic circuits. Nat. Commun. **3**, 1325 (2012)
27. F. Marsili, D. Bitauld, A. Fiore, A. Gaggero, F. Mattioli, R. Leoni, M. Benkahoul, F. Lévy, High efficiency NbN nanowire superconducting single photon detectors fabricated on MgO substrates from a low temperature process. Opt. Express **16**, 3191 (2008)
28. A.J. Kerman, E.A. Dauler, W.E. Keicher, J.K.W. Yang, K.K. Berggren, G. Gol'tsman, B. Voronov, Kinetic-inductance-limited reset time of superconducting nanowire photon counters. Appl. Phys. Lett. **88**, 111116 (2006)
29. S. Miki, M. Fujiwara, M. Sasaki, B. Baek, A.J. Miller, R.H. Hadfield, S.W. Nam, Z. Wang, Large sensitive-area NbN nanowire superconducting single-photon detectors fabricated on single-crystal MgO substrates. Appl. Phys. Lett. **92**, 061116 (2008)
30. M.G. Tanner, C.M. Natarajan, V.K. Pottapenjara, J.A. O'Connor, R.J. Warburton, R.H. Hadfield, B. Baek, S. Nam, S.N. Dorenbos, E.B. Urena, T. Zijlstra, T.M. Klapwijk, V. Zwiller, Enhanced telecom wavelength single-photon detection with NbTiN superconducting nanowires on oxidized silicon. Appl. Phys. Lett. **96**, 221109 (2010)
31. S. Miki, T. Yamashita, H. Terai, Z. Wang, High performance fiber-coupled NbTiN superconducting nanowire single photon detectors with Gifford-McMahon cryocooler. Opt. Express **21**, 10208 (2013)
32. B. Baek, A.E. Lita, V. Verma, S.W. Nam, Superconducting a-W_xSi_{1-x} nanowire single-photon detector with saturated internal quantum efficiency from visible to 1850 nm. Appl. Phys. Lett. **98**, 251105 (2011)
33. A. Engel, A. Aeschbacher, K. Inderbitzin, A. Schilling, K. Il'in, M. Hofherr, M. Siegel, A. Semenov, H.W. Hubers, Tantalum nitride superconducting single-photon detectors with low cut-off energy. Appl. Phys. Lett. **100**, 062601 (2012)
34. Y.P. Korneeva, M.Y. Mikhailov, Y.P. Pershin, N.N. Manova, A.V. Divochiy, Y.B. Vakhtomin, A.A. Korneev, K.V. Smirnov, A.G. Sivakov, A.Y. Devizenko, G.N. Goltsman, Superconducting single-photon detector made of MoSi film. Supercond. Sci. Technol. **27**, 095012 (2014)
35. V.B. Verma, B. Korzh, F. Bussières, R.D. Horansky, S.D. Dyer, A.E. Lita, I. Vayshenker, F. Marsili, M.D. Shaw, H. Zbinden, R.P. Mirin, S.W. Nam, High-efficiency superconducting nanowire single-photon detectors fabricated from MoSi thin-films (2015). arXiv:1504.02793
36. V.B. Verma, A.E. Lita, M.R. Vissers, F. Marsili, D.P. Pappas, R.P. Mirin, S.W. Nam, Superconducting nanowire single photon detectors fabricated from an amorphous $Mo_{0.75}Ge_{0.25}$ thin film. Appl. Phys. Lett. **105**, 022602 (2014)
37. S.N. Dorenbos, P. Forn-Diaz, T. Fuse, A.H. Verbruggen, T. Zijlstra, T.M. Klapwijk, V. Zwiller, Low gap superconducting single photon detectors for infrared sensitivity. Appl. Phys. Lett. **98**, 251102 (2011)
38. H. Shibata, Fabrication of a MgB_2 nanowire single-photon detector using Br_2 -N_2 dry etching. Appl. Phys. Ex. **7**, 103101 (2014)
39. H. Shibata, T. Akazaki, Y. Tokura, Fabrication of MgB_2 nanowire single-photon detector with meander structure. Appl. Phys. Express **6**, 023101 (2013)
40. A.E. Lita, A.J. Miller, S.W. Nam, Counting near-infrared single-photons with 95% efficiency. Opt. Express **16**, 3032 (2008)
41. A.D. Semenov, G.N. Gol'tsman, A.A. Korneev, Quantum detection by current carrying superconducting film. Physica C **351**, 349 (2001)
42. A. Semenov, A. Engel, H.W. Hübers, K. Il'in, M. Siegel, Spectral cut-off in the efficiency of the resistive state formation caused by absorption of a single-photon in current-carrying superconducting nano-strips. Eur. Phys. J. B **47**, 495 (2005)
43. A.D. Semenov, P. Haas, H.W. Hübers, K. Ilin, M. Siegel, A. Kirste, T. Schurig, A. Engel, Vortex-based single-photon response in nanostructured superconducting detectors. Physica C **468**, 627 (2008)
44. L.N. Bulaevskii, M.J. Graf, V.G. Kogan, Vortex-assisted photon counts and their magnetic field dependence in single-photon superconducting detectors. Phys. Rev. B **85**, 014505 (2012)

45. J.J. Renema, R. Gaudio, Q. Wang, Z. Zhou, A. Gaggero, F. Mattioli, R. Leoni, D. Sahin, M.J.A. de Dood, A. Fiore, M.P. van Exter, Experimental Test of Theories of the Detection Mechanism in a Nanowire Superconducting Single Photon Detector. Phys. Rev. Lett. **112**, 117604 (2014)
46. J.K.W. Yang, A.J. Kerman, E.A. Dauler, V. Anant, K.M. Rosfjord, K.K. Berggren, Modeling the electrical and thermal response of superconducting nanowire single-photon detectors. IEEE Trans. Appl. Supercond. **17**, 581 (2007)
47. E.A. Dauler, M.E. Grein, A.J. Kerman, F. Marsili, S. Miki, S.W. Nam, M.D. Shaw, H. Terai, V.B. Verma, T. Yamashita, Review of superconducting nanowire single-photon detector system design options and demonstrated performance. Opt. Eng. **53**, 081907 (2014)
48. A.J. Kerman, E.A. Dauler, J.K.W. Yang, K.M. Rosfjord, V. Anant, K.K. Berggren, G.N. Gol'tsman, B.M. Voronov, Constriction-limited detection efficiency of superconducting nanowire single-photon detectors. Appl. Phys. Lett. **90**, 101110 (2007)
49. A.J. Kerman, J.K.W. Yang, R.J. Molnar, E.A. Dauler, K.K. Berggren, Electrothermal feedback in superconducting nanowire single-photon detectors. Phys. Rev. B **79**, 100509 (2009)
50. F. Marsili, F. Najafi, C. Herder, K.K. Berggren, Electrothermal simulation of superconducting nanowire avalanche photodetectors. Appl. Phys. Lett. **98**, 093507 (2011)
51. V.B. Verma, F. Marsili, S. Harrington, A.E. Lita, R.P. Mirin, S.W. Nam, A three-dimensional, polarization-insensitive superconducting nanowire avalanche photodetector. Appl. Phys. Lett. **101**, 251114 (2012)
52. F. Marsili, V. Verma, M.J. Stevens, J.A. Stern, M.D. Shaw, A. Miller, D. Schwarzer, A. Wodtke, R.P. Mirin, S.W. Nam, Mid-Infrared Single-Photon Detection with Tungsten Silicide Superconducting Nanowires, *CLEO: 2013* (Optical Society of America, California, 2013), p. CTu1H.1
53. F. Najafi, A. Dane, F. Bellei, Q. Zhao, K.A. Sunter, A.N. McCaughan, K.K. Berggren, Fabrication Process Yielding Saturated Nanowire Single-Photon Detectors With 24-ps Jitter. IEEE Journal of Selected Topics in Quantum Electronics **21**, 3800507 (2015)
54. F. Najafi, F. Marsili, E. Dauler, A.J. Kerman, R.J. Molnar, K.K. Berggren, Timing performance of 30-nm-wide superconducting nanowire avalanche photodetectors. Appl. Phys. Lett. **100**, 152602 (2012)
55. F. Marsili, F. Najafi, E. Dauler, R.J. Molnar, K.K. Berggren, Afterpulsing and instability in superconducting nanowire avalanche photodetectors. Appl. Phys. Lett. **100**, 112601 (2012)
56. V. Anant, A.J. Kerman, E.A. Dauler, J.K.W. Yang, K.M. Rosfjord, K.K. Berggren, Optical properties of superconducting nanowire single-photon detectors. Opt. Express **16**, 10750 (2008)
57. F. Najafi, J. Mower, N.C. Harris, F. Bellei, A. Dane, C. Lee, X. Hu, P. Kharel, F. Marsili, S. Assefa, K.K. Berggren, D. Englund, On-chip detection of non-classical light by scalable integration of single-photon detectors. Nat. Commun. **6**, 5873 (2015)
58. Q. Zhao, A.N. McCaughan, A.E. Dane, F. Najafi, F. Bellei, D. De Fazio, K.A. Sunter, Y. Ivry, K.K. Berggren, Eight-fold signal amplification of a superconducting nanowire single-photon detector using a multiple-avalanche architecture. Opt. Express **22**, 24574 (2014)
59. E.A. Dauler, A.J. Kerman, B.S. Robinson, J.K.W. Yang, B. Voronov, G. Goltsman, S.A. Hamilton, K.K. Berggren, Photon-number-resolution with sub-30-ps timing using multi-element superconducting nanowire single photon detectors. J. Mod. Opt. **56**, 364 (2009)
60. M.J. Stevens, B. Baek, E.A. Dauler, A.J. Kerman, R.J. Molnar, S.A. Hamilton, K.K. Berggren, R.P. Mirin, S.W. Nam, High-order temporal coherences ofchaotic and laser light. Opt. Express **18**, 1430 (2010)
61. F. Marsili, D. Bitauld, A. Fiore, A. Gaggero, R. Leoni, F. Mattioli, A. Divochiy, A. Korneev, V. Seleznev, N. Kaurova, O. Minaeva, G. Gol'tsman, Superconducting parallel nanowire detector with photon number resolving functionality. J. Mod. Opt. **56**, 334 (2009)
62. F. Marsili, D. Bitauld, A. Gaggero, S. Jahanmirinejad, R. Leoni, F. Mattioli, A. Fiore, Physics and application of photon number resolving detectors based on superconducting parallel nanowires. New J. Phys. **11**, 045022 (2009)
63. Z. Zhou, S. Jahanmirinejad, F. Mattioli, D. Sahin, G. Frucci, A. Gaggero, R. Leoni, A. Fiore, Superconducting series nanowire detector counting up to twelve photons. Opt. Express **22**, 3475 (2014)

64. S. Jahanmirinejad, A. Fiore, Proposal for a superconducting photon number resolving detector with large dynamic range. Optics Express **20**, 5017 (2012)
65. P.K. Day, H.G. LeDuc, B.A. Mazin, A. Vayonakis, J. Zmuidzinas, A broadband superconducting detector suitable for use in large arrays. Nature **425**, 817 (2003)
66. S. Miki, T. Yamashita, Z. Wang, H. Terai, A 64-pixel NbTiN superconducting nanowire single-photon detector array for spatially resolved photon detection. Opt. Express **22**, 7811 (2014)
67. D. Rosenberg, A.J. Kerman, R.J. Molnar, E.A. Dauler, High-speed and high-efficiency superconducting nanowire single photon detector array. Opt. Express **21**, 1440 (2013)

Chapter 2
Superconducting Transition Edge Sensors for Quantum Optics

Thomas Gerrits, Adriana Lita, Brice Calkins and Sae Woo Nam

Abstract High efficiency single-photon detectors allow novel measurements in quantum information processing and quantum photonic systems. The photon-number resolving transition edge sensor (TES) is known for its near-unity detection efficiency and has been used in a number of landmark quantum optics experiments. We review the operating principle of the optical superconducting TES, its use for quantum optics and quantum information processing and review its recent implementation in integrated photonic platforms.

2.1 Introduction

The superconducting transition edge sensor (TES) is an exquisitely sensitive device which exploits the abrupt change in resistance at the onset of the superconducting transition. Since its demonstration in 1998 [1] the optical TES has been improved and utilized by groups around the world and tailored for optical photon detection [2–6]. Owing to its intrinsic energy resolution capability, it is perfectly suited for the detection of optical pulses stemming from quantum optical processes in which the number of photons per pulse is needed to determine the state of the optical quantum system. Since interaction with the environment degrades every quantum system, detection of

T. Gerrits (✉) · A. Lita · B. Calkins · S.W. Nam
National Institute of Standards and Technology, 325 Broadway,
Boulder, CO 80305, USA
e-mail: thomas.gerrits@nist.gov

A. Lita
e-mail: adriana.lita@nist.gov

B. Calkins
e-mail: brice.calkins@nist.gov

S.W. Nam
e-mail: nams@boulder.nist.gov

© Springer International Publishing Switzerland 2016

R.H. Hadfield and G. Johansson (eds.), *Superconducting Devices in Quantum Optics*, Quantum Science and Technology,
DOI 10.1007/978-3-319-24091-6_2

optical quantum states must be done very efficiently, i.e. with as little loss as possible added by the detector. Thus the optical TES was continuously improved to achieve the highest detection efficiency possible, with efficiencies nowadays routinely in excess of ninety percent for fiber-coupled devices [7].

The detection of visible to near-infrared light is an important enabling technology for many emerging applications [8]. High optical intensities can easily be detected with room temperature devices, such as silicon photodiodes. These detectors can have high detection efficiencies close to 100 % and can detect light levels from tens of milliwatts down to picowatts (millions of photons per second). However, these devices are not capable of registering single photons. Photo-multiplier tubes (PMTs) had been and continue to be the workhorse for single-photon optical detection. However, their efficiencies are relatively low in the visible [8], and even lower in the telecom band at ~1550 nm, where much of the current quantum information and communication technology is being developed. InGaAs avalanche photodiodes (InGaAs APDs) offer higher efficiencies (~20 %) at telecom wavelengths. However, their dark count rates may be in the 100s of kilohertz. To mitigate this effect, their bias voltage is typically gated and synchronized with the expected optical input signal. This method of course precludes continuous quantum-state production techniques, and even then only offers incremental improvements in the effective detection efficiency and dark count rate. Currently the only available single-photon detectors offering high (near unity) detection efficiency paired with low dark count rates at telecom wavelengths are low-temperature superconducting devices. One of these is the superconducting nanowire single-photon detector (SNSPD). Najafi et al. give an introduction to SNSPD technology in Chap. 1 of this book; Chaps. 3, 4 and 5 expand on implementations and applications of SNSPD devices in quantum optics, information processing and communications.

This chapter gives an overview of the development and applications of superconducting transition edge sensors tailored for optical wavelengths. This work was principally carried out by the authors at the US National Institute of Standards and Technology (NIST) in Boulder, Colorado. We discuss the TES, its applications and future perspectives for single-photon and photon-number-resolved detection. We first review the operating principle of the TES, discuss techniques to improve the performance of these devices from the first demonstration, and describe techniques to characterize the performance of TES detector systems. Next we review important quantum optics experiments that benefited from the use of the TES. We also review our recent efforts on integration of TESs into optical waveguide circuits that will be used in future applications of TES detectors.

2.2 The Optical Transition Edge Sensor

2.2.1 TES Operation

2.2.1.1 Basic Operating Principle

Superconducting transition-edge sensors (TES) are highly-sensitive microcalorimeters that are used as microbolometers to detect radiation from sub-mm wavelengths to gamma-rays. They typically consist of an absorber, a sensitive thermometer, and a weak thermal link to a thermal bath [9]. The optical TES is a superconducting sensor measuring the amount of heat absorbed from an optical photon with energy on the order of 1 eV. When an optical photon is absorbed by the sensor, the associated photon energy is transformed into a measurable temperature change of the sensor. The most successful optical TESs operate at temperatures below 1 K [7, 10, 11]. However, higher operating temperatures are possible by reducing detector size (to maintain equal heat capacity of the electron system), even though light may miss the detector, reducing the system detection efficiency [12]. For devices operating below 1 K, the thermal isolation required is usually provided through the Coupling, electron-phonon within the superconductor itself, which can be weak at low temperatures, especially with certain superconducting metals. Tungsten is an example of a material which achieves the necessary weak coupling below its thin-film superconducting critical temperature T_c, typically $\sim$100 mK.

Due to the low photon energies involved, detection of single photons at optical and near-infrared wavelengths requires very low heat capacity and extremely sensitive thermometry. For optical TES detectors based on tungsten [7], the absorber is the electron system in the metal; the thermometer is the superconducting-resistive transition in the metal; the weak thermal link is the weak electron-phonon coupling in the metal itself; the thermal bath is the tungsten phonon system, strongly coupled to the lattice of the silicon substrate and, in turn, to the cryostat cold-plate onto which the detector chip is mounted [13].

2.2.1.2 Thermal Response

In our tungsten-TES detectors, after an optical photon is absorbed by an electron, the hot electron ($\sim$11, 000 K) scatters with neighboring electrons at a length scale equal to the mean free path. The thermal link between the electron and phonon system in tungsten is small compared to other metals, enabling strong non-equilibrium effects, in particular allowing for the electron system to remain hot for several hundreds of nanoseconds [14]. By contrast, a strong thermal link between the tungsten phonon system and the surrounding thermal bath allows immediate dissipation of the thermal energy into the thermal bath, a process much faster than the electron-phonon coupling, as shown in Fig. 2.1a. Thus the heat dissipation process can generally be

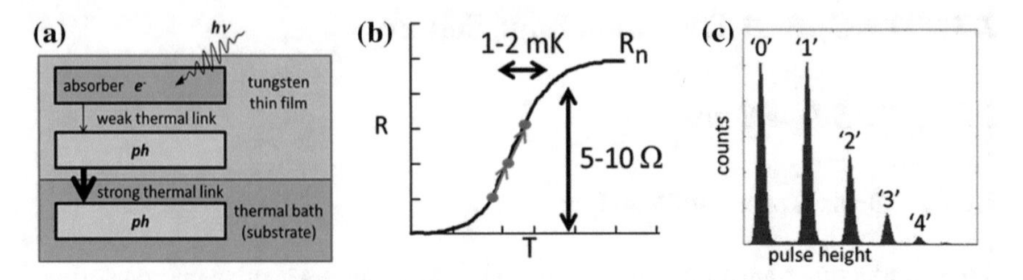

Fig. 2.1 The fundamentals of TES operation: **a** Principle of the TES as a microcalorimeter; **b** Resistance versus temperature "Transition edge" of a typical TES. Energy resolution is achieved by careful design of the system parameters; **c** Typical TES photon number histogram of amplitudes of the output response when illuminating the sensor with a weak coherent state

understood by a simple two-body model treating the phonon system of the tungsten and the thermal bath as one body:

$$C_e(T_e)\frac{dT_e}{dt} = \Sigma_{e-p}V\left(T_e^5 - T_{bath}^5\right) + P_J + \eta_\gamma \delta(t) P_\gamma, \tag{2.1}$$

where $C_e(T_e)$ is the temperature-dependent heat capacity of the electron system ($C_e = \gamma_e \cdot V \cdot T_e$); γ_e is the specific electron heat capacity coefficient, V is the absorber volume, T_e, and T_{bath} are the electron and thermal bath temperatures, Σ_{e-p} is the electron-phonon thermal coupling parameter, P_J is the Joule heating power due to the TES voltage bias, $\delta(t)P_\gamma$ is the absorbed optical power incident at the TES at initial time $t = 0$, and η_γ is the fraction of the optical energy transferred to the electron system after the optical pulse is absorbed with remainder being transferred to phonon system, i.e. η_γ is the energy collection efficiency. For tungsten, typical values are $\Sigma_{e-p} \sim 0.4\,\text{nW}/(\mu\text{m}^3\text{K}^5)$, $\gamma_e \sim 140\,\text{aJ}/(\mu\text{m}^3\text{K}^2)$ and $\eta_\gamma \sim 0.50$. During operation at the transition edge, the electron temperature is kept near the transition temperature via voltage biasing [15], i.e. $T_e = T_c$. The voltage bias causes Joule heating that raises the electron temperature of the detector to a temperature in the narrow superconducting-to-normal transition region, as shown in Fig. 2.1b. The transition edge is about 1–2 mK wide with a resistance change of several ohms. In steady state operation, the power dissipation of the device (Joule heating of the slightly resistive device) equals the power flow from the electrons in the tungsten to the phonons in the metal: $\Sigma_{e-p}V\left(T_e^5 - T_{bath}^5\right) = P_J = V^2/R(T)$. The positive temperature coefficient of resistance ($\text{d}R/\text{d}T > 0$), combined with the voltage bias, results in a negative electrothermal feedback that keeps the detector stably biased in the narrow superconducting transition. For example, if the temperature increases (decreases), the resistance increases (decreases), resulting in a smaller (larger) amount of Joule heating. With the detector thus equilibrated at a very steep point in its superconducting-to-normal transition, the absorption of a single photon results in a relatively large instantaneous change in the electrical resistance that can be measured with a Superconducting quantum interference device (SQUID) electronics described in the following section.

Figure 2.1b shows an example of a resistance versus temperature curve for a TES. Devices fabricated at NIST have a typical volume of 12.5 μm^3 (25 μm × 25 μm × 0.02 μm) and a transition temperature of ≈100 mK. A photon with an energy of 0.8 eV heats the electron system by ≈0.05 mK, resulting in a measurable resistance change. Due to the shape of the transition edge, an optical wave packet with some few numbers of photons will result in an increase in the resistance change of the device until the device approaches its normal resistance value. The response of the device to many such wave packets is shown in Fig. 2.3, where clear separation between different waveforms, corresponding to different numbers of absorbed photons, can be seen. A typical histogram of the different response pulse heights is shown in Fig. 2.1c. When operating the TES in the linear regime, the energy resolution determines the photon-number resolving capability of the detector (a monochromatic source is assumed). The energy resolution of the transition edge sensor is limited by the noise associated with the Johnson noise of the intrinsic resistance of the device, the noise associated with the statistical process of the electron-phonon thermal link [16] and the readout noise of the electronics. The TESs fabricated at NIST have a typical energy resolution of less than 0.25 eV in the linear operating regime [7].

2.2.1.3 SQUID Readout

The optical transition edge sensor readout is typically accomplished by use of superconducting quantum interference devices (SQUIDs) [17], which serve as low noise amplifiers of the current flowing through the TES. The SQUIDs are used to measure current flowing through coils that are inductively coupled to the SQUID loop (Fig. 2.2a shows the schematic of the TES readout). There are two coils. The one on the left couples signals from the TES (input coil), and the one on the right is used to optimize the output of the SQUID (feedback coil). Generally NIST optical

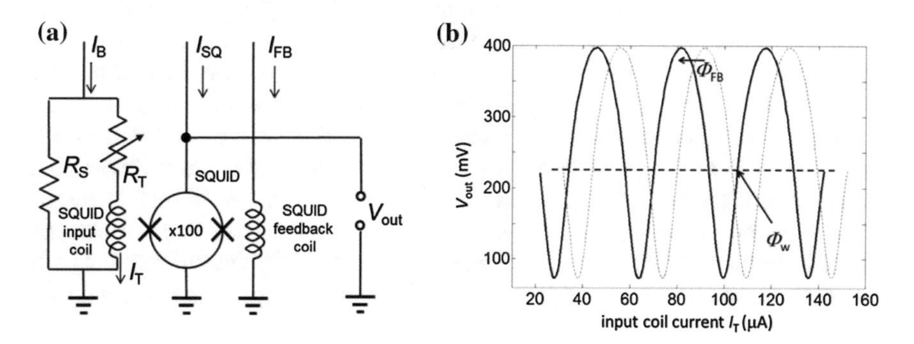

Fig. 2.2 TES readout via SQUID: **a** Schematic of the SQUID readout; **b** measured SQUID output voltage response (after x100 room temperature amplifier) as a function of input coil current (I_T) for $I_{FB} = 0$ (*dashed line*) and I_{FB} set to some value to maximize the TES response at it quiescent operating point (*solid line*)

TESs use a series array of SQUIDs, typically 100 (denoted by x100 in Fig. 2.2a), operating coherently [18] to produce signals large enough to be easily amplified at a later stage using simple room-temperature electronics. Because of the low operating temperature, and the need for a relatively stiff voltage bias, a small shunt resistor (R_s) is placed in parallel with the TES (the TES is drawn as a variable resistor (R_T) in the schematic). The value of R_s is chosen to be much smaller than the TES operating resistance when the TES is biased to detect photons ($R_s \sim 10\,\mathrm{m}\Omega$). The SQUID input coil inductively couples the TES current (I_T) via a magnetic flux (Φ_B) to the SQUID and therefore allows direct equivalent current readout.

Figure 2.2b shows a measured SQUID response as a function of feedback coil current. The period of oscillation is about 36 μA, revealing a SQUID sensitivity of about $36\,\mu\mathrm{A}/\Phi_0$, where Φ_0 is the magnetic flux quantum. The amplitude of the oscillation is determined by the geometry and material of the SQUID's Josephson junctions. In this case, a x100 room temperature amplifier along with a x100 SQUID array was used. Consequently, the magnitude of the voltage swing of one SQUID itself is ≈32 μV. The y-axis-offset originates from the resistance of the lead cable that supplies the bias current to the SQUID. For maximum signal-to-noise, the SQUID readout should operate around the maximum slope of the SQUID response, the working point. To accomplish this proper "flux bias," another coil is added to the SQUID, which inductively couples I_{FB} inducing flux Φ_{FB} to the SQUID. This added flux allows tuning of the overall flux through the SQUID, hence tuning the SQUID to the optimal working point: $\Phi_W = \Phi_B + \Phi_{FB}$.

2.2.1.4 TES Output

Figure 2.3 shows typical waveforms for a TES optimized for light at ≈800 nm wavelength. The TES was illuminated with weak coherent state pulses with a mean photon number per pulse of ≈2. The photon energy was 1.55 eV (800 nm). Detection efficiency of this particular device was found to be ≈94 % [13]. The figure shows 1024 individual waveforms of the TES response. Clearly, photon-number resolution is present, and up to 4 photons can easily be distinguished from the detector output. When no photon is present the signal consists entirely of the readout noise of the electronics, the Johnson noise of the detector itself and the fluctuations of the thermal bath [9]. The clear separation of the no-photon waveforms from those due to a photon means that the TES does not have any dark counts (false positives). There are however background counts. These background counts are real detections, which can be due to spurious visible wavelength photons coupled to the detector from the lab environment, spurious black-body photons or even high energy cosmic ray muons.

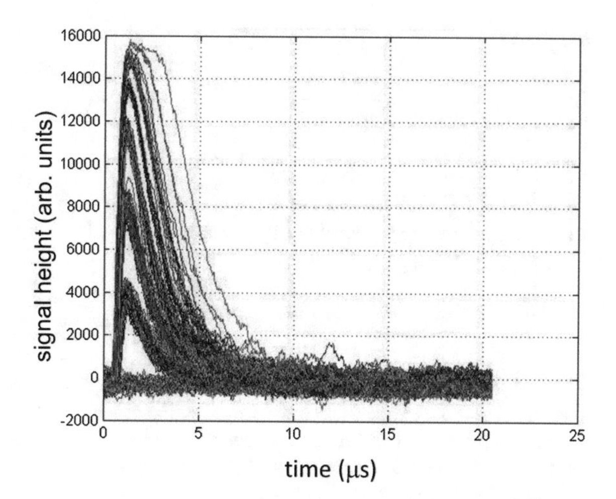

Fig. 2.3 Typical TES output $V_{out}(t)$ response waveforms under illumination with 1.55 eV (800 nm wavelength) photons

2.2.2 *TES Optimization*

2.2.2.1 Optical Stack Design

Absorption of optical photons in metals is generally dominated by the metal's interband transitions [19]. The interband transitions originate from the band structure of the metal. In principle, the optical absorption coefficient of a metal will be altered when the metal undergoes a superconducting transition due to a change of its band structure [20]. However, since this effect is small in tungsten, our studies show no significant change in the absorption coefficient when the tungsten undergoes the superconducting transition. A bare superconducting film is neither a perfect reflector nor highly absorptive at optical frequencies. For optical photons, the optical properties of the material can be well described using a complex index of refraction. At NIST, using the complex index of refraction for tungsten we have been able to embed a tungsten-TES in a dielectric stack with a quarter-wave back-short reflector, analogous to a microwave "backshort", to impedance-match the TES to free-space. The thickness of the dielectric layers can be varied to optimize the absorption at particular wavelengths. In general, it is possible to absorb >95 % of normal incident photons over a wide range of wavelengths from the UV to the mid-IR using variations of the dielectric stack layers optimized for a specific wavelength. The bandwidth of the absorption depends on the details of the stack design [7] (see Fig. 2.4a). A typical dielectric stack consists of a highly reflective mirror, a dielectric spacer, the active detector film and an anti-reflection (AR) coating.

A critical step in designing a multilayer structure that maximizes the absorption in a metal film is to determine the indices of refraction of all component layers. The optical properties of thin films differ from the bulk tabulated values, so it is necessary to measure the actual indices of refraction. The indices of refraction can be extracted from reflectance and transmittance measurements of thin-film samples deposited

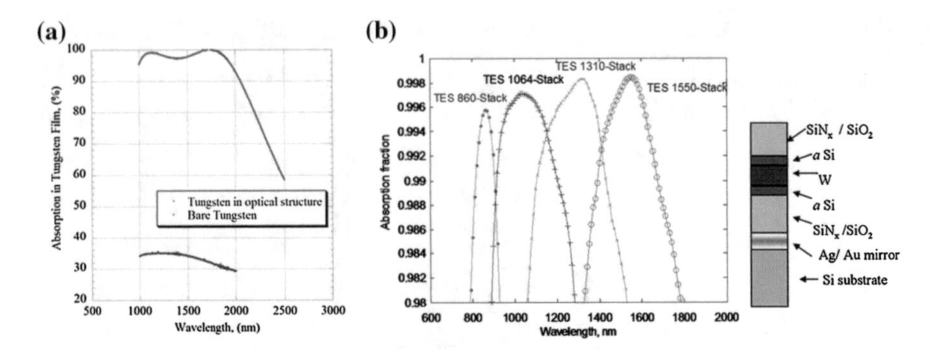

Fig. 2.4 Maximizing optical TES performance via optical cavity design: **a** Spectrophotometer data taken at room temperature indicating significant improvement in device absorption at 1550 nm when bare tungsten is embedded between appropriate dielectric layers; **b** Calculated absorption for tungsten (W) embedded in optical multilayer structure designed to optimize absorption at different wavelengths. (**a**) Reproduced with permission from Ref. [7]. Copyright 2008, Optics Express (OSA); (**b**) Reproduced with permission from Ref. [21]. Copyright 2010, Proceedings of SPIE

onto quartz or silicon wafers. Film thickness strongly influences the refractive index calculation and so must be measured with high accuracy. Using these measured indices of refraction for all layers in a thin-film modeling program, we designed structures that enhance absorption in the active device material. An optimization algorithm varies thicknesses for a fixed sequence of materials, and calculates the expected performance of the optical stack in terms of reflectance, transmittance, and absorption in the active device material. The component layer thicknesses giving maximum absorption in the active device material are used as target thicknesses when fabricating the optimized multilayer structure. The optical stacks for tungsten TESs designed at NIST are similar for all wavelength-specific designs (see Fig. 2.4b). The differences are in the layer thicknesses and the materials choice of non-absorbing dielectrics and/or top-layer AR coatings. Our simulations indicate greater than 99 % absorption in the tungsten layer at each target wavelength [21].

The bandwidth of the absorption depends on the details of the stack design. There are several important parameters to consider when designing optical stacks: thickness variation from design values to deposited layers in the structure, as well as interface intermixing or material reactivity can affect film optical constants and will shift the absorption of the multilayer stack from the design target value. Figure 2.5 shows how thickness variation from target values (output of the optimization algorithm) may affect the overall reflectance of the multilayer structure for a TES optical stack optimized for absorption at 810 nm. By employing a Monte-Carlo search algorithm randomly varying layer thicknesses within a standard deviation of 5 % from the target values, a better estimate of the actual measured absorption is obtained. The altered simulation curve represents the expected reflectance corresponding to slightly different layers thicknesses than designed, in this case 1 % to 7 % thickness variations for 4 out of 6 layers in the structure.

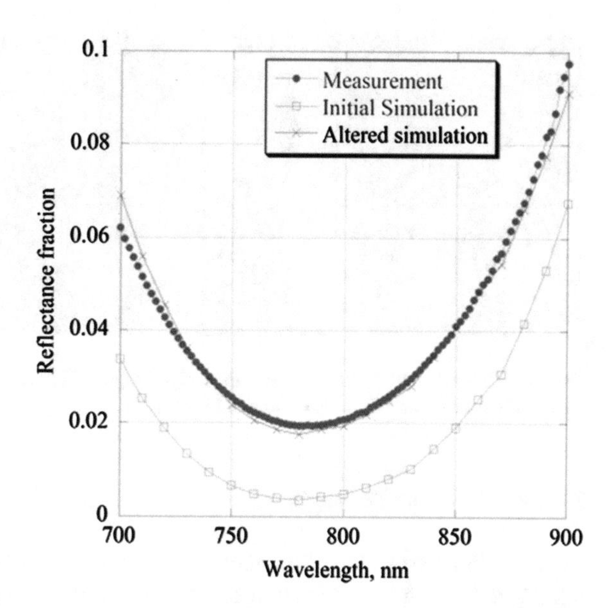

Fig. 2.5 The tungsten TES optical structure consisting of 6 layers is optimized for absorption at 810 nm wavelength. Reflectance curves for initial simulation (*open circles*), measurement (*solid circles*), and Monte Carlo altered simulation (*crosses*) are displayed. Reproduced with permission from Ref. [21]. Copyright 2010, Proceedings of SPIE

Compatibility of layers in the optical structure is also important, especially since stresses will arise from the differences in coefficients of thermal expansion when cooling down to the cryogenic operating temperatures, 100 mK to 200 mK. An example of incompatibility between layers is seen when hafnium (Hf) is used as a TES. When embedding Hf in a multilayer structure (mirror of gold, dielectric spacer Si_3N_4, Hf active device material, Si_3N_4 as antireflection coating), absorption in the active device material can be greater than 99 % at 860 nm [22]. However, stress induces a broadening in the superconducting transition of Hf in the multilayer structure. That broadening results in a significant increase in TES recovery time (20 μs for Hf in multilayer compared to ~500 ns for the bare Hf) [22]. For tungsten TESs in multilayer structures, the tungsten is deposited as part of an *in situ* trilayer: amorphous-silicon / tungsten / amorphous-silicon, which was found to enhance and stabilize the tungsten T_c against thermal-stress-induced suppression [23]. Titanium (Ti), is another material used for optical TES. By embedding Ti in an optical structure consisting of successive layers of dielectric deposited by ion beam assisted sputtering (IBS), high detection efficiencies at 850 nm (98 %) and 1550 nm (84 %) with fast response and 25 ns timing jitter were reported [10]. However, oxidation of Ti and subsequent oxygen migration (even when Ti is embedded in the optical structure) results in degradation of superconductivity and changes in refractive index. By adding an *in-situ* Au layer (10 nm) on top of Ti (30 nm), simulations have shown that absorption larger than 99 % is possible when using an antireflection coating consisting of 11 dielectric layers [24].

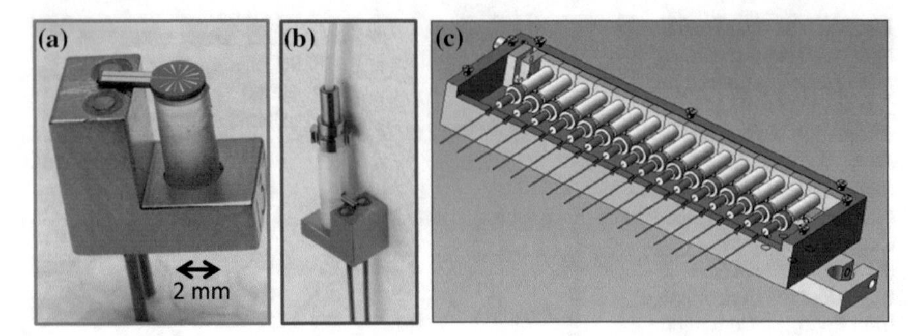

Fig. 2.6 TES packaging: **a** TES in self-aligned package; **b** TES package with optical fiber attached; **c** 16 channel detector mount box. Adapted from [13]

2.2.2.2 Detector Packaging

Reproducible and stable alignment of the optical fiber with the TES is crucial for reliably achieving high detection efficiencies. Since the optical-mode field diameter in the optical fiber is $\approx 10\,\mu\text{m}$, and the size of the TES is $25\,\mu\text{m} \times 25\,\mu\text{m}$, slight misalignment of the optical fiber with respect to the TES will result in decreased collection of optical photons onto the active area of the TES, and hence decreased overall detection efficiency. Also, a mechanically stable mount is required to allow for robust fiber-to-detector alignment that can survive multiple thermal cycling without drifting. Miller et al. [13] presented the self-alignment mount shown in Fig. 2.6. This shows the TES chip mounted on top of a sapphire rod held by a gold-plated oxygen-free, high heat-conductivity copper mount. The TES is located at the center of the 2.5 mm disk. Bond pads allow wire bonding the TES to gold pins integrated in the copper mount. A commercial zirconia fiber sleeve is put around the sapphire rod and TES chip (Fig. 2.6b). The TES chip and sapphire rod are precision machined to a diameter of 2.499 ± 0.0025 mm, such that the commercial fiber sleeve locates the TES precisely in the center of the sleeve. A fiber ferrule is inserted into the fiber sleeve. Both the fiber ferrule and sleeve have a combined specified tolerance of $\approx 1\,\mu\text{m}$. Along with the precision micro-machining of the TES chip this ensures fiber-to-detector alignment precision of about $3\,\mu\text{m}$ [13]. Many of these self-aligned detectors can be compactly mounted together in a single unit (shown in Fig. 2.6c), which can easily be mounted in a cryogenic system.

2.2.2.3 Refrigeration of the TES

Since the TES are required to operate at temperatures below the temperature of liquid helium, elaborate cooling systems are required to operate these devices. The workhorse for operating the TES at around 100 mK temperatures is the adiabatic demagnetization refrigerator (ADR). The ADR relies on the magnetocaloric effect, which manifests itself by cooling a paramagnetic material upon adiabatic demagnetization of that material, achieving temperatures of about 50 mK. In our case, the demagnetization process is initiated at temperatures around 4 K, given by the lowest temperature achievable by standard commercially available cryogenic cooling techniques. Since the total energy stored in the paramagnetic material is limited (about 1 Joule), the ADR only has a limited cooling capacity (or hold time). Due to the heat load via wires and optical fibers connected to the detectors, our ADRs generally have hold times of about 8–12 hours, depending on the number of TESs present. After the energy is depleted, the ADR temperature can be recycled, a process taking about 2 hours.

2.2.2.4 Speed Improvements and Timing Jitter

The recovery rate of the optical transition edge sensor is limited by the thermal link between the electron and phonon system. While a weak thermal link between the electron and phonon system is required for a photon to be converted to a detectable electrical signal, the weakness of this link can lead to long recovery times compared to other single-photon detectors such as superconducting nanowire single-photon detectors (SNSPDs) and single-photon avalanche photodiodes [8, 25]. The thermal recovery time of the TESs fabricated at NIST can be as short as 1–2 μs. During the thermal recovery time, the TES is still able to receive and detect a photon, and therefore has no dead time, which we note is unique among photon counting detectors. However, if an absorption event occurs during the recovery of the detector, signal pile-up will occur and will make the detector response characterization complicated. Thus, one seeks short recovery times to allow for high-repetition rate experiments. Calkins et al. have recently demonstrated such engineering of the thermal coupling between the electron and phonon system of the TES [26] achieving a thermal recovery time of less than half a microsecond.

Figure 2.7a shows a micrograph of a device fabricated with two 2 μm × 10 μm × 0.115 μm gold bars deposited on the edge of the tungsten TES. Since the thermal

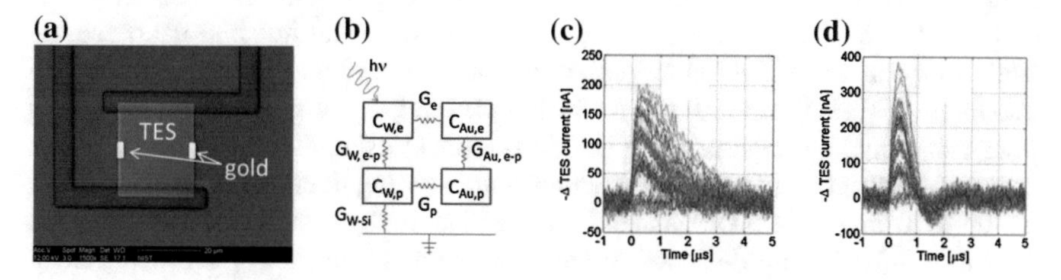

Fig. 2.7 **a** Micrograph of the fabricated TES with gold bars deposited on top of the tungsten; **b** Sensor electron-phonon coupling with the addition of gold bars; **c** TES response without gold bars; **d** TES response with added gold bars. Reproduced with permission from Ref. [26]. Copyright 2011, AIP Publishing LLC

electron-phonon coupling parameter of gold is ≈10x larger than for tungsten, one can engineer the thermal coupling by adding a controlled volume of gold to the TES. Figure 2.7b shows the thermal model of the TES with the gold bars added. The gold's electron system is strongly coupled to the tungsten's electron system. A strong thermal link $\Sigma_{\text{Au,e-p}}$ between the gold's electron and phonon system allows for faster thermalization to the thermal bath, hence a modified thermal coupling constant of the whole system. In addition to the thermal coupling parameter, the strength of thermal coupling improvement depends on the gold's volume and its specific heat capacity coefficient γ_{Au} [26]. The difference in TES response with (Fig. 2.7d) and without (Fig. 2.7c) gold bars under illumination with an attenuated pulsed laser is easily seen. The TES with the gold bars reduced the thermal recovery time to ≈460 ns, an improvement of a factor of 4 without affecting the energy resolution [26].

In addition to faster recovery times, low timing jitter is another desirable parameter for most single-photon experiments [10, 27]. The timing jitter is the uncertainty in identifying the arrival time of the photon. It is important to realize that this is separate from the 'latency time' of the detector. Latency is the time delay from when the photon was absorbed by the detector to the time when the system indicates that a photon has been detected and is largely due to propagation delays in cables and amplifiers. The main contribution to timing jitter in TESs is noise in the identification of the time associated with the threshold crossing of electrical output pulse originating from a photon absorption event. The timing jitter (Δt_σ) of such an electrical signal can be approximated by:

$$\Delta t_\sigma = \frac{\sigma}{\frac{dA}{dt}\big|_t} \approx \frac{\sigma}{A_{\max}} \tau_{\text{rise}}, \tag{2.2}$$

where σ is the root-mean-square (RMS) noise of the electrical signal, $\mathrm{d}A/\mathrm{d}t$ is the slope of the signal at a given time t, t_{rise} is the rise time and $A_{\max}$ is the maximum amplitude. The slope ($\sim\tau_{\text{rise}}/A_{\max}$) of the electrical signal is directly proportional to the timing uncertainty of the electrical output. The rise time of the electrical signal for TESs used at NIST is generally limited by the overall inductance in the readout circuit and the resistance of the TES. In most of the experiments performed with NIST TESs, the combination of SQUID input inductance and wiring inductance leads to an inductance of a few hundred nH. This inductance along with the low TES resistance (≈1 Ω) will limit the signal rise time $\tau_{\text{rise}}\sim L/R$ to several hundred nanoseconds. With a typical x100 SQUID amplifier, we measure a timing jitter full-width half maximum (FWHM) of ≈50–100 ns for the TES output, far greater than the timing jitter that would be required for a gated experiment using a Ti:Sapphire oscillator. Lamas-Linares et al. have recently demonstrated timing jitter FWHM values of 4.1 ns for 1550 nm single photons and 2.3 ns for 775 nm by directly wire-bonding the TES to a SQUID chip designed to have lower input inductance, and using high-bandwidth room-temperature amplification [27]. The lower timing jitter for higher energy photons stems from the larger amplitude of the TES output waveform.

2.2.3 Detector Characterization

2.2.3.1 Photon-Number Resolution

The optical TES is one of the few detectors with intrinsic photon-number-resolving capability. Other detection strategies exist where detectors with no number-resolving capability, 'click/no-click' detectors,[1] are multiplexed to achieve a quasi-photon number resolution. However, in those cases the fidelity of the measured state is always degraded compared to true/intrinsic photon-number-resolving detection. In many applications, however, multiplexed click detectors are sufficient to achieve high fidelity state characterization [28–30]. In contrast to simple 'click/no-click' detectors, the TES output contains information about the number of photons absorbed.

Figure 2.8a illustrates the photon-number-resolving capability of the TES. The TES was illuminated with weak coherent state pulses with a mean photon number per pulse of ≈ 2. The photon energy was 1.55 eV (800 nm). Figure 2.8a also shows a ≈ 2 μm wavelength black-body photon detection at ≈ 10 μs after the weak coherent pulses. When TESs are coupled to typical telecom optical fibers, background photons such as this result from the section of the high-energy tail of the room-temperature black-body emission spectrum that falls below the long cutoff wavelength of the fiber around 2000 nm [31]. However, when the energy of the signal photons is large enough, the black-body photons can easily be separated out by pulse height analysis. Generally, our TESs coupled to telecom optical fibers detect black-body photons at a rate of 500–1000 photons/sec^{-1}, which has a negligible effect on pulsed-light measurements and minimal effect on measurements of continuous sources in the near-IR down to the several femtowatt level. As can be seen in Fig. 2.8a, at higher photon numbers, the TES response enters the non-linear regime close to the normal conducting regime, and the photon-number resolution capability degrade.

To maximize the signal to noise ratio, post-processing of the output waveforms can be done. One fast, reliable method is optimum filtering using a Wiener filter [16]. Even though the optimum filtering method using Wiener filters only works in the case of white Gaussian noise (the TES response is not white due to the temperature dependent Johnson noise in the transition), the TES pulse heights can be reconstructed reliably with improved signal to noise ratio [7]. A measured coherent state with mean photon number of ≈ 5 is shown in Fig. 2.8b. In this case, the optimum filter analysis was used to determine the pulse-height histogram. Since the optimum filter makes use of a known linearly-scalable detector response, the method fails when the assumed detector response does not match the actual response, i.e. when entering the non-linear region. This effect is evidenced by the reduced visibility of the pulse height peaks for photon numbers larger than 4, shown in Fig. 2.8b. To improve the photon-number resolution at higher photon numbers in the non-linear region, one can find the most optimum representation of each photon-number response [32]. Also, linearization of the non-linear detector response can be accomplished to enable better

[1] We refer detectors with no photon-number-resolving capability as 'click/no-click' detectors, i.e. the detector cannot discriminate between the absorption of one or more photons.

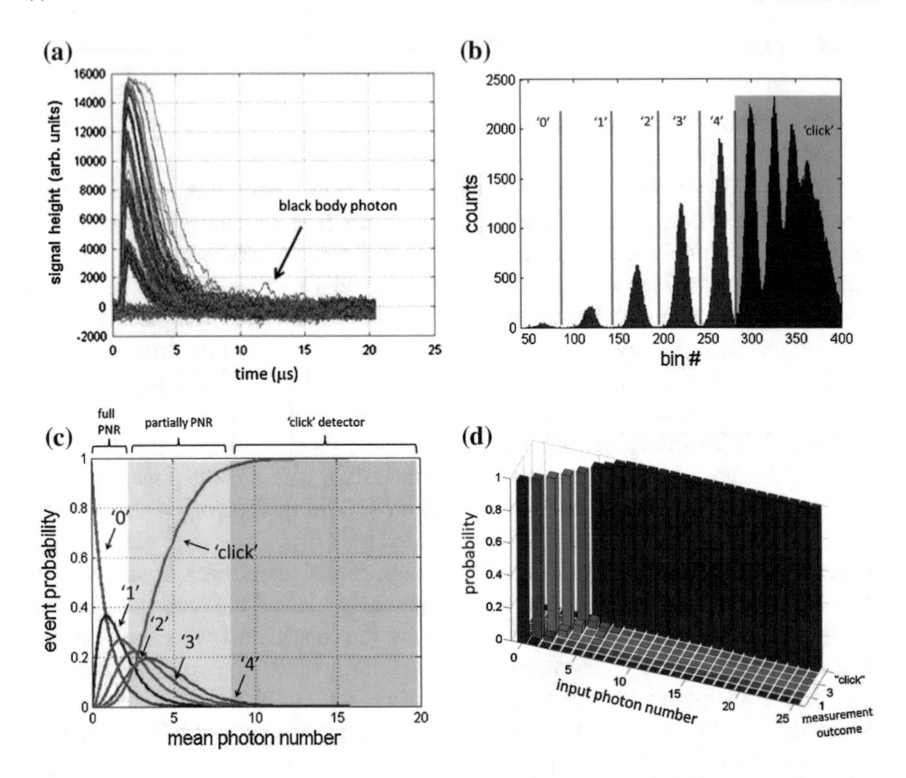

Fig. 2.8 Detector tomography results **a** Many TES output waveforms; **b** TES output pulse height histogram shown with boundaries dividing one detected number from the next; **c** Photon-number detection event probability as a function of input mean photon number; **d** Reconstructed POVMs for 0 through 4 photons and the "click" event

photon-number resolution close to or beyond the normal-conducting region [33]. Another method, allowing for analyzing detector responses far beyond the normal conducting regime was presented in Ref. [34] and will be presented below.

When characterizing the detector outcome for a photon-number-resolving detector one seeks to determine the probability of a detector response given a certain number of photons on the input. This probability is equivalent to the positive-operator-valued measure (POVM). The POVM characterizes the detector outcomes (in photon number k) for any given input photon number Fock state. Since pure photon number states at any value of k are still hard to realize in the laboratory, one employs a tomographic reconstruction of the POVMs based on measurement outcomes from excitation with weak coherent states, which are easily generated by attenuating a coherent laser beam [28, 35, 36]. Given a set of m coherent probe states C and a detector POVM Π, the measurement result R is given by: $R = C\Pi$. In general, POVMs are phase-dependent. However, in the case of TESs one can assume no phase dependence to the detector outcomes. Thus the POVM matrix is diagonal and can easily be represented by a vector. Also, in practice the Hilbert space needs to be truncated at some photon number k_{max}. We chose detector outcomes 0 through 4 photons and define the 'click' outcome for photon numbers greater than 4. Figure 2.8c shows the detector outcome probability as a function of input coherent state mean photon number. For a given input mean photon number, we can identify three regimes: 1) Fully photon-number-resolving, where the probability of a click outcome is small; 2) Partially photon-number-resolving where a significant number of detector outcomes are 'click' outcomes; 3) 'click' detector where the majority of detector outcomes are above 4 photons. Figure 2.8d shows the POVMs reconstructed by the maximum likelihood reconstruction. These POVMs fully characterize the phase-insensitive detector response under the illumination with a k photon number state. In the case of the 1-photon outcome POVM, the POVM value at $k = 1$ yields the detection efficiency of the TES. A more complete tomographic reconstruction of the detector outcomes not relying on continuous POVMS instead of partitioning the outcomes for specific photon numbers, considers was recently presented in Ref. [37].

2.2.3.2 Extending the TES Response to the Normal-Conducting Regime

Transition edge sensors typically resolve photons over a range of 1 to 10 or 20, depending on the wavelength, heat capacity of the device, and steepness of the superconducting resistance transition. The severely reduced variation of resistance with temperature above the superconducting transition of the detector makes it more difficult to extract (or assign) photon numbers from detection events that cause saturation in the pulse height as illustrated in Fig. 2.8. Despite the saturation in the pulse height, one can still extract some information about the number of photons detected. The uncertainty in the number of photons detected may exceed one photon, but the number may be below the shot noise limit of the pulsed input coherent state. This sub-shot noise performance still could be useful in applications requiring weak light detection.

Figure 2.9a shows a TES response under illumination with a coherent state pulse with a mean photon number of 4.8×10^6 photons at a wavelength of 1550 nm [34]. The rise at $t = 0$ shows the initial heating of the electron system, and the TES immediately enters the normal-conductive regime, evidenced by the flat response out to

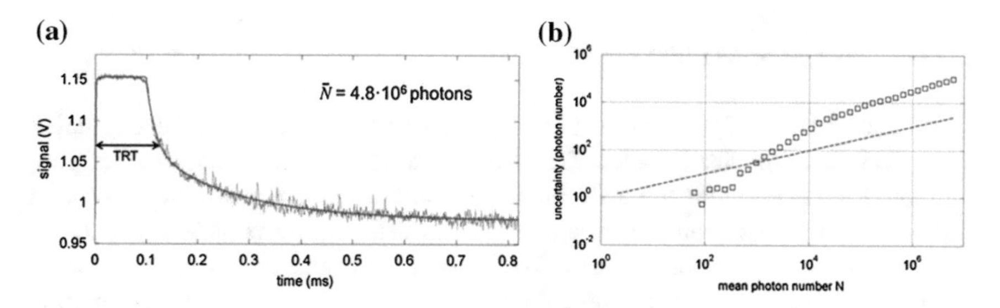

Fig. 2.9 **a** TES output waveform after excitation with about 4.8×10^6 photons in a single pulse; **b** uncertainty in photon number (*black squares*) after subtracting the photon source shot noise (*dotted line*) as a function of input mean photon number. Reproduced with permission from Ref. [34]. Copyright 2011, Optics Express (Optical Society of America)

0.1 ms. We can estimate the time it takes for the TES to return to its transition region (thermal relaxation time [TRT]) and use it to determine the initial heating of the electron system. A full fit to the detector response (dark solid line) yields the thermal relaxation time. Figure 2.9b shows the uncertainty in units of photon number as a function of input mean photon number derived from the TRTs. The black squares show the outcome uncertainty after subtracting the photon source shot noise. The uncertainties were obtained by determining the variance of the fitting parameters obtained from 20,480 waveforms for each input mean photon number. The dashed line shows the photon source shot noise. The results show that detection of $\approx$1,000 photons below the shot-noise is possible. Beyond 1,000 photons ($\approx$1 keV) the uncertainty in the detector response rises above the input photon shot noise. The detector can reliably produce an outcome up to a mean photon number of $\approx$10,000 as outlined in ref. [34]. Beyond $\approx$10 keV the detector response deviates from the theoretical prediction due to cumulative heating of the detector substrate at the repetition rate of 1 kHz used in those measurements.

2.3 Applications of the Optical Transition Edge Sensor

2.3.1 Key Experiments in Quantum Optics

In this section we review experiments that have benefited from the high-detection efficiency and photon-number resolving capabilities of transition edge sensors (TESs). We review the first experiment using the TESs measuring the output photon-number statistics from a Hong-Ou-Mandel (HOM) interference experiment [38], describe how the TESs enabled us to generate high-fidelity coherent state superpositions (optical Schrödinger cat states) [3], and how subtraction of photons from a thermal state affects the photon-number distribution of the heralded state [39]. Lastly, we review fundamental tests of quantum non-locality that require high total detection efficiencies—detection-loophole-free Bell tests based on optical photons [40, 41].

2.3.1.1 Hong-Ou-Mandel Interference

HOM interference [42] lies at the heart of quantum interference and has been studied extensively over the last decades. The effect exploits the bosonic character of photons and forbids two indistinguishable single photons that enter two different input ports of a semitransparent beam splitter to exit different ports—the two photons will always exit either port together. To demonstrate or use this quantum interference, experiments usually rely on coincidence counting of single photons exiting the two output ports of the beam splitter. When full HOM interference exists (that is the two incident photons are completely indistinguishable), the coincidence rate should drop to zero. Using TESs, however, Di Giuseppe et al. were able to directly

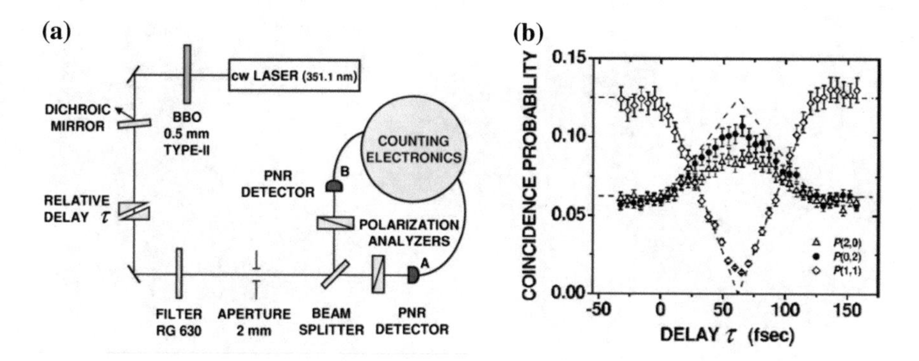

Fig. 2.10 **a** Setup for determining photon-number statistics from a Hong-Ou-Mandel experiment; **b** Measured Hong-Ou-Mandel interference dip and photon bunching. Reproduced with permission from Ref. [38]. Copyright 2003, Physical Review A (American Physical Society)

measure the photon-number statistics from a HOM beam splitter for the first time [38]. Figure 2.10a shows their experiment utilizing TESs for determining the output photon number statistics after HOM interference of two indistinguishable photons generated during a spontaneous parametric down-conversion event. The photons ($\lambda \approx 700$ nm) are delayed with respect to one another to observe the photon-number statistics as a function of the distinguishability of the photons' paths. Figure 2.10b shows the observed HOM interference dip (open diamonds), i.e. the probability of seeing one photon in one and one photon in the other output port, $P(1,1)$. As expected $P(1,1)$ approaches zero when both photons' paths become nearly indistinguishable. The open triangles and solid circles show the probability of two photons exiting one port, and no photon exiting the other port, $P(2,0)$, and vice versa, $P(0,2)$. In both cases the probabilities increase as the photon delay nears zero. This result showed the transition of a binomial (classical) distribution to a bosonic distribution of the two-photon state after interference on the beam splitter due to photon-number resolving capability of the TES.

2.3.1.2 Photon-Number Subtracted States

Photon-number subtraction is a useful technique for applications in quantum optics. One example of the use of photon subtraction is the generation of a coherent state superposition (CSS) state. Such states are often called Schrödinger cat states when each subsystem of the superposition contains a macroscopic number of photons. The use for high quality CSSs ranges from super-resolution metrology to quantum computing [43]. High fidelity CSSs are required to minimize the overlap between the two superimposed states and to achieve fault-tolerant quantum computing [44]. A number of experiments that aimed to generate such optical CSSs have been performed [45–48], two of which used TESs to herald the presence of the CSS [3, 48]. To create the CSSs a photon-subtraction scheme depicted in Fig. 2.11a is utilized, and

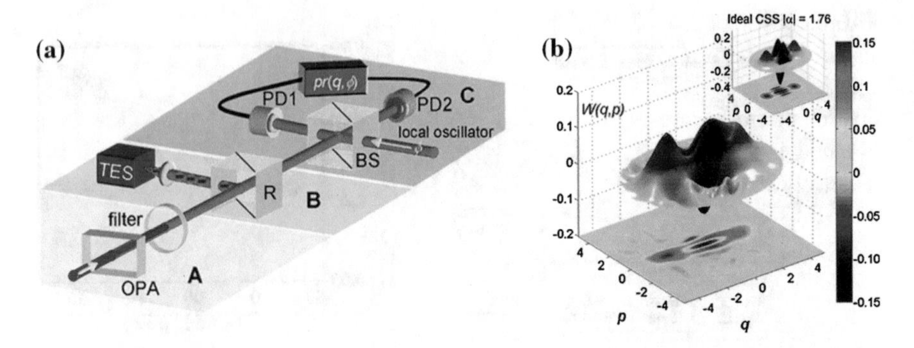

Fig. 2.11 **a** Scheme for optical Schrödinger cat state generation. An up-converted laser pulse enters an optical parametric amplifier (OPA) to create a squeezed vacuum state. After spectral filtering, photons are probabilistically subtracted via a weakly reflecting beam splitter and detected with a TES. Quadratures of the heralded state are measured by homodyne detection; **b** The reconstructed and theoretical cat state. Reproduced with permission from Ref. [3]. Copyright 2010, Physical Review A (American Physical Society)

we will concentrate in the following on the experiment performed in Ref. [3]. The scheme, in principle allows for large amplitude and high fidelity CSSs by heralding on the measurement of multiple photons subtracted from a squeezed vacuum [49, 50]. A squeezed vacuum state is prepared and sent through a weakly reflecting beam splitter. Reflected photons that are detected herald an approximate CSS in the transmitted beam. The TES used in this experiment had an efficiency of 85 % and was able to resolve up to 10 photons at a wavelength of 860 nm. The experiment aimed at generating the largest CSS with the highest fidelity. Figure 2.11b shows the theoretical and reconstructed Wigner functions of the CSS obtained after subtraction of three photons. The measured CSS had a mean photon number of 2.75 and a fidelity of 0.59, which remains the largest CSS generated from squeezed-vacuum photon-subtraction to date. Whilst the demonstration of large and high fidelity CSS was achieved, the rate at which these states were generated was low and made scaling impossible. However, recent work has demonstrated the potential for high-rate, high-fidelity approximations to CSS, through the use of photon replacement operations implemented with number-resolved detection [51].

Continuous variable (CV) quantum states, such as the CSS described above, are measured by use of optical-homodyne detection and reconstructed by maximum-likelihood estimation [52]. In this case, a strong local oscillator is used to determine the quadrature of the weak quantum field. Since the TES can resolve up to 20 photons, another method of measuring the quadrature of CV states is by performing optical homodyne measurements in the weak local oscillator regime, where the overall signal strength is on the order of the weak quantum state. A number of experiments have already been presented in the literature showing the capability of weak-field homodyne detection [53–55], one of which utilized the TES [56]. Combined with more efficient state generation schemes and because of the small amount of energy absorbed by the detector, weak-field homodyne detection may become an important resource in the future when generating these states on-chip and mating them with high-efficiency superconducting integrated detectors [57–60], where low power dissipation platforms are a key challenge.

Another example of a photon-subtracted state was recently presented by Zhai et al. [39]. Up to eight photons were subtracted from a thermal state of light obeying Bose-Einstein statistics. A laser pulse illuminated a spinning disk of ground glass. Collection of the transmitted beam with a single-mode optical fiber selects a single-mode thermal state. By use of TESs, Zhai et al. reconstructed the photon-number statistics and showed that the photon number of the photon-subtracted state increases linearly with the number of subtracted photons. This seemingly counterintuitive result is explained by the noting that because thermal light is bunched, when sampling of a portion the light pulse made via the beam splitter finds a high number of photons, it means that the other portion of the light pulse (the other port of the beam splitter) is likely to also have a high number of photons. It is also important to note that these states, conditioned on the presence of a photon number larger than zero in the subtracted path, are necessarily rare because the photon-number probability always falls with increasing photon number. Given the measured photon number distributions, Zhai et al. also demonstrated that when using TESs, second- and higher-order correlation functions can easily be derived from the measured photon-number statistics [39, 61], and the thermal Bose-Einstein distribution approached a Poisson distribution representing the subtracted state when the number of subtracted photons is large.

2.3.1.3 Fundamental Tests of Quantum Non-locality

In classical physics, locality refers to the measurement outcome of an object that is influenced by its local interaction with the environment. To explain observed quantum mechanical correlations, an unknown theory that could describe the observed correlations at a local level may exist. Such a theory is referred to generically as a local hidden variable theory. Quantum Non-locality (QN) refers to the description of observed quantum mechanical correlations that cannot be mimicked by local hidden variable theories.

In 1964 Bell [62] developed an inequality showing that no local hidden variable theory can reproduce certain predictions made by using quantum mechanics. Bell's inequalities constrain correlations that systems governed by local hidden variable theories can exhibit, but some entangled quantum systems exhibit correlations that violate the inequalities. Thus, testing an inequality with a quantum system experimentally could falsify the hypothesis that the system is governed by a local hidden variable. However in all previous experiments, loopholes exist that a local hidden variable model can exploit to mimic the observed outcomes that violate the tested inequalities. These loopholes make an experimental test of local hidden variables a challenging endeavor. The three most prominent loopholes are the freedom of choice, locality, and detection (also called fair-sampling) loophole. While all loopholes have now been closed individually [40, 41, 63–66] with photons, an experiment addressing all of these loopholes simultaneously is still lacking. Here, we summarize experiments relying on the TES's high detection efficiency to close the fair-sampling loophole.

A variation of Bell's inequality based on a maximally polarization entangled state was introduced in 1970 by Clauser et al. [67], and an upper limit of ≈88 % for the overall detection efficiency of the entangled pairs necessary to close the fair-sampling loophole was introduced.[2] Even with the best detectors, such high overall system detection efficiency is difficult to achieve in photonic systems, since the entangled photons generally have to couple into single-mode optical fibers. This coupling can be lossy, although as it has recently been shown to be quite efficient (theoretically reaching almost 100 %) when considering the correct geometric coupling properties [68, 69]. Eberhard came to the conclusion that lowering the amount of entanglement of polarization entangled states, allows for lower overall system detection efficiencies, as low as 2/3 [70]. This absolute lower limit assumes no detector dark counts or degradation of the entanglement visibility. Thus, a system efficiency of >70 % is a more realistic efficiency to aim for.

Recently two groups succeeded in closing the fair-sampling loophole by use of photons and TESs [40, 41]. Both experiments made use of entangled photon pairs generated in a spontaneous parametric down-conversion (SPDC) process and a variation of the Bell inequality, the so-called CH Bell inequality [71]. In both cases a clear violation of the CH Bell inequality was shown with an efficiency allowing for detector background counts and entanglement visibility degradation. Christensen et al. employed a non-collinear entanglement scheme at a down-conversion center wavelength of ≈710 nm. The TES optimized for this study had a detection efficiency of ≈95 %. The symmetric single-mode fiber coupling efficiency of the entangled photons was > 90 %. Additional loss of ≈5 % was introduced at the single-mode fiber/telecom fiber interface splice (the telecom fiber is generally used for coupling photons to the optical TES). Further, spectral filtering added ≈7 % loss to the overall system performance, leading to an overall system efficiency of ≈75 %. Giustina et al. used an entangled SPDC source based on a Sagnac configuration and achieved an overall system efficiency of ≈73 % and 78 % in the two collection modes. The TESs in this study were optimized for a center wavelength of 810 nm and had a detection efficiency of ≈95 %. The latter experiment was performed using a continuous wave excitation of the SPDC process.

Progress continues toward implementing an experiment using photons that closes all loopholes simultaneously. In these experiments, both high-efficiency detection and good timing resolution is needed. Further analysis of data obtained in Ref. [41] showed that because of the continuous excitation of the SPDC process, the results are in principle subject to the coincidence-time loophole [72]. Even though a re-analysis of the data revealed that the experiment was not subject to the coincidence-time

[2] Note that overall detection efficiency is a product of all optical losses in transferring the photons from where they are generated to where they are detected and the detector's detection efficiency (see Sect. 1.1.1).

loophole [73], future loophole-free experiments will use pulsed sources to define every trial of the Bell test. Also, an improved TES timing jitter results in better rejection of background light. Better rejection of background light lowers the needed overall source-to-detector coupling efficiencies needed to perform a loophole-free experiment.

2.4 Integration of Optical TES on Waveguide Structures

We now present our recent efforts towards integrating optical TESs into optical on-chip architectures for scalable quantum information processing. We review the results presented in two publications [57, 58]. Integrated optics has become one of the leading technologies for generating and manipulating complex quantum states of light, since mode-matching, low-loss and small device footprints allow for more complex structures. A recent example of photon routing using integration to build complex networks is Boson sampling [74–77]. In these experiments the photon detection was done off-chip, which severely limited the number of modes that could feasibly be measured due to interface losses. Therefore, high-efficiency on-chip detection via evanescent coupling of the optical mode is key to realizing such large-scale optical quantum information and communication applications. Currently, a number of groups are pursuing integration of superconducting single-photon detectors on scalable platforms. Most efforts center around superconducting nanowire single-photon detectors [59, 60, 78–80]—two such approaches are reviewed in Chaps. 3 and 4 of this volume. Our method exploits integrated TESs which combine both high efficiency and intrinsic number resolution. Any integrated detector will avoid loss originating from coupling off-chip. In addition, such a detector can in principle be placed at arbitrary locations within an optical waveguide structure to allow optical mode probing, and allow in principle fast feed-forward operations, if all detection, decision, and photon routing is performed on chip. As we show, such a detector will also enable mode-matched photon-number subtraction without the need for complicated optical setups. However, one challenge remains when integrating source, circuit and detector on-chip. Since the number of pump photons required to generate single-photon states is large, a small fraction of the pump photons may leak into the single photon detector, resulting in unwanted detection events degrading the fidelity of the desired state. Therefore, pump filtering and pump rejection is currently one of the major challenges towards realization of a fully-integrated device.

2.4.1 On-Chip Transition Edge Sensor

For integration of our TES, we use a silica-on-silicon planar structure and UV laser writing to define the optical waveguide [81] fabricated at the University of Southampton (U.K.). Figure 2.12a shows the schematic of this structure. A thermal

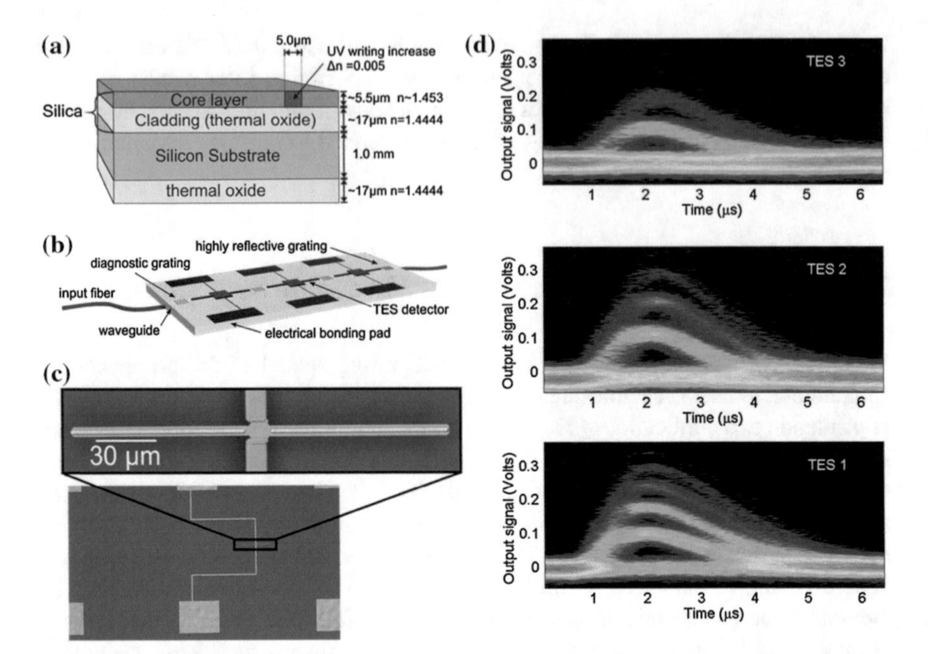

Fig. 2.12 **a** Schematic of the silica waveguide structure; **b** Schematic of the multiplexed long absorber evanescently coupled TES; **c** Micrograph of one fabricated long absorber TES; **d** Electrical output waveforms for all three multiplexed TESs. (**a**) Courtesy James Gates, University of Southampton, U.K. (**b**)–(**d**) Reproduced with permission from Ref. [58]. Copyright 2013, Optics Express (Optical Society of America)

oxide cladding layer is deposited on a silicon substrate. The waveguide core layer is a germanium-doped silica layer and deposited via Flame Hydrolysis Deposition (FHD). UV laser writing into the photosensitive core layer achieved an index contrast change of $\approx$0.3 % and results in a waveguide mode of Gaussian profile [82]. After UV writing, the optical waveguide is $\approx$5 μm wide. As a proof-of-principle we embedded a 25 μm × 25 μm × 0.04 μm TES on top of such a silica waveguide structure [57]. The detection efficiency of a photon inside the waveguide was found to be $\approx$7 %. We found that when aiming for close-to-unity detection efficiency in such a silica waveguide structure, the evanescent coupling of the absorber has to be on the order of the length of 1 mm due to the small optical mode overlap with the absorber. In contrast, a silicon waveguide device with mode-field dimensions a factor of ten smaller than those of our silica device, only require $\approx 20 - 30$ μm absorber length to achieve high absorption of the optical mode [60].

To improve the overall absorption and therefore the detection efficiency of our devices, we increased the absorber length and multiplexed three devices. Figure 2.12b shows the schematic of an integrated high-efficiency TES presented in Ref. [58]. The device consists of three detector regions with absorber lengths of 210 μm each, and yielded an overall device efficiency of 79 % for a photon inside the waveguide. This approach used a separate absorber geometry increasing the interaction length with the optical waveguide mode. The total heat capacity of the absorber/sensor device was low enough to retain the photon-number resolution of the TES. A micrograph of the fabricated device is shown in Fig. 2.12c. The active TES is a 10 μm × 10 μm sensor made of tungsten, 40 nm thick operating at a transition temperature of ≈85 mK. Two 100 μm long tungsten absorbers are attached to the device and placed directly on top of the optical waveguide. To maintain an overall low heat capacity, the layer was 3.5 μm wide and 40 nm thick. A gold spine facilitates the heat conduction to the sensor area and allows the heat to reach the sensor region before being dissipated into the substrate. The gold layer was 2 μm wide and 80 nm thick. Simulations showed that for an absorber this length, the probability of absorbing a photon inside the waveguide should be ≈48 %. Simulations also revealed that the effective mode index mismatch at the absorber/no-absorber interface is on the order of 0.1 %. Therefore the optical mode is almost undisturbed, so reflections at the absorber/no-absorber interface are negligible. This allows multiplexing of several such detectors on a single waveguide to maximize the absorption of the optical mode. Calkins et al. also modeled the detector response due to the absorption of a photon along the long absorber [58]. When a photon is absorbed along the tails of these detectors, that energy must be transmitted to the sensor region before escaping via some other thermal path, e.g. through the phonon system into the substrate. This process is described by electron diffusion and depends on the thermal conductivity of the electron system and the amount of energy the electron system can carry. Calkins et al. found that a lower device temperature (lower T_c) and a large thermal conductivity are advantageous to facilitate such properties [58]. To maintain photon-number resolution, the maximum length of the absorber depends on the heat capacity, transition temperature, thermal conductivity, and electron-phonon coupling strength of the device. The latter will eventually leak heat out of the system into the substrate [58].

Figure 2.12d shows the waveforms from the device presented in Ref. [58]. A strongly attenuated pulsed 1550 nm laser was used to measure the device's response. The device consists of 3 individual detectors, TES 1 through TES 3. Clearly, photon-number resolution is well maintained for these long absorber detectors. Also, the measured detection efficiency was about 43 % per detector element, yielding an overall efficiency of ≈79 %. Figure 2.12d also shows the subsequent decay of photon-number amplitudes. Since all three detectors have the same length, each preceding detector absorbs part of the incoming photon state resulting in an exponential decrease in detected mean photon number with each additional detector. In this study all TESs were read out individually, each with a dedicated SQUID amplifier. This poses a scalability constraint when multiplexing many TESs on a single circuit.

2.4.2 A New Experimental Tool: On-Chip Mode-Matched Photon Subtraction

The experimental results, along with optical-mode modeling, allowed us to find the dependence of the detection efficiency as a function of the absorber length. We are now able to fabricate a detector with pre-determined absorption (detection efficiency). Since the silica platform requires relatively long absorbers per unit of detection efficiency, variation in the detection efficiency due to fabrication imperfections are considered to be small. This will allow us to implement the first quantum optics experiment utilizing the non-Gaussian operation of photon-subtraction on-chip, an experimental challenge when working with free-space optical components. Figure 2.13 shows a schematic of a heralded photon-subtraction experiment. This is similar to the scheme used for the generation of coherent state superpositions described above. A generated photon state is sent towards a beam splitter with known reflectivity. The beam splitter probabilistically reflects photons from the input state and sends it to a heralding photon detector. Measurement of a photon at this detector projects a subtracted photon-number state onto the input state. In this way, the transmitted (heralded) state can show non-classicality. The heralded state detection is accomplished by some other scheme such as homodyne detection or photon counting. The challenge in these kinds of experiments in free-space is to mode-match the photon subtraction (heralding) detector to the heralded state detection. The mode-matching must be done in the spatial, spectral and temporal regimes to achieve high-fidelity state generation. If the subtracted photon is not mode-matched to the heralded state detection, a vacuum state is detected, thus degrading the fidelity of the heralded state. The integrated approach, however, offers an almost perfect mode-matched heralding detector, since the optical mode transmitted matches the optical mode evanescently coupled to the heralding detector. Also, since the TES intrinsically has no dark counts, there should be no false heralding photon detection events. In addition, the TES has 100 % internal detection efficiency (probability of an absorbed photon yielding an output signal) for the subtracted photons. When combining the integrated photon subtraction scheme with weak-field homodyne detection, the integrated TES platform may become a powerful tool for the study and manipulation of optical quantum states.

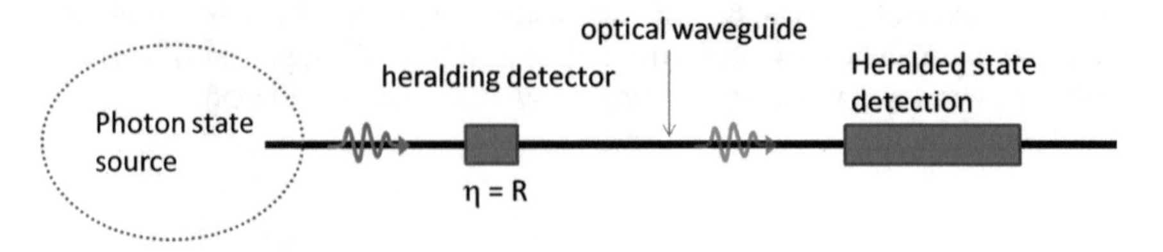

Fig. 2.13 Schematic of on-chip photon-number subtraction (heralding) for the experimental realization of non-Gaussian operations

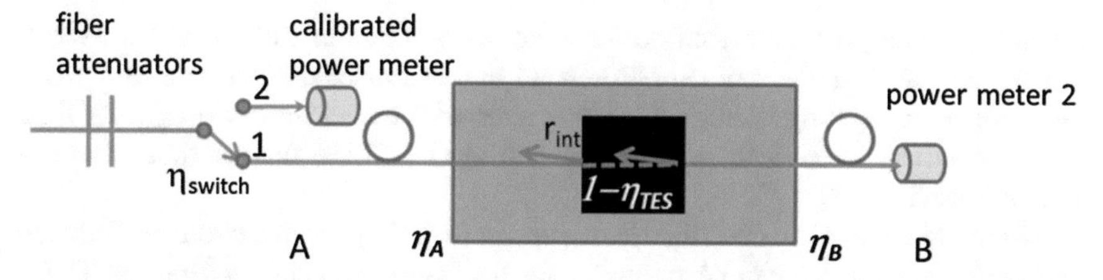

Fig. 2.14 Experimental scheme for measuring the coupling efficiency and detection efficiency of the TES detectors

2.4.3 *On-Chip Detector Calibration*

Determining the detection efficiency of a waveguide-integrated single-photon detector is a crucial aspect of studying the device performance. In contrast to fiber-coupled devices where the detection efficiency can be determined for photons inside the optical fiber, the overall detection efficiency of a waveguide-integrated detector also includes the fiber-to-waveguide loss, a non-negligible contribution to the overall device-loss. Here we show how we are able to discriminate between the fiber-to-waveguide coupling loss and the detection efficiency of a photon residing inside the waveguide, following the derivation found in the supplemental material of Ref. [57] and in [58]. Figure 2.14 shows the scheme for measuring the detection efficiency of the TES centered on top of the waveguide and simultaneously measuring the fiber-to-waveguide loss. We perform three independent measurements:

- We measure the overall transmission through the whole device. Under the assumption that fiber-to-waveguide coupling is single-mode, the coupling efficiencies at sides A and B are independent of the propagation direction of the photons. This assumption can easily be verified by measuring the transmission through the device in both directions: if $T = T_{\mathrm{AB}} = T_{\mathrm{BA}}$ then $T = \eta_{\mathrm{switch}}\eta_A(1 - r_{\mathrm{int}})(1 - \eta_{\mathrm{TES}})\eta_{\mathrm{B}}$, where η_{switch} is the switching ratio of the switch, η_{A} and η_{B} are the coupling efficiencies into the waveguide on side A and B, respectively. η_{TES} is the detection efficiency of the TES, and r_{int} is the reflection at the absorber/non-absorber interface. Since r_{int} is small ($\approx$0.1 %): $T \approx \eta_{\mathrm{switch}}\eta_A(1 - \eta_{\mathrm{TES}})\eta_{\mathrm{B}}$;
- We measure the overall system detection efficiencies including the fiber-to-waveguide coupling losses: $\eta_{sA/sB} \approx \eta_{\mathrm{A/B}}\eta_{\mathrm{TES}}$ (with $r_{\mathrm{int}} \ll 1$) for each propagation direction.

We can solve for η_{A}, η_{B} and η_{TES}. We can also estimate the uncertainty of the final outcome when including the calculated absorber/non-absorber boundary loss. It is important to note that this technique does not account for losses within the detection region, i.e. underneath the detector. These losses would be equivalent to a boundary reflection, and, if these are small they would only have small effects on the final detection efficiency outcome. We use two power meters to measure the overall transmission along both possible propagation directions (T_{AB} or T_{BA}). To achieve well-defined attenuated single-photon laser pulses, we calibrate optical fiber attenuators using the calibrated power meter as outlined in Ref. [13]. The switch imbalance (ratio) is determined beforehand and is modeled according to a weak laser pulse propagating through a beam splitter with transmissivity η_{switch}. By this, we obtained: $\eta_{\text{A}} = 39.8\,\%$, $\eta_{\text{B}} = 47.9\,\%$ and $\eta_{\text{TES}} = 7.2\,\%$ for our first proof-of-principle device [57].

Similar to the method described above and in Ref. [57], the detection and fiber-to-waveguide efficiencies can be estimated for the high-efficiency multiplexed TESs. Calkins et al. determined the detection efficiency for all three multiplexed detectors also by measuring the overall system efficiencies for both propagation directions. In addition, they were also able to deduce the waveguide mode absorption in their estimation. In general the method yields $2N + 1$ measurements for N multiplexed detectors, allowing determination of more system parameters than necessary to extract the sensor and fiber-to-waveguide coupling efficiencies ($N + 2$). In principle, this method can be extended to many low-efficiency detectors allowing the probing of crucial circuit parameters, such as scattering losses, beam splitter ratios and detection efficiencies.

2.5 Outlook

We have given an overview of the superconducting optical transition edge sensor (TES) and its use in quantum optics experiments. The high detection efficiency of the TES combined with photon number resolving capability are key requirements for experiments in quantum optics and quantum information. However, it should be noted that the TES can in principle have a large dynamic range owing to its thermal response with low uncertainty. This large dynamic range at the single-photon level may be advantageous for calibration of photon pulses that are in the mesoscopic regime between single-photon detection and conventional sensitivities of semiconductor photodiodes. The relatively slow response and recovery time of the TES, particularly compared to superconducting nanowire single-photon detectors (Chap. 1), preclude their use in some high-speed photon-counting applications, but when true photon-number resolution is required, the TES is the most efficient tool to measure photon states across an extremely broad wavelength range. The almost unity detection efficiency of the TES is still unmatched and is crucial for fundamental tests of quantum nonlocality. We have presented our efforts on integration of optical TES on optical waveguide structures and showed a route towards high-detection efficiency implementation. A full source-circuit-detector implementation is still missing due to the challenges of optical pump reduction to minimize both the heat load on the cryogenic system and unwanted detection of spurious pump photons. However, circuit implementation, along with high-efficiency TESs, is on its way also investigating the potential for mode-matched high-efficiency photon subtraction for future quantum information applications.

In the future, the dynamic range of the TES may be interesting to studies that aim at macroscopic entanglement [83], a developing route of research that requires single-photon resolution at levels of hundreds of photons. In principle, the energy resolution of the TES can be improved at the cost of speed, e.g. lower thermal heat escape mechanism. The combination of weak-field homodyne detection and integrated optics may lead to interesting, more complex experiments in the continuous variable regime of quantum optics.

Acknowledgments This work was supported by the Quantum Information Science Initiative (QISI) and the NIST *'Innovations in Measurement Science'* Program. The NIST authors thank all collaborators who enabled the joint experiments summarized in this chapter.

References

1. B. Cabrera et al., Detection of single infrared, optical, and ultraviolet photons using superconducting transition edge sensors. Appl. Phys. Lett. **73**, 735 (1998)
2. D. Rosenberg et al., Quantum key distribution at telecom wavelengths with noise-free detectors. Appl. Phys. Lett. **88**, 021108 (2006)
3. T. Gerrits et al., Generation of optical coherent-state superpositions by number-resolved photon subtraction from the squeezed vacuum. Phys. Rev. A **82**, 031802 (2010)
4. D. Rosenberg et al., Long-distance decoy-state quantum key distribution in optical fiber. Phys. Rev. Lett. **98**, 010503 (2007)
5. R.W. Romani et al., First astronomical application of a cryogenic transition edge sensor spectrophotometer. Astrophys. J. **521**, L153 (1999)
6. J. Burney et al., Transition-edge sensor arrays UV-optical-IR astrophysics. Nucl. Instrum. Methods Phys. Res. Sect. A: Accel., Spectrom., Detect. Assoc. Equip. **559**, 525–527 (2006)
7. A.E. Lita, A.J. Miller, S.W. Nam, Counting near-infrared single-photons with 95 % efficiency. Opt. Express **16**, 3032 (2008)
8. M.D. Eisaman et al., Invited review article: single-photon sources and detectors. Rev. Sci. Instrum. **82**, 071101 (2011)
9. K.D. Irwin, G.C. Hilton, *Transition-Edge Sensors, in Cryogenic Particle Detection* (Springer, Heidelberg, 2005)
10. D. Fukuda et al., Titanium superconducting photon-number-resolving detector. IEEE Trans. Appl. Supercond. **21**, 241 (2011)
11. L. Lolli, E. Taralli, M. Rajteri, Ti/Au TES to discriminate single photons. J. Low Temp. Phys. **167**, 803 (2012)
12. D.F. Santavicca, F.W. Carter, D.E. Prober, Proposal for a GHz count rate near-IR single-photon detector based on a nanoscale superconducting transition edge sensor. Proc. SPIE 8033, Adv. Photon Count. Tech. V **80330W**, (2011)
13. A.J. Miller et al., Compact cryogenic self-aligning fiber-to-detector coupling with losses below one percent. Opt. Express **19**, 9102 (2011)
14. B. Cabrera, Introduction to TES physics. J. Low Temp. Phys. **151**, 82 (2008)
15. K.D. Irwin, An application of electrothermal feedback for high resolution cryogenic particle detection. Appl. Phys. Lett. **66**, 1998 (1995)
16. D.J. Fixsen et al., Pulse estimation in nonlinear detectors with nonstationary noise. Nucl. Instrum. Methods Phys. Res. Sect. A: Accel., Spectrom., Detect. Assoc. Equip. **520**, 555 (2004)
17. R. Jaklevic et al., Quantum interference effects in Josephson tunneling. Phys. Rev. Lett. **12**, 159 (1964)
18. R.P. Welty, J.M. Martinis, A series array of DC SQUIDs. IEEE Trans. Magn. **27**, 2924 (1991)
19. C. Kittel, *Introduction to Solid State Physics* (Wiley, New York, 1956)
20. G. Dresselhaus, M. Dresselhaus, Interband transitions in superconductors. Phys. Rev. **125**, 1212 (1962)

21. A.E. Lita et al., Superconducting transition-edge sensors optimized for high-efficiency photon-number resolving detectors. Proc. SPIE 7681, Advanced Photon Count. Tech. IV **76810D** (2010)
22. A.E. Lita et al., High-efficiency photon-number-resolving detectors based on hafnium transition-edge sensors. AIP Conf. Proc. **1185**, 351 (2009)
23. A.E. Lita et al., Tuning of tungsten thin film superconducting transition temperature for fabrication of photon number resolving detectors. Appl. Supercond., IEEE Trans. **15**, 3528 (2005)
24. G. Fujii et al., Thin gold covered titanium transition edge sensor for optical measurement. J. Low Temp. Phys. **167**, 815 (2012)
25. R.H. Hadfield, Single-photon detectors for optical quantum information applications. Nat. Photon. **3**, 696 (2009)
26. B. Calkins et al., Faster recovery time of a hot-electron transition-edge sensor by use of normal metal heat-sinks. Appl. Phys. Lett. **99**, 241114 (2011)
27. A. Lamas-Linares et al., Nanosecond-scale timing jitter for single photon detection in transition edge sensors. Appl. Phys. Lett. **102**, 231117 (2013)
28. J.S. Lundeen et al., Tomography of quantum detectors. Nat. Phys. **5**, 27 (2009)
29. D. Achilles et al., Fiber-assisted detection with photon number resolution. Optics Lett. **28**, 2387 (2003)
30. T. Bartley et al., Direct observation of sub-binomial light. Phys. Rev. Lett. **110**, 173602 (2013)
31. D. Rosenberg et al., Noise-free high-efficiency photon-number-resolving detectors. Phys. Rev. A **71**, 061803 (2005)
32. Z.H. Levine et al., Algorithm for finding clusters with a known distribution and its application to photon-number resolution using a superconducting transition-edge sensor. J. Opt. Soc. Am. B **29**, 2066 (2012)
33. D.J. Fixsen et al., Optimal energy measurement in nonlinear systems: an application of differential geometry. J. Low Temp. Phys. **176**, 16 (2014)
34. T. Gerrits et al., Extending single-photon optimized superconducting transition edge sensors beyond the single-photon counting regime. Opt. Express **20**, 23798 (2012)
35. G. Brida et al., Ancilla-assisted calibration of a measuring apparatus. Phys. Rev. Lett. **108**, 253601 (2012)
36. B. Giorgio et al., Quantum characterization of superconducting photon counters. New J. Phys. **14**, 085001 (2012)
37. P.C. Humphreys et al., Tomography of photon-number resolving continuous output detectors. New J. Phys. **17**, 103044 (2015)
38. G. Di Giuseppe et al., Direct observation of photon pairs at a single output port of a beam-splitter interferometer. Phys. Rev. A **68**, 063817 (2003)
39. Y. Zhai et al., Photon-number-resolved detection of photon-subtracted thermal light. Opt. Lett. **38**, 2171 (2013)
40. B. Christensen et al., Detection-Loophole-free test of quantum nonlocality, and applications. Phys. Rev. Lett. **111**, 130406 (2013)
41. M. Giustina et al., Bell violation using entangled photons without the fair-sampling assumption. Nature **497**, 227 (2013)
42. C.K. Hong, Z.Y. Ou, L. Mandel, Measurement of subpicosecond time intervals between two photons by interference. Phys. Rev. Lett. **59**, 2044 (1987)
43. A. Gilchrist et al., Schrödinger cats and their power for quantum information processing. J. Opt. B: Quantum Semiclassical Opt. **6**, S828 (2004)
44. A.C. Lund, T.P. Ralph, H.L. Haselgrove, Fault-tolerant linear optical quantum computing with small-amplitude coherent states. Phys. Rev. Lett. **100**, 030503 (2008)
45. A. Ourjoumtsev et al., Generation of optical Schrödinger cats from photon number states. Nature **448**, 784 (2007)
46. A. Ourjoumtsev et al., Generating Optical Schrödinger Kittens for Quantum Information Processing. Science **312**, 83 (2006)
47. K. Wakui et al., Photon subtracted squeezed states generated with periodically poled $KTiOPO_4$. Opt. Express **15**, 3568 (2007)

48. N. Namekata et al., Non-Gaussian operation based on photon subtraction using a photon-number-resolving detector at a telecommunications wavelength. Nat. Photon. **4**, 655 (2010)
49. M. Dakna et al., Generating Schrödinger-cat-like states by means of conditional measurements on a beam splitter. Phys. Rev. A **55**, 3184 (1997)
50. S. Glancy, H.M. de Vasconcelos, Methods for producing optical coherent state superpositions. J. Opt. Soc. Am. B **25**, 712 (2008)
51. T.J. Bartley et al., Multiphoton state engineering by heralded interference between single photons and coherent states. Phys. Rev. A **86**, 043820 (2012)
52. A.I. Lvovsky, Iterative maximum-likelihood reconstruction in quantum homodyne tomography. J. Opt. B: Quantum Semiclassical Opt. **6**, S556 (2004)
53. K. Banaszek et al., Direct measurement of the Wigner function by photon counting. Phys. Rev. A **60**, 674 (1999)
54. K. Laiho et al., Probing the negative Wigner function of a pulsed single photon point by point. Phys. Rev. Lett. **105**, 253603 (2010)
55. G. Donati et al., Observing optical coherence across Fock layers with weak-field homodyne detectors. Nat. Commun. **5**, 5584 (2014)
56. N. Sridhar et al., Direct measurement of the Wigner function by photon-number-resolving detection. J. Opt. Soc. Am. B **31**, B34 (2014)
57. T. Gerrits et al., On-chip, photon-number-resolving, telecommunication-band detectors for scalable photonic information processing. Phys. Rev. A **84**, 060301 (2011)
58. B. Calkins et al., High quantum-efficiency photon-number-resolving detector for photonic on-chip information processing. Opt. Express **21**, 22657 (2013)
59. J.P. Sprengers et al., Waveguide superconducting single-photon detectors for integrated quantum photonic circuits. Appl. Phys. Lett. **99**, 181110 (2011)
60. W.H.P. Pernice et al., High-speed and high-efficiency travelling wave single-photon detectors embedded in nanophotonic circuits. Nat. Commun. **3**, 1325 (2012)
61. T. Gerrits et al., Generation of degenerate, factorizable, pulsed squeezed light at telecom wavelengths. Opt. Express **19**, 24434 (2011)
62. J.S. Bell, On the Einstein-Podolsky-Rosen paradox. Physics **1**, 195 (1964)
63. M.A. Rowe et al., Experimental violation of a Bell's inequality with efficient detection. Nature **409**, 791 (2001)
64. M. Ansmann et al., Violation of Bell's inequality in Josephson phase qubits. Nature **461**, 504 (2009)
65. J. Hofmann et al., Heralded entanglement between widely separated atoms. Science **337**, 72 (2012)
66. T. Scheidl et al., Violation of local realism with freedom of choice. Proc. Natl. Acad. Sci. **107**, 19708 (2010)
67. J. Clauser et al., Proposed experiment to test local hidden-variable theories. Phys. Rev. Lett. **23**, 880 (1969)
68. P.B. Dixon et al., Heralding efficiency and correlated mode coupling of near-IR fiber-coupled photon pairs. Phys. Rev. A **90**, 043804 (2014)
69. R.S. Bennink, Optimal collinear Gaussian beams for spontaneous parametric down-conversion. Phys. Rev. A **81**, 053805 (2010)
70. P. Eberhard, Background level and counter efficiencies required for a loophole-free Einstein-Podolsky-Rosen experiment. Phys. Rev. A **47**, R747 (1993)
71. J. Clauser, M. Horne, Experimental consequences of objective local theories. Phys. Rev. D **10**, 526 (1974)
72. J.Å. Larsson, R.D. Gill, Bell's inequality and the coincidence-time loophole. EPL (Europhysics Letters) **67**, 707 (2004)
73. J.-Å. Larsson et al., Bell-inequality violation with entangled photons, free of the coincidence-time loophole. Phys. Rev. A **90**, 032107 (2014)
74. M.A. Broome et al., Photonic Boson sampling in a tunable circuit. Science **339**, 794 (2013)
75. A. Crespi et al., Integrated multimode interferometers with arbitrary designs for photonic boson sampling. Nat. Photon. **7**, 545 (2013)

76. J.B. Spring et al., Boson sampling on a photonic chip. Science **339**, 798 (2013)
77. M. Tillmann et al., Experimental boson sampling. Nat. Photon. **7**, 540 (2013)
78. G. Reithmaier et al., On-chip time resolved detection of quantum dot emission using integrated superconducting single photon detectors. Sci. Rep. **3**, 1901 (2013)
79. F. Najafi et al., On-chip detection of non-classical light by scalable integration of single-photon detectors. Nat. Commun. **6**, 5873 (2015)
80. C. Schuck et al., Optical time domain reflectometry with low noise waveguide-coupled superconducting nanowire single-photon detectors. Appl. Phys. Lett. **102**, 191104 (2013)
81. A.S. Webb et al., MCVD planar substrates for UV-written waveguide devices. Electron. Lett. **43**, 517 (2007)
82. D. Zauner et al., Directly UV-written silica-on-silicon planar waveguides with low insertion loss. Electron. Lett. **34**, 1582 (1998)
83. V. Vedral, Quantifying entanglement in macroscopic systems. Nature **453**, 1004 (2008)

Chapter 3
Waveguide Superconducting Single- and Few-Photon Detectors on GaAs for Integrated Quantum Photonics

Döndü Sahin, Alessandro Gaggero, Roberto Leoni and Andrea Fiore

Abstract Integrated quantum photonics offers three principal advantages over bulk optics—low loss, simplicity and scalability. Quantum photonic integrated circuits show promise for on-chip generation, manipulation and detection of tens of single photons for quantum information processing. For quantum photonic integration, gallium arsenide is one of the most promising material platforms as full integration of all active and passive circuit elements can be obtained.

3.1 Introduction

Quantum information processing with photons relies on single-photon sources, passive circuit elements and single-photon detectors. In order to take advantage of quantum physics in advanced quantum technologies such as quantum simulation and quantum computing, tens of photons must be generated, manipulated and detected. However, when the number of photons exceeds a few, bulk optics becomes complex and difficult to scale [1]. Integrated quantum photonics offers a solution to these formidable challenges. Quantum photonic integrated circuits (QPICs) may enable the scalable generation, manipulation and detection of single photons on a chip,

D. Sahin (✉)
Centre for Quantum Photonics, H. H. Wills Physics Laboratory,
University of Bristol, Tyndall Avenue, Bristol, BS8 1TL, UK
e-mail: d.sahin@bristol.ac.uk

A. Gaggero · R. Leoni
Istituto di Fotonica e Nanotecnologie, Consiglio Nazionale delle Richerche (CNR), Via Cineto Romano 42, 00156 Roma, Italy
e-mail: alessandro.gaggero@ifn.cnr.it

R. Leoni
e-mail: roberto.leoni@ifn.cnr.it

A. Fiore
COBRA Research Institute, Eindhoven University of Technology,
PO Box 513, 5600 MB Eindhoven, The Netherlands
e-mail: a.fiore@tue.nl

© Springer International Publishing Switzerland 2016

R.H. Hadfield and G. Johansson (eds.), *Superconducting Devices in Quantum Optics*, Quantum Science and Technology,
DOI 10.1007/978-3-319-24091-6_3

thereby opening the way to quantum information processing at the level of tens of qubits. To date, spontaneous parametric down-conversion, nitrogen-vacancy centres in diamond, and semiconductor systems such as quantum dots (QDs) have been explored as integrated single-photon sources. QDs made of III-V semiconductors, such as InP and GaAs, have attracted considerable attention for their potential to be integrated with semiconductor photonic circuits and particularly for their efficiency when combined with photonic crystal structures. Superconducting nanowire single-photon detectors (referred to as SSPDs or SNSPDs in the literature), on the other hand, are a promising enabling technology for single-photon detection due to their fast response, low dark count rates, low jitter, and scalability [3]. The integration of superconducting nanowire detectors with III-V semiconductor waveguides is therefore of primary importance for the realisation of a scalable and fully-integrated quantum photonic technology. In this chapter, we will first describe the deposition of NbN thin films on GaAs, the nanofabrication of detector structures and the measurement techniques needed to characterise waveguide detectors. We then discuss the design of waveguide single-photon detectors and their electro-optical characteristics. We show a key functionality for the characterisation of quantum light, i.e. the on-chip measurement of the second order intensity correlation function, $g^{(2)}(\tau)$. In the last part of the chapter, we discuss integrated photon-number resolving detectors able to resolve the photon number in a given pulse. The combination of these functionalities makes waveguide photon detectors on the GaAs platform strong candidates for integrated quantum information processing with photons.

3.1.1 Integrated Quantum Photonics

Quantum information processing (QIP), which enables new protocols and functionalities in communication and computing, is on the verge of a transformation from a scientific field to a research-usable technology. The simplest example of a commercial application is quantum key distribution, based on the exchange of a cryptographic key by coding the information onto single photons. The security of the exchange is guaranteed by the laws of quantum mechanics, enabling a physically-secure form of communication [4]. Furthermore, the tantalizing prospect of computational speed-up for a number of important problems has motivated research in the direction of quantum computers, with many experimental platforms being investigated. Single photons are one of the preferred implementations of quantum bits, since they are relatively easily generated, detected and manipulated, and can be transmitted over long distances with low loss and low decoherence. For these reasons, they are used in QKD and will form an integral part of any future QIP system where quantum information needs to be transferred between different nodes or processors. While photons hardly interact with each other, the nonlinearity associated with photon detection can be used to realize quantum gates [5] and in principle even a scalable optical quantum computer [6, 7]. Photonic quantum simulation appears a particularly promising mid-term goal, as it requires moderate numbers of photons (~50–100) (see review in Ref. [8]). However, the production, processing and detection of single photons is still mostly realized using discrete, free-space or fiber-optic devices, resulting in unacceptable complexity and cost when circuits with several photons are built. The development of an integrated quantum photonic circuit is therefore needed to advance the state-of-the-art of photonic QIP in both science and applications. This has motivated a strong recent interest for integrated quantum photonics. Starting from the seminal demonstration of an optical CNOT quantum gate in a silica-on-silicon waveguide circuit [9], outstanding progress has been achieved in controlling photonic qubits on a planar chip by combining waveguides, splitters and phase modulators, based on position-coding [10]. Three-dimensional quantum circuits using polarisation-coding have also been demonstrated [11]. Integrated quantum photonics based on this simple set of components has already equalled or surpassed the state-of-the-art of table-top or fiber-optic quantum optics, leading for example to the demonstration of quantum walks of entangled photons [12] and to several experimental demonstrations of boson sampling [13–16]. However, in view of scaling up the number of photons to few tens, the on-chip integration of sources and detectors is also needed. First demonstrations of the integration of single-photon sources in waveguide circuits have been reported, based either on artificial two-level systems in optical semiconductors [17] or on parametric down-conversion [18–20]. On the detector side, single-photon detectors on waveguides have been reported almost simultaneously by three groups, based on transition-edge sensors (TES) on silica waveguides [21], superconducting nanowires on GaAs/AlGaAs ridge waveguides [22] and superconducting nanowires on Si/SiO_2 waveguides [23]. In this chapter we describe the fabrication and characterization of different detector structures integrated on GaAs waveguides, while waveguide TES and nanowire detectors on Si are discussed in Chaps. 2 and 4 respectively.

3.1.2 GaAs-Based Quantum Photonic Integrated Circuits

III-V semiconductors, and GaAs in particular, are strong candidates for the implementation of a dense QPIC. Single photons and entangled photon pairs can be produced from the radiative recombination from excitonic complexes in quantum dots (see a review in Ref. [17]). QDs are semiconductor nanostructures where carriers are confined in three dimensions, giving rise to energy quantization and a discrete, atom-like density of states. Similarly to atoms, the spontaneous emission from an excited energy state results in the emission of a single photon, while a cascade recombination from a biexcitonic state can produce a pair of entangled photons. The collection of spontaneously emitted photons into waveguides is normally very inefficient, but it can be optimized by controlling the optical density of states around the emitter (for example using photonic crystal structures), so that most photons are emitted into a given mode (see recent reviews in Refs. [25, 26]). The channeling of single photons into photonic crystal waveguides with probabilities above 98 % has been reported [27], and efficient coupling of single photons from photonic crystal to ridge waveguides has been shown [28]. Methods for actively tuning the spectrum of photons emitted from distinct, waveguide-coupled QDs are also being investigated [29, 30], which is promising in view of the fabrication of arrays of sources of indistinguishable photons.

GaAs is also a suitable material for waveguiding and passive circuits. GaAs/AlGa As waveguides with propagation loss of 0.2 dB/cm were reported [31] and directional couplers and multimode-interference (MMI) couplers can be realized, similarly to other material platforms. As compared to Si photonics, a key advantage is the presence of a second-order optical nonlinearity, which enables phase shifting via the Pockels effect, at very low power consumption. Quantum photonic circuits based on GaAs, including phase modulation and directional couplers, were recently demonstrated [32]. As will be shown later in this chapter, single-photon and photon-number detection is possible by integrating superconducting nanowires on GaAs ridge waveguides, completing the suite of functionalities needed e.g. for boson sampling. In the long term, coupling of single photons to spins in waveguide systems [33] may open the way to integrated quantum memories.

3.1.3 Nanowire Detectors on GaAs

Nanowire detectors on GaAs exploit the same photon detection mechanism as free-space coupled SNSPDs [3] (as extensively discussed in Chap. 1). Upon absorption of a photon, quasi-particles are created around the vicinity of absorption point. Through a complex microscopic process involving diffusion of quasi-particles and vortex crossing (see Ref. [34] and references therein), and depending on the bias current, this local perturbation can lead to the creation of a resistive region across the cross-section of the wire. The resistive section further grows due to Joule heating, resulting in a resistance of the order of $1\,\mathrm{k}\Omega$. Therefore, the applied DC bias current to the circuit is diverted to the load resistance producing an electrical pulse which can be amplified with room temperature amplifiers and detected.

Nanowire detectors were initially fabricated on sapphire [3] and MgO [35]. In 2008, they were realised on Si [36]. Later in 2009, a numerical study showed that when nanowires are placed on top of Si_3N_4 waveguide with an inverse tapering (for adiabatic coupling), the absorptance on the nanowires would increase up to 76 % for about $\geq$50 μm–long wires [37]. Nearly simultaneously, nanowire detectors were demonstrated for the first time on GaAs/AlGaAs Bragg mirrors [38] in a free-space-coupled configuration, with a motivation of integrating the active and passive optical elements on a single chip. Then in 2011, the first nanowire detectors integrated on a III-V waveguide were shown, achieving a high efficiency of $\sim$20 % with a waveguide length of only 50 μm [22]. It was also theoretically calculated that an efficiency close to 100 % can be obtained due to high absorptance via evanescent-wave coupling, as discussed in the next section.

Recently, further progress in functionality and integration has been demonstrated by several groups, such as the on-chip measurement of photon statistics (covered in this chapter) [39], the integration of InAs QDs with waveguide detectors and the corresponding time-resolved characterization of QD emission [40, 41]. Besides, in

Ref. [41], the generation and detection of single photons have been demonstrated by using superconducting NbN nanowires on GaAs multimode ridge waveguide. All these recent works demonstrate that the III-V system is highly promising for the realisation of fully-integrated QPICs, potentially paving the way to scalable quantum photonic applications.

3.2 Fabrication of Nanowire Waveguide Detectors on GaAs

3.2.1 Deposition of NbN Thin Films on GaAs

The fabrication of nanowire detectors is extremely challenging due to their low tolerance to variations of the wire properties. As high quantum efficiency with NbN wires can only be obtained if the wire is biased very close to its critical current, any imperfections locally limiting the I_c strongly affect the efficiency of the whole nanowire. This in practice requires a uniform NbN film with thickness in the 3–5 nm range and high critical current. We note that the use of lower-gap superconducting materials such as WSi results in a less critical dependence on the bias current [42, 43] and thereby potentially higher yield. However, waveguide nanowire detectors based on WSi have not yet been demonstrated and are not discussed in what follows. The technique most commonly used for depositing 2D NbN films is DC reactive magnetron sputtering. This method, widely used for depositing metals, alloys and nitrides, exploits the energy and momentum transfer from the ionized sputtering gas, argon (Ar) and nitrogen (N), to a solid target, niobium in our case. For NbN films with an intended thickness of 3.5–4 nm, the superconducting transition temperatures, T_c, as high as 12 K are reported. Among them, NbN films deposited on MgO [44], MgO buffer layers [45] and sapphire substrates [46] exhibit higher T_c than those sputtered on Si and GaAs. In fact, sapphire has a lattice constant closer to NbN, allowing the growth of a polycrystalline film with larger grains. However, as we have discussed in the previous sections, for the realization of a fully integrated QPIC, direct-gap III-V semiconductors, and particularly GaAs, are very promising as they enable the integration of single-photon sources. The use of GaAs substrates poses significant challenges for the nanowire detector fabrication:

(a) the mismatch between the substrate and NbN film lattice parameters is higher ($\sim$23 %) than with other substrates such as MgO ($\sim$4 %), sapphire ($\sim$9 %) or even Si ($\sim$20 %);
(b) the best quality of NbN films is usually obtained with deposition temperatures (800 °C) incompatible with GaAs processing; and
(c) the relatively high atomic density of GaAs (compared with sapphire, MgO or Si) makes high-resolution electron beam lithography (EBL) more challenging because of the stronger proximity effect.

In the following, we show that high-quality NbN films and nanowire devices can be obtained on GaAs despite these problems.

The magnetron sputtering technique allows deposition at low plasma pressure (2–3 mTorr) by confining the plasma only around the cathode. Therefore, the sputtered particles retain most of their kinetic energy because the low pressure strongly reduces the collisions between sputtered particles and the inert gas (argon, Ar). The Ar ions are accelerated towards the cathode, they are practically not affected by the magnetic field, impinge without collision on the target and give rise to Nb sputtering. To deposit nitride compounds (NbN in this case), a partial pressure of nitrogen gas (referred to as reactive gas) is introduced together with Ar in the chamber. As nitrogen reacts with the sputtered Nb, it forms NbN, which grows on the substrate. The Nb/N ratio in the films depends on the balance between the sputtering and nitridization rates of the target. In order to change the film composition and to influence the superconducting properties in a reproducible way it is then important to have control over the target sputtering/nitridization balance. Depositions have been performed in constant current mode as in this mode of operation the nitrogen partial pressure uniquely determines the nitridization state of the target [47], i.e. the composition of the deposited NbN films, which makes it possible to fabricate films with good reproducibility. We use a vacuum chamber equipped with two electrodes, facing each other: the cathode, electrically coupled with the solid Nb target, and the anode, on which the substrate is mounted. A heater sets the substrate temperature during the deposition. The optimum deposition temperature is determined by varying it between 300–600 °C, and it is found to be in the range 350–410 °C, which corresponds to the best compromise between low surface roughness and high T_c [38, 48].

Indeed, high substrate temperatures $T_S \geq 800\,°C$ are used on sapphire substrates [3], to promote the surface diffusion of the sputtered particles, resulting in films with high crystalline quality. We previously developed a low-temperature deposition process ($T_S = 400\,°C$) on MgO, which yielded high quality NbN thin films [35]. However, films deposited on GaAs at elevated temperatures did not reach the same quality as the films on MgO [35]. We have found that the reason for this degradation is as follows: at 400 °C oxides (AsO and As_2O_3) evaporate from the surface of the GaAs substrate during the baking and the deposition procedure, resulting in poor morphology of the substrate surface and thus of the deposited films. Indeed, evaporation of AsO starts at 150 °C and between 300 and 400 °C various chemical reactions take place, leading to the formation of a very stable Ga_2O_3 oxide and evaporation of As and Ga from the substrate, which, due to the masking effect of the oxide, enhances surface roughness [49]. We investigated the surface morphology of NbN/GaAs films by an atomic force microscope (AFM). A 5 nm thick NbN film grown at 400 °C shows a very large granularity with a grain size of about 100 nm [38]. From the roughness analysis we obtained a root mean square roughness $R_{\mathrm{rms}} \sim 10.9$ nm and a mean peak to peak distance of ~45 nm while the peak to peak maximum value is ~75 nm. We attribute this granularity to As-oxide desorption during baking, resulting in the formation of Ga or Ga_2O_3 droplets [49] as confirmed by a baking test performed on a GaAs substrate without Nb. The substrate granularity results in disordered superconducting films with poor superconducting properties, i.e.

low critical temperature ($T_c = 6.8$ K) and wide transition width ($\Delta T_c = 2.4$ K), where the localization of charge carriers by Coulomb interaction and the corresponding enhancement of quantum fluctuations of the phase of the superconductor order parameter induces the superconductor–insulator transition [50]. Lowering the baking and deposition temperature leads to a dramatic improvement of the surface quality and superconducting properties. In fact 4.5 nm NbN film deposited at 350 °C shows an $R_{\mathrm{rms}} \sim 0.1$–0.2 nm and a mean peak to peak value 0.61 nm with the maximum peak to peak value of about 1.1 nm [38]. This improvement in the microstructure has a direct effect on the superconducting transition. Indeed the transition width for 4.5 nm thick NbN film is very narrow with $\Delta T_c = 0.7$ K and the critical temperature is quite high, $T_c = 10.3$ K. Besides, as expected, for a slightly thicker 5 nm NbN film, the transition width is further decreased and the critical temperature is higher [51].

3.2.2 Detector Nanofabrication

For the integration of nanowire detectors on ridge waveguides, NbN films (thickness 4–6 nm) are deposited on top of a GaAs (300 nm or 350 nm) /$Al_{0.75}Ga_{0.25}As$ (1.5 μm) heterostructure grown by molecular beam epitaxy on an undoped GaAs (001) substrate. After NbN deposition, there are four steps of EBL for the fabrication of the superconducting nanowire waveguide single-photon detectors (WSPDs) Waveguide-integrated SNSPD (see Fig. 3.1). The EBL is the main top-down pattern-generation technique in nanotechnology because it provides arbitrary patterning capabilities at the sub-micro-meter scale. EBL is therefore increasingly important for the fabrication of novel photonic, magnetic, and nanoelectronic devices. The EBL working principle is relatively simple: a focused beam of electron is scanned across a substrate covered by an electron-sensitive material (resist) that changes its solubility properties according to the energy deposited by the electron beam. Areas exposed (or not exposed, depending on the tone of the resist) are removed by developing.

In the first EBL step, the electrical contact pads Ti(10 nm)/Au(60 nm) (see Fig. 3.1a) and alignment markers are fabricated by electron beam evaporation and

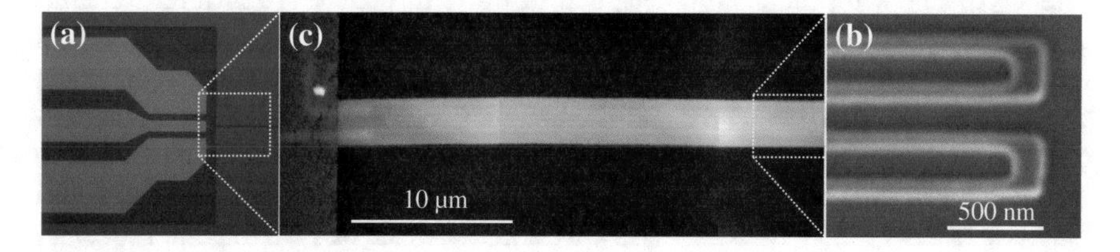

Fig. 3.1 **a** An optical microscope image of the Ti/Au electric contacts, waveguide and detector of a WSPD. **b** Atomic force microscope image of a 30 μm long nanowire detector on top of a 1.85 μm wide GaAs waveguide. **c** Scanning electron micrograph of a 100 nm wide meandered NbN nanowire (Both the detector and the waveguide are still covered by the HSQ etching mask in (**b**) and (**c**))

lift-off using a polymethyl methacrylate (PMMA- a positive tone electronic resist) stencil mask.

In the second EBL step, the meandered nanowire (100 nm-wide and 60–200 μm-long nanowires with 250 nm pitch) is defined on a 180 nm thick hydrogen silsesquioxane (HSQ) resist mask and then transferred to the NbN film with a (CHF_3+SF_6+Ar) reactive ion etching (RIE) (see, Fig. 3.1). The RIE is used to remove unwanted areas of the NbN material, not covered by the HSQ resist, using a reactive gas (SF_6, sulfur hexafluoride in this case or CF_4, carbon tetrafluoride). A high radio frequency (RF) voltage is applied between the anode and the cathode to create the plasma. Highly volatile compounds of Nb-F are formed during the reaction and pumped away, leaving the masked part of the NbN film untouched.

In the third step, another HSQ-mask is patterned by carefully realigning this layer with the previous one in order to define 1.85 μm-wide waveguides, as shown in Fig. 3.1b. This layer also protects the Ti/Au pads against the subsequent reactive etching process. Successively, a 250–300 nm thickness of the underlying GaAs layer is etched by a Cl_2 + Ar electron cyclotron resonance RIE to obtain a ridge waveguide. Finally, for the electrical wiring to the TiAu pads, holes through the remaining HSQ-mask are opened using a PMMA mask and RIE in a CHF_3plasma.

The waveguide photon-number-resolving detectors [39] described below require two extra EBL steps for the parallel integrated resistors [52, 53]. First, Ti(5 nm)/Au (20 nm) electrical connection pads for the parallel resistors are written and lifted-off by using polymethyl methacrylate as second step before the patterning of the nanowires on HSQ. The AuPd resistors, on the other hand, are defined right before the waveguide mask patterning. The Ti(10 nm)/AuPd(50 nm) resistances are fabricated by lift-off of a PMMA stencil mask. Each resistance is 500 nm wide and 3.5 μm long corresponding to a design value of $R_p = 49\ \Omega$ (See Fig. 3.5a for the sketch of parallel resistances, R_p).

3.3 Measurement Setup for Waveguide Detectors

The WSPDs were characterized in a continuous-helium-flow micro-probe station. This cryogenic setup is custom-designed in order to couple the infrared photons into the waveguides using a polarisation-maintaining lensed fibre and to contact the detectors with RF-probes *in situ*. The electro-optical response of the detectors was measured at 1310 nm by the end-fire coupling technique. The fiber and the RF-probe were located on XYZ piezo stages and thermally anchored to the base plate to maintain a low base temperature of below 4 K. The lensed fiber has a nominal spot diameter of 2.5 ± 0.5 μm at $1/e^2$ and a working distance of 14 ± 2 μm. The count rate of the detectors was observed to be extremely sensitive to the fiber-waveguide alignment as well as to their distance, confirming that the detector responds to guided photons and not to stray light propagating along the surface or in the substrate. The output signals, measured with RF-probes, were amplified using a set of low-noise, negative-gain amplifiers from Mini-Circuits (with a bandwidth of 20–6000 MHz).

The count rate was measured to be proportional to the laser power, proving operation in the single-photon regime.

The *on-chip* detection efficiency (OCDE) was derived by dividing the number of counts (after subtracting the dark counts) by the number of photons coupled to the input waveguide. The system detection efficiency (SDE) was derived by dividing the number of counts (after subtracting the dark counts) by the number of photons coupled to the system. Therefore, these two figures are related by SDE = OCDE$\times \eta_c$, where η_c is the coupling efficiency from the lensed fibre into the waveguide. The coupling efficiency was determined by a transmission measurement on GaAs/AlGaAs ridge waveguides without NbN wires and contact pads. The transmission was measured with a tuneable laser operating around 1310 nm, using two nominally identical lensed fibers positioned at the two end facets of the waveguide. Therefore, we assumed that the coupling efficiencies at the two facets were equal. Then the η_c was derived from the measured Fabry-Pérot fringes in the transmission spectra through the waveguide [48]. From standard Fabry-Pérot theory, the transmittance (T) through the waveguide is calculated as

$$T = \eta_c^2 \frac{e^{-\alpha l}}{1 + \mathrm{Re}^{-\alpha l} cos(2kl)} \tag{3.1}$$

where α, l and R stand for propagation loss coefficient, waveguide length and facet reflectance, respectively. Using this expression, the loss coefficient and coupling efficiency can be derived from the measured minimum and maximum transmittance values, using the reflectance value calculated from the simulated effective modal index.

3.4 Waveguide Single-Photon Detectors

3.4.1 Design

Here, two different designs for WSPDs integrated on GaAs waveguides are discussed. All designs are based on 5 nm-thick and 100 nm-wide NbN nanowires niobium nitride (NbN). The first design is based on a ridge-waveguide geometry [22] and the second design is based on a nanobeam structure [28]. Whilst only experimental results based on the former geometry will be discussed later in this chapter, the latter is considered as a promising approach for improving the efficiency.

In the former design, a U-bend NbN nanowire (pitch $p = 250$ nm) is placed on top of a GaAs(300 nm)/$Al_{0.75}Ga_{0.25}As$(1.5 μm) multimodal ridge-waveguide. The ridge waveguide is $w = 1.85$ μm-wide and $h = 250$ nm-deep and provides a strong 2D confinement of the fundamental mode [22]. In all our calculations, we consider a 100 nm-thick silicon oxide (SiO_x) layer on top of the waveguide, remaining of HSQ after processing of electron-beam resist (discussed in the previous section). By using

Table 3.1 Refractive index of niobium nitride (NbN), gallium arsenide (GaAs), aluminum gallium arsenide (AlGaAs) and silicon oxide (SiOx) at a wavelength of 1310 nm

Material	Refractive index
NbN	ñ = 4.35–i4.65
GaAs	ñ = 3.39
$Al_{0.75}Ga_{0.25}As$	ñ = 3.07
SiOx	ñ = 1.46

The GaAs and AlGaAs indices are valid at 10 K, while for NbN and SiOx room temperature values are used, due to the unavailability of ellipsometry data for low-temperature

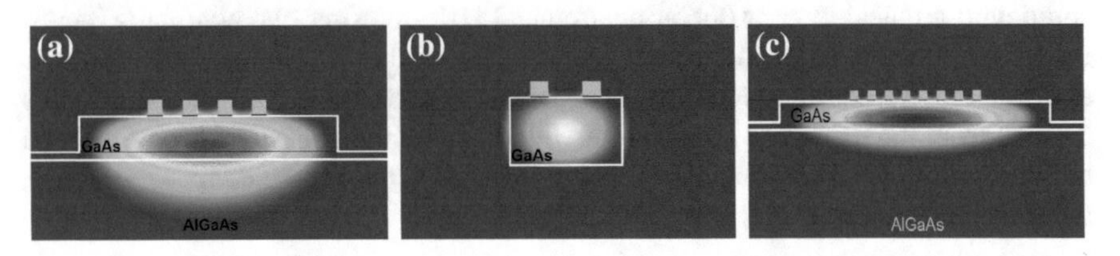

Fig. 3.2 Electric field distribution of the fundamental quasi-TE mode for **a** a ridge-waveguide single-photon detector (core height $t = 300$ nm, ridge height $h = 250$ nm and width $w = 1.85$ μm), **b** a nanobeam single-photon detector ($t = 300$ nm and $w = 500$ nm) and **c** a ridge-waveguide photon-number-resolving detector ($t = 350$ nm, $h = 260$ nm and $w = 3.85$ μm). The white contour is drawn around GaAs and nanowires and SiO_x are coloured with red and grey, respectively for clarity

the refractive indices in Table 3.1 at $\lambda = 1310$ nm, a modal absorption coefficient of $\alpha_{abs}^{TE} = 370\ cm^{-1}$ ($\alpha_{abs}^{TM} = 347\ cm^{-1}$) is calculated for the quasi-transverse electric, TE, (-transverse magnetic, TM) polarisation with a finite element mode solver (Comsol Multiphysics). That provides a total absorptance *(A)* of about 79 and 86 % for a 50 μm-long waveguide in TE and TM polarisations, respectively. Figure 3.2a shows the simulated electric field distribution of the fundamental TE mode. While a high absorptance for both TE and TM modes is obtained with this design, the first quasi-TM mode has a complex polarization profile and low in-/out-coupling efficiency for this design [48]. However, considering the fact that InAs QDs in waveguides emit in the TE polarisation, the design is fully compatible with QPICs. Furthermore, increasing the GaAs core layer thickness by 50 nm (to 350 nm core), a well-confined fundamental TM mode is supported in the waveguide as well. This modified design maintains high absorptance for both polarisations $A_{TE} = 79\,\%$ and $A_{TM} = 86\,\%$ for 50 μm-long waveguide detectors with corresponding modal absorption coefficients of $\alpha_{abs}^{TE} = 313\,cm^{-1}$ and $\alpha_{abs}^{TM} = 391\,cm^{-1}$ [22]. This design was implemented for waveguide autocorrelators [54]. The air/GaAs/AlGaAs waveguide geometry therefore allows high performance nanowire detectors at relatively short lengths. Moreover, when the waveguide length is long enough to satisfy nearly-unity absorptance on NbN nanowires, nearly polarisation-insensitive devices can be obtained.

The second design exploits a nanobeam structure as studied in Ref. [28]. This design is optimised to provide a high absorptance for a short nanowire on a suspended waveguide, i.e. nanobeam. The design is motivated by the fact that long nanowires are more sensitive to inhomogeneities in the NbN film and therefore present a lower internal QE (i.e. probability of detection once a photon is absorbed) [55–58]. This ultimately limits the yield as well and is a potential limitation for the realisation of QPICs featuring tens to hundreds of integrated detectors. The nanowire is again folded into a U-bend meander (the bend is required for electrical contacting both sides of the wire) on top of a 500 nm-wide and 300 nm-thick GaAs waveguide which is suspended in air (nanobeam) as presented in Fig. 3.2b. The symmetric index contrast of nanobeam (air/GaAs/air) improves the overlap between nanowire and the guided light by pushing the mode up into the NbN. That provides modal absorption coefficients as high as $\alpha_{\text{abs}}^{\text{TE}} = 957\,\text{cm}^{-1}$ and $\alpha_{\text{abs}}^{\text{TM}} = 3151\,\text{cm}^{-1}$ at 1310 nm for the fundamental quasi-TE and -TM modes, respectively (by using the refractive indices in Table 3.1). That allows 90 % absorptance for a 24 μm-long nanobeam for the TE polarisation and for a 7 μm-long nanobeam for the TM polarisation [48]. An efficient coupling between nanobeam and ridge waveguide can also be achieved as demonstrated in Ref. [28], which makes the nanobeam design promising for QPICs.

3.4.2 Results

Experimental results obtained using the WSPD design (Fig. 3.2a), representing the first demonstration of GaAs WSPDs, are summarized here. Nanowire WSPDs show the characteristic current-voltage (*IV*) behavior of nanowire detectors (see Chap. 1). A critical current $I_c = 16.9$ μA was measured for a 50 μm-long WSPD (total nanowire length is 4x50 μm). The on-chip detection efficiency (OCDE) measured at 1310 nm for TE polarisation is plotted in Fig. 3.3 (left axis) as a function of the normalized bias current I_b/I_c. The SDE (diamonds) reaches 3.4 % for a 50 μm-long device. A transmission measurement was performed on a sample containing only 3 mm-long ridge waveguides. From the measured Fabry-Pérot fringes with $T_{\text{max}} = 6.1$ % and $T_{\text{min}} = 1.8$ % in the TE polarisation, and a calculated facet

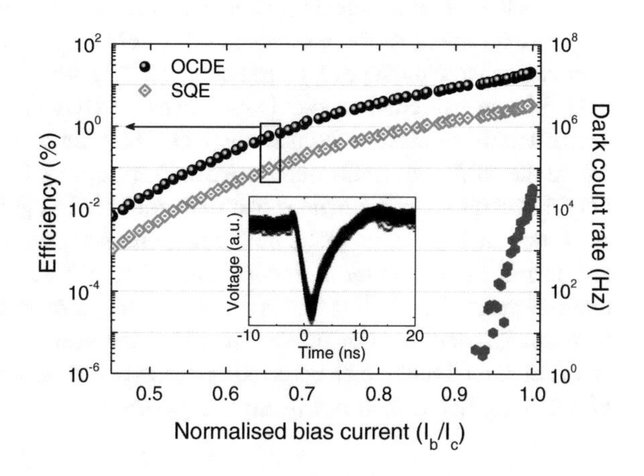

Fig. 3.3 *Left axis* System quantum efficiency (*diamonds*) and on-chip detection efficiency (*dots*) of a superconducting waveguide single-photon detector. *Right axis* Dark count rate as a function of normalized bias current. *Inset* Electrical output pulse of a WPNRD showing 1/e decay time of 3.6 ns

reflectance $R = 27\,\%$, the propagation loss over a 3 mm-long waveguide was calculated as negligible. Moreover, from Eq. (3.1), a coupling efficiency of 17.4 % was derived for the TE polarisation. The corresponding OCDE is plotted (dots) in Fig. 3.3, and reaches 19.7 % at the maximum bias current, $I_b = I_c$. While this value is similar to other works reported on GaAs in the literature on 10 nm-thick wires at 950 nm [40], it is still lower than the calculated absorptance, 84 % in the 50 μm-long WSPD. As discussed earlier in this chapter, uniform NbN film deposition is relatively difficult on GaAs [38] compared to the traditional substrates Al_2O_3 [3] and MgO [35]. This arises from the lower deposition temperature (as limited by the occurrence of surface roughness on the GaAs surface), which limits the surface diffusion of the sputtered atoms on the substrate [38, 48]. Therefore, we attributed the difference between measured OCDE and calculated absorptance to a limited internal quantum efficiency. Further improvements of film quality by increasing the film thickness (and correspondingly reducing the width) may improve the device OCDE.

Another potential cause for limited efficiency can be the modal reflection due to the nanowires, which is however calculated to be very small (in the order of $10^{-5}\,\%$). The dark count rate (DCR) was measured in another cryostat with no optical window at 1.2 K. The DCR is presented on the right axis of Fig. 3.3, showing the usual exponential dependence as a function of the bias current. For a dark count as low as 100 Hz, an OCDE above 10 % is possible.

The inset in Fig. 3.3 shows an electrical output pulse of a NbN nanowire WSPD. A pulse duration (at full-width-half-maximum, FWHM) of 3.2 ns was measured. Furthermore, the 1/e decay time of 3.6 ns corresponds very well to the expected time constant of $\tau_{1/e} = L_{kin}/R = 3.6$ ns, where R is the load resistance, $R = 50\,\Omega$, and the kinetic inductance of the wire was assumed to be $L_{kin} = 180$ nH, as calculated from the kinetic inductance per square for similar NbN wires, $L_{\square} = 90$ pH/□ [59]. A maximum counting rate ∼100 MHz was calculated from the decay time. Furthermore, the detector jitter was measured while the device is illuminated with a pulsed diode laser, and a value of 61 ps at FWHM was derived for the intrinsic detector jitter.

After the demonstration of WSPDs, more complex functionalities were realised based on the integrated detector concept. One key functionality for QPICs is the on-chip measurement of the second-order correlation function, $g^{(2)}(\tau)$. Two different approaches were pursued. The first one, as a very straightforward implementation of the $g^{(2)}(\tau)$ measurements on-chip, was to pattern two distinct WSPDs on each output arm of a multi-mode interference coupler (MMI), acting as an integrated beamsplitter, where the MMI was optimised for a 50:50 splitting [60]. However, here we discuss only a second approach where two detectors are integrated on a single waveguide as shown in Fig. 3.4a so that both sense the same waveguide mode. This device is referred as *waveguide autocorrelator* [54]. While both approaches enable the measurement of the $g^{(2)}(\tau)$ for single-photons on a *compact*, integrated photonic chip, waveguide autocorrelator further simplifies the design and the footprint. Moreover, it was observed that the close packing of the two detectors increased the probability of obtaining two detectors with similar performances.

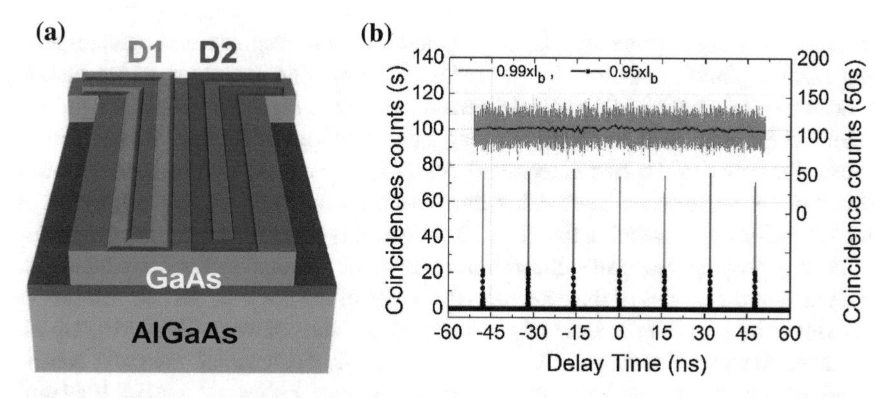

Fig. 3.4 **a** Schematic of the integrated autocorrelator with two, electrically separated nanowire detectors, D1 and D2 on a GaAs waveguide. **b** Coincidence rate under illumination with continuous wave with an excitation power of 77 pW (*right axis*, integration time of 50s) and pulsed laser with 34 pW average power (*left axis*, integration time of 1 s). The detectors are biased at 95 % (*dotted-line*) and 99 % (*line*) of their bias current. The *black line* embedded in CW illumination is the averaging over 1 ns data range

The detectors of a waveguide autocorrelator featured very similar *IV* curves at $T_{base} = 2.1$ K with a critical current $I_c = 23\,\mu$A [54]. The electro-optical response to a continuous wave (CW) 1310 nm diode laser was measured for each element of the autocorrelators. In order to prove that the detectors responded to the single photons, the count rate was measured to be proportional to the laser power as mentioned earlier in this chapter. A 1/e pulse decay time of 1.5 ns was measured for a 50 μm-long detector [54], approximately in agreement with the calculated value $\tau_{1/e} = L_{kin}/R_{out} = 1.8$ ns (the small difference can be due to the different thickness and kinetic inductance as compared to Ref. [59]). The short recovery time of the waveguide autocorrelator with a potential maximum count rate of >200 MHz is a great advantage of integrating two wires on top of the same waveguide.

The OCDE of both detectors was measured using a continuous-wave laser at 1300 nm in the TE polarisation [54]. The two detectors performed very similarly in terms of efficiency, with a peak OCDE value reaching 0.5 % (D1) and 0.9 % (D2) at $I_b = 0.99 I_c$. This value has been derived by taking into account the measured coupling efficiency $\eta_c = 17\,\%$ and 19 % for the fundamental quasi-TE and quasi-TM modes, respectively. The relatively low value of the OCDE for both detectors, despite the high absorptance, was also attributed to the limited internal quantum efficiency as also indicated by the unsaturated bias dependence of the OCDE. This was probably related to the film thickness ($t = 5.9$ nm), larger than the conventional thickness used in nanowire detectors (4–5 nm). Moreover, the quality of this NbN film was also not optimized (as indicated by the $T_c = 10.1$ K, lower than previous demonstrations of NbN nanowire detectors on GaAs [38]. It was anticipated that the OCDE may also be increased using narrower wires [61, 62].

In view of the application to autocorrelation measurements, a detailed study of the mutual coupling of the nanowires was performed. In an ideal autocorrelator, detectors work independently, causing no modified/false response on a detector arising from a photon detection by the other [63, 64]. The detectors on waveguide autocorrelator were closely packed to ensure an equal, high coupling to the guided light, which may lead to electrical, magnetic or thermal coupling, referred to as crosstalk, between the two detectors. Crosstalk is expected to introduce spurious correlations at and around zero delay and therefore would affect the measurement of the second-order correlation function. Therefore, it is of utmost importance to investigate any possible crosstalk-related limitation of the waveguide autocorrelators. All the relevant timescales of the detectors (the decay of the hotspot, the detector recovery time, the propagation time of photons and phonons between the wires) are all expected to be within a few ns. To assess the possible existence of crosstalk, coincidence counts of a continuous wave (1310 nm) and pulsed laser (62.5 MHz, 1064 nm) were measured and are presented in Fig. 3.4b at bias current as high as $0.99 \times I_b$. Indeed, a laser source is expected to present no peak or dip in the correlation counts at zero delay. Within the shot noise limit and our experimental uncertainty of 4 %, no crosstalk could be identified.

Coincidence measurements on a pulsed laser (λ= 1064 nm, ~6 ps-wide pulse width) at zero delay were also used to determine the timing resolution of the autocorrelator. The measured 125 ps of Gaussian distribution at a full-width-half-maximum (FWHM) corresponded to the jitter of two detectors as well as the jitter of the electronics, laser and the correlation card. As two detectors were nominally the same, considering an equal timing jitter for each detector, a jitter of 88 ps (FWHM) was calculated for one detector (amplifiers and the cabling are included).

3.5 Waveguide Photon-Number-Resolving Detectors on GaAs

3.5.1 Photon-Number-Resolving (PNR) Detectors Using Superconducting NbN Nanowires

A single-photon detector is a binary detector and responds to one or more photons arriving onto the detector at the same time or within the dead time in the same way. On the other hand, a photon-number-resolving detector (PNRD) can resolve the number of photons in a single optical pulse. PNRDs are beneficial in quantum key distribution, for example to detect the photon-number splitting attacks to the transmission line, and generally in photonic quantum information processing [6].
Recently, there has been a considerable effort to realise PNRDs either by multiplexing single-photon detectors or by using intrinsically photon-number-resolving detectors. The charge integration photon detectors [65] and superconducting transition edge sensors (TESs) [21, 66, 67] have been shown to resolve the photon number intrinsically. To date, of those only TESs are reported to be integrated with silica waveguides [21, 67]. As TESs require superconducting quantum interference

devices for the small signal amplification with low noise, they are limited in speed. Therefore, TESs are unsuited for high-speed QIP. Moreover, recently it has been shown that an avalanche photodetector can discriminate the photon numbers when the signal is measured at the very early stage of the avalanche, [68], but there has been no demonstration of integrated single-photon avalanche photodetectors. On the other hand, using the multiplexing technique, avalanche photo-diode detectors [69, 70] and SNSPDs [52, 53, 71–74] have been used for resolving the number of photons in an optical pulse. Among those, SNSPDs are outstanding devices as they are suitable for an integrated circuitry and they show great promise for scalability.

For any PNRD, fidelity is an important figure of merit. It defines the accuracy in the photon number measurement, i.e. the probability of measuring n photons when n photons are incident on the detector, $P(n|n)$. The low OCDE, the limited number of elements, the low signal-to-noise ratio (SNR) and the crosstalk represent potential limits to the fidelity of a PNRD. For typical OCDE values $\eta < 90\,\%$, and when $n << N$ (N being the number of elements of the multiplexed detector), the fidelity is mainly limited by the low OCDE rather than the number of elements [39, 72]. This is intuitively understood: a low OCDE implies that often at least one photon will not be detected, leading to a wrong photon number measurement. The increase in the OCDE increases the fidelity dramatically while with a low OCDE, increasing N has no significant effect. However, once the OCDE is high and $n \sim N$, the number of elements will be a limiting factor for the fidelity of a PNRD. Indeed, if the incoming photons are spread over the N elements, there is a chance that two of them will be absorbed by the same element, which produces an error. Moreover, a high SNR is needed for well-separated signal levels, which leads to a clear discrimination of the photon levels. The last limitation can be the crosstalk, which is a spurious firing of a detector after photon absorption in the other one. As it is not a real detection event, it is expected to limit the fidelity as well. Besides those factors, PNRDs based on time-multiplexing are affected from the transmission losses in the delay line(s). They are modelled as a multi-element detector with different efficiencies.

PNRDs based on superconducting nanowires connected in parallel [71] and in series [53] have been demonstrated. In particular, the series configuration offer the best potential for scalability to large photon numbers [52]. Four-element series PNRDs [39, 53] as well as twelve-element [74] and twenty-four-element [75] series PNRDs based on NbN nanowires have been demonstrated with a good SNR. Here, only series-nanowire PNRDs integrated with GaAs ridge waveguides are presented. These devices are called waveguide photon-number-resolving detectors, or WPNRDs. The WPNRDs based on nanowires are promising due to their relatively high operation temperature and speed, as compared to TESs.

3.5.2 *Design of WPNRDs*

The detector was based on four NbN superconducting nanowires on top of GaAs/$Al_{0.75}Ga_{0.25}As$ (0.35 $\mu m/1.5 \mu m$) ridge waveguide [39]. As for WSPDs, the

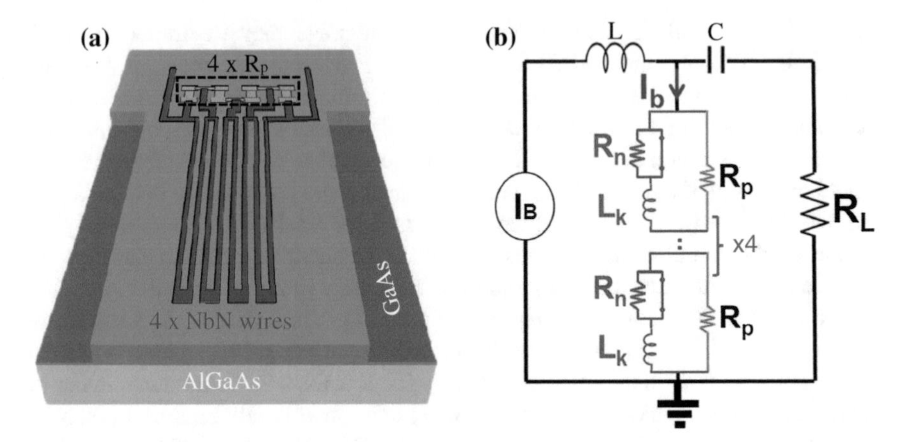

Fig. 3.5 **a** Schematic diagram of a waveguide photon-number-resolving detector (WPNRD). Four NbN wires (*red*), connected in series, are located on top of a GaAs waveguide and each is in parallel to an integrated resistance (*dark yellow*, R_p). **b** The equivalent circuit diagram of the WPNRD. L and C are the inductance and the capacitance of the bias-tee and R_n and L_k are the resistance (in the normal phase) and inductance of each wire, respectively

wires are 5 nm-thick and 100 nm-wide with a pitch of 250 nm. The waveguide length is 30 μm and each wire is 60 μm long. Figure 3.5a depicts a sketch of WPNRD, based on the series connection of four wires with their own integrated shunt resistances (R_p) [52]. The photon detection mechanism in each wire is the same as in nanowire detectors ([3], Chap. 1). The circuit diagram in Fig. 3.5b shows the equivalent electrical circuit (only two wires are shown for simplicity). Similarly to SNSPDs [76], each nanowire was modelled with a variable resistance (R_n) and an inductance (kinetic inductance, L_k). For WPNRD, additionally each wire was connected in parallel to an integrated resistance (R_p) while connected in series with the other nanowires. The nanowires are biased with a current (bias current, I_b) close to their critical currents (I_c). Upon absorption of a single-photon, if a resistive region is formed across the nanowire, the I_b is diverted to its own resistance integrated in parallel to the wire and a voltage pulse is produced. If several wires switch simultaneously, a voltage approximately proportional to the number of switching wires is measured on the load resistance. The concept and expected performance of the series nanowire detectors have been discussed in detail in Ref. [52].

The wider geometry as compared to WSPDs [22] was enforced by the increased number of wires (four wires) along the lateral direction of the waveguide. The width is optimised to minimize the difference in the modal absorption coefficients due to the central and lateral wires while still maintaining the high absorptance of the TE and TM modes. The four nanowires on top of a 3.85 μm-wide and 350 nm-thick ridge GaAs waveguide, etched by 260 nm, was simulated. The total absorptance for the fundamental TE and TM modes was calculated as $A_{TE} = 63\,\%$ and $A_{TM} = 72\,\%$ along a 30 μm-long ridge waveguide with the respective modal absorption coefficients of $\alpha_{tot}^{TE} = 327\,\text{cm}^{-1}$ and $\alpha_{tot}^{TM} = 429\,\text{cm}^{-1}$. The modal absorption coefficients

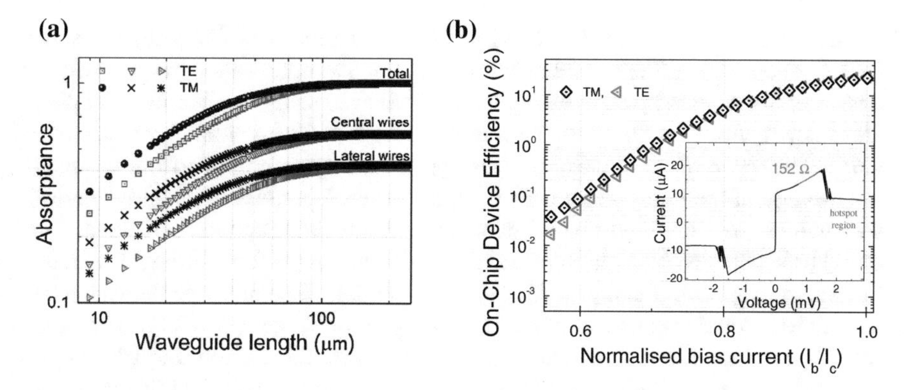

Fig. 3.6 **a** Absorptance calculation for a WPNRD for TE (*red*) and TM (*blue*) polarisations, for central and lateral wires as well as all the wires. **b** On-chip detection efficiency (OCDE) of a WPNRD for TE and TM polarisations. *Inset* Experimental current versus characteristic of a series superconducting photon-number-resolving detector, with $4 \times R_p = 152\,\Omega$

by only the two central wires, $\alpha^{TE}_{cent} = 193\,cm^{-1}$ and $\alpha^{TM}_{cent} = 251\,cm^{-1}$, were higher than the corresponding absorption by the two lateral wires $\alpha^{TE}_{lat} = 136\,cm^{-1}$ and $\alpha^{TM}_{lat} = 182\,cm^{-1}$ for both polarisations due to the confinement profile of the mode as shown in Fig. 3.2c. In order to understand the effect of this unbalance to the PNRD fidelity, the probability of absorption after propagating over a length l, $P(l)$, for central(lateral), $P_{cent(lat)}(l)$, wires was calculated from

$$P_{cent(lat)}(l) = \frac{\alpha_{cent(lat)}}{\alpha_{tot}}(1 - e^{\alpha_{tot} l}) \tag{3.2}$$

and it is plotted for both TE and TM polarisations for the two central (circles) and lateral (diamonds) wires in Fig. 3.6a. The situation is analogous to an unbalanced N-port splitter [77] followed by single-photon detectors. It was calculated that the corresponding unbalance in detection probability does not significantly limit the fidelity of the PNR measurement [39]. Moreover, the design is tolerant to the variation of the etching depth between 250 and 300 nm.

3.5.3 Experimental Results on WPNRDs

The inset of Fig. 3.6b presents the characteristic *IV* response of a four-wire WPNRD [39], following the design discussed above. A critical current $I_c = 10\,\mu A$ was measured. The linear slope observed in the *IV* curve after reaching I_c was related to the series connection of the four resistances, $4 \times R_p = 152\,\Omega$ (38 Ω /each). Then, a hotspot plateau region was observed after all four wires switch to their resistive state (see Fig. 3.6b, inset).

The SDE was measured by using a continuous-wave (CW) laser attenuated to the single-photon level at 1310 nm. It reached 4 % in the TE polarisation and 3.3 %

in the TM polarisation. The OCDE of a WPNRD is plotted in Fig. 3.6b, where the OCDE reaches $24 \pm 2\,\%$ and $22 \pm 1\,\%$ for TE and TM polarisation, respectively, at a bias current $I_b = 9.3\,\mu$A. These numbers are derived using the measured coupling efficiency η_c from the fiber into the waveguide, $\eta_c^{TE} = 17 \pm 1\,\%$ and $\eta_c^{TM} = 14.8 \pm 0.6\,\%$. The value of η_c in this case was approximately determined from the spectral average of the Fabry-Pérot (FP) fringes measured on four, nominally identical waveguides (with no wires on top) by using a tunable laser around 1310 nm and its error bar is obtained from the standard deviation among the four waveguides. In the transmission spectra, a complex fringe pattern for the TE polarisation motivated the use of the spectral average, instead of the Fabry-Pérot fringes (Eq. (3.1)). The value of η_c is confirmed to correspond well to the one determined from the fringe contrast ($\eta_c^{TM} = 14 \pm 1\,\%$) for the TM polarisation. While this was the highest OCDE reported for superconducting nanowire detectors on GaAs, the OCDE did not reach unity [39]. That was attributed to two reasons. First, as discussed above, the internal quantum efficiency may be lower than one [38, 48]. Second, the absorptance of the 30 μm-long waveguides was calculated as $A_{TE} = 63\,\%$ and $A_{TM} = 72\,\%$. We anticipated that longer wires may lead to a higher device OCDE for high quality film. A change in the ratio of the TE and TM efficiencies was observed at low bias current, which seems to indicate a polarisation dependent internal quantum efficiency, as also observed also in Ref. [78] under free-space illumination.

The temporal response of the WPNRD was probed with a TE polarized pulsed laser-diode (10 MHz) at 1310 nm using a sampling oscilloscope with the detector biased at $I_b = 8.8\,\mu$A. A 1/e decay time, $\tau_{1/e} = 6.2$ ns is calculated from the photoresponse pulse, corresponding to four-photon absorption [39]. That value agreed well with the value of $\tau_{1/e} = 5.6$ ns obtained from the simulation using the electrothermal model [52] and plotted in the same graph (red line). From the decay time, a maximum count rate of >50 MHz was estimated.

The PNR capability of WPNRDs was tested under illumination of an attenuated pulsed laser-diode (~100 ps pulse width, 2 MHz repetition rate), where the photon statistics can be assumed to be Poissonian. The photoresponse of the detector was measured with a 40 GHz bandwidth sampling oscilloscope and Fig. 3.7a shows an example of a typical measurement. The measurement was conducted in TE polarisation for an average number of detected photons $\bar{\mu} \approx 2.3$ /pulse at $I_b = 8.8\,\mu$A. Five distinct detection levels correspond to the detection of 0-, 1-, 2-, 3- and 4-photon events. A low-pass filter (LPF: DC - 80 MHz), added to the circuit to remove the high frequency noise, was responsible for the slow rise time, as compared to the measurement shown in the inset of Fig. 3.7b.

Moreover, the count rate of a WPNRD was measured at a fixed bias current, $I_b = 8.8\,\mu$A, as a function of the threshold voltage (V_{th}) of a frequency counter for TE polarisation. At different laser powers (with 12 MHz repetition rate), when the threshold levels in the counter were set to the different levels, the detection probability corresponding to ≥1-, ≥2- , ≥3- and ≥4-photon detection events was measured as given in Fig. 3.7a [39]. The results were in a good agreement with the expected detection probability for a Poissonian source in the regime where detected average

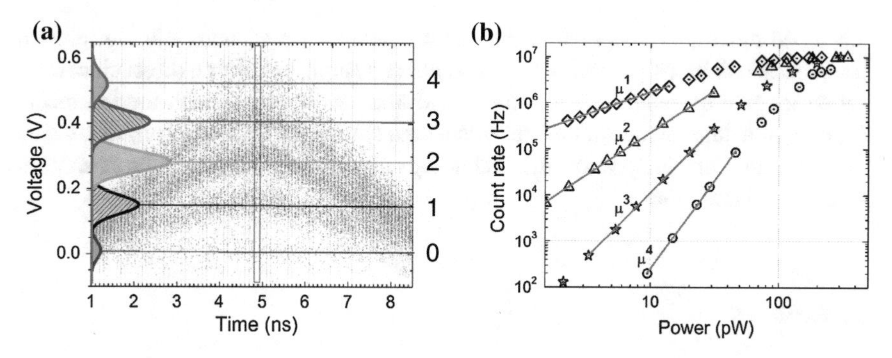

Fig. 3.7 **a** Photodetection of a WPNRD on ersistence mode of sampling oscilloscope. The shown area in the *centre* of the plot, used to reproduce the pulse height distribution on the *left* axis corresponding to 0-, 1- , 2- , 3- and 4-photon detections, shows about 100 ps time window. **b** Count rate measured with a pulsed laser with a repetition rate of 12 MHz. Photon counting levels are shown for corresponding >1- (*diamond*), >2- (*triangle*), >3- (*star*) and >4-photon (*circle*) photon absorption

photon number was $\bar{\mu} << 1$, as shown by the μ^x fits (black lines with $x = 1$–4) in the figure for each photon level.

The excess noise in these WPNRDs was analyzed and it was concluded that serially connected WPNRDs are a promising approach for scaling the number of photons beyond a few photons [39, 74]. Furthermore, we studied the potential factors affecting the fidelity (as discussed in Sect. 5.2) for the two-photon case, i.e. two photons incident on the four-wire WPNRD. With the reported efficiency value, the main limitation of the fidelity is represented by the low device OCDE. However, once the high efficiency detectors are realised, the limited number of wires will limit the highest fidelity to 0.75 for 2 photons on a four-element PNRD. On the other hand, the unbalanced absorptance between the central and lateral wires will only affect the fidelity marginally, decreasing it to 0.73. More detailed study on the excess noise and the fidelity calculations can be found in Ref. [51].

3.6 Conclusions

An approach to integrated quantum photonics, based on the GaAs material system, has been discussed in this chapter. The GaAs platform is very promising for this application, since it enables the monolithic integration of all the active and passive circuit elements. By using NbN superconducting nanowires, the successful integration of the key photon detection functionalities (single-photon detection, autocorrelation, photon-number measurement) has been demonstrated. The WPNRD technology benefits from high efficiency while preserving all the benefits of nanowire detectors in terms of low dark counts, small jitter, high count rate and ungated operation. Simple waveguide engineering also allows the design of polarisation-independent detectors.

Advanced functionalities, such as the measurement of the second-order correlation function on a chip, can be realized in a compact design, and offer a direct characterisation method of quantum light on chip. Moreover, WPNRDs with four elements allow photon number measurements with a high count rate. Due to the promise GaAs holds in monolithic integration, this technology may pave the way for future quantum photonic technologies.

References

1. X.-C. Yao, T.-X. Wang, P. Xu, H. Lu, G.-S. Pan, X.-H. Bao et al., Observation of eight-photon entanglement. Nat. Photon. **6**, 225–228 (2012)
2. J.L. O'Brien, A. Furusawa, J. Vučković, Photonic quantum technologies. Nat. Photonics **3**, 687–695 (2009)
3. G.N. Gol'tsman et al., O. Okunev, G. Chulkova, A. Lipatov, A. Semenov, K. Smirnov et al., Picosecond superconducting single-photon optical detector. Appl. Phys. Lett. 79, 705-707 (2001)
4. N. Gisin, G. Ribordy, W. Tittel, H. Zbinden, Quantum cryptography. Rev. Mod. Phys. **74**, 96–145 (2002)
5. J.L. O'Brien, G.J. Pryde, A.G. White, T.C. Ralph, D. Branning, Demonstration of an all-optical quantum controlled-NOT gate. Nature **426**, 264–267 (2003)
6. E. Knill, R. Laflamme, G.J. Milburn, A scheme for efficient quantum computation with linear optics. Nature **409**, 46–52 (2001)
7. R. Raussendorf, H.J. Briegel, A one-way quantum computer. Phys. Rev. Lett. **86**, 5188–5191 (2001)
8. A. Aspuru-Guzik, P. Walther, Photonic quantum simulators. Nat. Phys. **8**, 285–291 (2012)
9. A. Politi, M.J. Cryan, J.G. Rarity, S.Y. Yu, J.L. O'Brien, Silica-on-silicon waveguide quantum circuits. Science **320**, 646–649 (2008)
10. P.J. Shadbolt, M.R. Verde, A. Peruzzo, A. Politi, A. Laing, M. Lobino et al., Generating, manipulating and measuring entanglement and mixture with a reconfigurable photonic circuit. Nat. Photon. **6**, 45–49 (2012)
11. A. Crespi, R. Ramponi, R. Osellame, L. Sansoni, I. Bongioanni, F. Sciarrino et al., Integrated photonic quantum gates for polarization qubits. Nat. Commun. **2**, 566 (2011)
12. A. Peruzzo, M. Lobino, J.C.F. Matthews, N. Matsuda, A. Politi, K. Poulios et al., Quantum walks of correlated photons. Science **329**, 1500–1503 (2010)
13. S. Aaronson, A. Arkhipov, The computational complexity of linear optics, in Stoc 11: Proceedings of the 43rd ACM Symposium on Theory of Computing, pp. 333-342 (2011)
14. A. Crespi, R. Osellame, R. Ramponi, D.J. Brod, E.F. Galvao, N. Spagnolo et al., Integrated multimode interferometers with arbitrary designs for photonic boson sampling. Nat. Photon. **7**, 545–549 (2013)
15. J.B. Spring, B.J. Metcalf, P.C. Humphreys, W.S. Kolthammer, X.M. Jin, M. Barbieri et al., Boson sampling on a photonic chip. Science **339**, 798–801 (2013)
16. M. Tillmann, B. Dakic, R. Heilmann, S. Nolte, A. Szameit, P. Walther, Experimental boson sampling. Nat. Photon. **7**, 540–544 (2013)
17. A.J. Shields, Semiconductor quantum light sources. Nat. Photon **1**, 215–223 (2007)
18. A. Martin, O. Alibart, M.P.D. Micheli, D.B. Ostrowsky, S. Tanzilli, A quantum relay chip based on telecommunication integrated optics technology. New J. Phys. **14**, 025002 (2012)
19. N. Matsuda, H. Le Jeannic, H. Fukuda, T. Tsuchizawa, W.J. Munro, K. Shimizu et al., A monolithically integrated polarization entangled photon pair source on a silicon chip. Sci. Rep. 2 (2012)

20. J.W. Silverstone, D. Bonneau, K. Ohira, N. Suzuki, H. Yoshida, N. Iizuka et al., On-chip quantum interference between silicon photon-pair sources. Nat. Photon. **8**, 104–108 (2014)
21. T. Gerrits, N. Thomas-Peter, J.C. Gates, A.E. Lita, B.J. Metcalf, B. Calkins et al., On-chip, photon-number-resolving, telecommunication-band detectors for scalable photonic information processing. Phys. Rev. A **84**, 060301 (2011)
22. J.P. Sprengers, A. Gaggero, D. Sahin, S. Jahanmirinejad, G. Frucci, F. Mattioli et al., Waveguide superconducting single-photon detectors for integrated quantum photonic circuits. Appl. Phys. Lett. **99**, 181110 (2011)
23. W.H.P. Pernice, C. Schuck, O. Minaeva, M. Li, G.N. Goltsman, A.V. Sergienko et al., High-speed and high-efficiency travelling wave single-photon detectors embedded in nanophotonic circuits. Nat. Commun. **3**, 1325 (2012)
24. O. Benson, C. Santori, M. Pelton, Y. Yamamoto, Regulated and entangled photons from a single quantum dot. Phys. Rev. Lett. **84**, 2513–2516 (2000)
25. S. Buckley, K. Rivoire, J. Vuckovic, Engineered quantum dot single-photon sources. Rep. Prog. Phys. **75**, 126503 (2012)
26. P. Lodahl, S. Mahmoodian, S. Stobbe, Interfacing single photons and single quantum dots with photonic nanostructures. Rev. Mod. Phys. **87**, 347 (2015)
27. M. Arcari, I. Söllner, A. Javadi, S. Lindskov Hansen, S. Mahmoodian, J. Liu et al., Near-unity coupling efficiency of a quantum emitter to a photonic crystal waveguide. Phys. Rev. Lett. **113**, 093603 (2014)
28. S.F. Poor, T.B. Hoang, L. Midolo, C.P. Dietrich, L.H. Li, E.H. Linfield et al., Efficient coupling of single photons to ridge-waveguide photonic integrated circuits. Appl. Phys. Lett. **102**, 131105 (2013)
29. T.B. Hoang, J. Beetz, M. Lermer, L. Midolo, M. Kamp, S. Hofling et al., Widely tunable, efficient on-chip single photon sources at telecommunication wavelengths. Opt. Express **20**, 21758–21765 (2012)
30. L. Midolo, F. Pagliano, T.B. Hoang, T. Xia, F.W.M. van Otten, L.H. Li et al., Spontaneous emission control of single quantum dots by electromechanical tuning of a photonic crystal cavity. Appl. Phys. Lett. **101**, 091106 (2012)
31. A.D. Ferguson, A. Kuver, J.M. Heaton, Y. Zhou, C.M. Snowden, S. Iezekiel, Low-loss, single-mode GaAs/AlGaAs waveguides with large core thickness. IEE Proc. Optoelectron. **153**, 51–56 (2006)
32. J. Wang, A. Santamato, P. Jiang, D. Bonneau, E. Engin, J.W. Silverstone et al., Gallium arsenide (GaAs) quantum photonic waveguide circuits. Opt. Commun. **327**, 49–55 (2014)
33. I.J. Luxmoore, N.A. Wasley, A.J. Ramsay, A.C.T. Thijssen, R. Oulton, M. Hugues et al., Interfacing spins in an in GaAs quantum dot to a semiconductor waveguide circuit using emitted photons. Phys. Rev. Lett. **110**, 037402 (2013)
34. J.J. Renema, R. Gaudio, Q. Wang, Z. Zhou, A. Gaggero, F. Mattioli et al., Experimental test of theories of the detection mechanism in a nanowire superconducting single photon detector. Phys. Rev. Lett. **112**, 117604 (2014)
35. F. Marsili, D. Bitauld, A. Fiore, A. Gaggero, F. Mattioli, R. Leoni et al., High efficiency NbN nanowire superconducting single photon detectors fabricated on MgO substrates from a low temperature process. Opt. Express **16**, 3191–3196 (2008)
36. S.N. Dorenbos, E.M. Reiger, U. Perinetti, V. Zwiller, T. Zijlstra, T.M. Klapwijk, Low noise superconducting single photon detectors on silicon. Appl. Phys. Lett. **93**, 131101 (2008)
37. H. Xiaolong, C.W. Holzwarth, D. Masciarelli, E.A. Dauler, K.K. Berggren, Efficiently coupling light to superconducting nanowire single-photon detectors. IEEE Trans. Appl. Supercond. **19**, 336–340 (2009)
38. A. Gaggero, S.J. Nejad, F. Marsili, F. Mattioli, R. Leoni, D. Bitauld et al., Nanowire superconducting single-photon detectors on GaAs for integrated quantum photonic applications. Appl. Phys. Lett. **97**, 151108 (2010)
39. D. Sahin, A. Gaggero, Z. Zhou, S. Jahanmirinejad, F. Mattioli, R. Leoni et al., Waveguide photon-number-resolving detectors for quantum photonic integrated circuits. Appl. Phys. Lett. **103**, 111116 (2013)

40. G. Reithmaier, S. Lichtmannecker, T. Reichert, P. Hasch, K. Muller, M. Bichler et al., On-chip time resolved detection of quantum dot emission using integrated superconducting single photon detectors. Sci. Rep. **3**, 1901 (2013)
41. G. Reithmaier, M. Kaniber, F. Flassig, S. Lichtmannecker, K. Müller, A. Andrejew, J. Vuckovic, R. Gross, J. Finley, On-chip generation, routing and detection of quantum light. Nano Lett. **15**, 5208 (2015)
42. B. Baek, A.E. Lita, V. Verma, S.W. Nam, Superconducting a-W_xSi_{1-x} nanowire single-photon detector with saturated internal quantum efficiency from visible to 1850 nm. Appl. Phys. Lett. **98**, 251105 (2011)
43. F. Marsili, V.B. Verma, J.A. Stern, S. Harrington, A.E. Lita, T. Gerrits et al., Detecting single infrared photons with 93% system efficiency. Nat. Photon. **7**, 210–214 (2013)
44. Z. Wang, A. Kawakami, Y. Uzawa, B. Komiyama, Superconducting properties and crystal structures of single-crystal niobium nitride thin films deposited at ambient substrate temperature. J. Appl. Phys. **79**, 7837–7842 (1996)
45. Y.B. Vachtomin, M.I. Finkel, S.V. Antipov, B.M. Voronov, K.V. Sminov, N.S. Kaurova, et al., Gain bandwidth of phonon-cooled HEB mixer made of NbN thin film with MgO buffer layer on Si, in Proceedings of the 13th International Symposium on Space Terahertz Technology (unpublished),Cambridge, MA, (2002) p. 259
46. F. Marsili, D. Bitauld, A. Fiore, A. Gaggero, F. Mattioli, R. Leoni et al., High efficiency NbN nanowire superconducting single photon detectors fabricated on MgO substrates from a low temperature process. Opt. Express **16**, 3191–3196 (2008)
47. P. Yagoubov, G. Gol'tsman, B. Voronov, L. Seidman, V. Siomash, S. Cherednichenko, et al., The bandwidth of HEB mixers employing ultrathin NbN films on sapphire substrate, in Presented at the 7th Int. Symp. on Space Terahertz Tech., Charlottesville, VA (1996) pp. 290-302
48. D. Sahin, A. Gaggero, J.W. Weber, I. Agafonov, M.A. Verheijen, F. Mattioli et al., Waveguide nanowire superconducting single-photon detectors fabricated on GaAs and the study of their optical properties. IEEE J. Sel. Top. Quantum Electron. **21**, 1–10 (2015)
49. A. Guillén-Cervantes, Z. Rivera-Alvarez, M. López-López, E. López-Luna, I. Hernández-Calderón, GaAs surface oxide desorption by annealing in ultra high vacuum. Thin Solid Films **373**, 159–163 (2000)
50. Y. Dubi, Y. Meir, Y. Avishai, Nature of the superconductor-insulator transition in disordered superconductors. Nature **449**, 876–880 (2007)
51. D. Sahin, PhD Thesis, Waveguide single-photon and photon-number-resolving detectors, Chapter 5, Eindhoven University of Technology. ISBN: 978-90-386-3537-8 (2014)
52. S. Jahanmirinejad, A. Fiore, Proposal for a superconducting photon number resolving detector with large dynamic range. Opt. Express **20**, 5017–5028 (2012)
53. S. Jahanmirinejad, G. Frucci, F. Mattioli, D. Sahin, A. Gaggero, R. Leoni et al., Photon-number resolving detector based on a series array of superconducting nanowires. Appl. Phys. Lett. **101**, 072602 (2012)
54. D. Sahin, A. Gaggero, T.B. Hoang, G. Frucci, F. Mattioli, R. Leoni et al., Integrated autocorrelator based on superconducting nanowires. Opt. Express **21**, 11162–11170 (2013)
55. A.J. Kerman, E.A. Dauler, J.K.W. Yang, K.M. Rosfjord, V. Anant, K.K. Berggren et al., Constriction-limited detection efficiency of superconducting nanowire single-photon detectors. Appl. Phys. Lett. **90**, 101110 (2007)
56. J.A. O'Connor, P.A. Dalgarno, M.G. Tanner, R.J. Warburton, R.H. Hadfield, B. Baek et al., Nano-optical studies of superconducting nanowire single photon detectors, in *Quantum Communication and Quantum Networking*, ed. by A. Sergienko, S. Pascazio, P. Villoresi (Springer, Berlin, 2010), pp. 158–166
57. F. Mattioli, R. Leoni, A. Gaggero, M.G. Castellano, P. Carelli, F. Marsili et al., Electrical characterization of superconducting single-photon detectors. J. Appl. Phys. **101**, 054302 (2007)
58. R. Gaudio, K.P.M. op 't Hoog, Z. Zhou, D. Sahin, A. Fiore, Inhomogeneous critical current in nanowire superconducting single-photon detectors, Appl. Phys. Lett. **105**, 222602 (2014)
59. F. Marsili, D. Bitauld, A. Gaggero, S. Jahanmirinejad, R. Leoni, F. Mattioli et al., Physics and application of photon number resolving detectors based on superconducting parallel nanowires. New J. Phys. **11**, 045022 (2009)

60. A. Gaggero, D. Sahin, P. Jiang, F. Mattioli, R. Leoni, J. Beetz, et al., Fully-integrated Hanbury Brown and Twiss interferometer, under preparation (2015)
61. F. Marsili, F. Bellei, F. Najafi, A.E. Dane, E.A. Dauler, R.J. Molnar et al., Efficient single photon detection from 500 nm to 5 μm wavelength. Nano Lett. **12**, 4799 (2012)
62. C. Schuck, W.H.P. Pernice, H.X. Tang, Waveguide integrated low noise NbTiN nanowire single-photon detectors with milli-Hz dark count rate. Sci. Rep. **3**, 1893 (2013)
63. E.A. Dauler, B.S. Robinson, A.J. Kerman, J.K.W. Yang, K.M. Rosfjord, V. Anant et al., Multi-element superconducting nanowire single-photon detector. IEEE Trans. Appl. Supercond. **17**, 279–284 (2007)
64. T. Yamashita, S. Miki, H. Terai, K. Makise, Z. Wang, Crosstalk-free operation of multielement superconducting nanowire single-photon detector array integrated with single-flux-quantum circuit in a 0.1 W Gifford-McMahon Cryocooler. Opt. Lett. **37**, 2982–2984 (2012)
65. M. Fujiwara, M. Sasaki, Direct measurement of photon number statistics at telecom wavelengths using a charge integration photon detector. Appl. Opt. **46**, 3069–3074 (2007)
66. D. Rosenberg, A.J. Kerman, R.J. Molnar, E.A. Dauler, High-speed and high-efficiency superconducting nanowire single photon detector array. Opt. Express **21**, 1440–1447 (2013)
67. B. Calkins, P.L. Mennea, A.E. Lita, B.J. Metcalf, W.S. Kolthammer, A. Lamas-Linares et al., High quantum-efficiency photon-number-resolving detector for photonic on-chip information processing. Opt. Express **21**, 22657–22670 (2013)
68. B.E. Kardynal, Z.L. Yuan, A.J. Shields, An avalanche-photodiode-based photon-number-resolving detector. Nat. Photon. **2**, 425–428 (2008)
69. M.J. Fitch, B.C. Jacobs, T.B. Pittman, J.D. Franson, Photon-number resolution using time-multiplexed single-photon detectors. Phys. Rev. A **68**, 043814 (2003)
70. E. Pomarico, B. Sanguinetti, R. Thew, H. Zbinden, Room temperature photon number resolving detector for infrared wavelengths. Opt. Express **18**, 10750–10759 (2010)
71. A. Divochiy, F. Marsili, D. Bitauld, A. Gaggero, R. Leoni, F. Mattioli et al., Superconducting nanowire photon-number-resolving detector at telecommunication wavelengths. Nat. Photon **2**, 302–306 (2008)
72. E.A. Dauler, A.J. Kerman, B.S. Robinson, J.K.W. Yang, B. Voronov, G. Goltsman et al., Photon-number-resolution with sub-30-ps timing using multi-element superconducting nanowire single photon detectors. J. Mod. Opt. **56**, 364–373 (2009)
73. D. Sahin, A. Gaggero, G. Frucci, S. Jahanmirinejad, J.P. Sprengers, F. Mattioli, et al., Waveguide superconducting single-photon autocorrelators for quantum photonic applications, (2013) Proc. SPIE Opto. pp. 86351B-86351B-6
74. Z. Zhou et al., S. Jahanmirinejad, F. Mattioli, D. Sahin, G. Frucci, A. Gaggero et al., Superconducting series nanowire detector counting up to twelve photons. Opt. Express 22, 3475-3489 (2014)
75. F. Mattioli, Z. Zhou, A. Gaggero, R. Gaudio, S. Jahanmirinejad et al., Photon-number-resolving superconducting nanowire detectors. Superconductor Science and Technology **28**(10), 104001 (2015)
76. A.J. Kerman, E.A. Dauler, W.E. Keicher, J.K.W. Yang, K.K. Berggren, G. Gol'tsman et al., Kinetic-inductance-limited reset time of superconducting nanowire photon counters. Appl. Phys. Lett. **88**, 111116 (2006)
77. T. Huang, X. Wang, J. Shao, X. Guo, L. Xiao, S. Jia, Single event photon statistics characterization of a single photon source in an imperfect detection system. J. Lumin. **124**, 286–290 (2007)
78. V. Anant, A.J. Kerman, E.A. Dauler, J.K.W. Yang, K.M. Rosfjord, K.K. Berggren, Optical properties of superconducting nanowire single-photon detectors. Opt. Express **16**, 10750–10761 (2008)

Chapter 4
Waveguide Integrated Superconducting Nanowire Single Photon Detectors on Silicon

Wolfram H.P. Pernice, Carsten Schuck and Hong X. Tang

Abstract Superconducting nanowire single-photon detectors integrated with nanophotonic waveguides hold tremendous potential for the development of silicon based quantum photonic devices. In this chapter we present an overview of recent efforts using scalable fabrication procedures to realize waveguide-coupled single-photon detectors. We will show how high detection efficiency, low noise and high timing resolution are achieved simultaneously over a large range of wavelengths. These features can be exploited, for example, in photon buffering and optical time domain reflectometry.

4.1 Introduction

Integrated photonic circuits allow for complex optical functionality by following a design methodology originally developed for the realization of integrated electrical circuits. While such photonic circuits have been extensively studied in the classical domain at relatively high intensities, they also hold tremendous potential for quantum optical applications [1]. For integrated quantum photonic circuits not only passive components such as waveguides and optical elements to shape the spectral response are required, but also active devices including non-classical light sources and single photon detectors. In the telecommunication wavelength range, which is primarily explored with silicon photonic devices, efficient single photon detectors are difficult to realize with approaches traditionally used in free-space quantum optics [2]. Instead, silicon based quantum photonic devices will benefit from a fully integrated

W.H.P. Pernice (✉)
Institute of Physics, University of Muenster, Heisenbergstr. 11, 48149 Muenster, Germany
e-mail: wolfram.pernice@uni-muenster.de

C. Schuck · H.X. Tang
Department of Electrical Engineering, Yale University, New Haven, CT 06511, USA
e-mail: carstenschuck@gmail.com

H.X. Tang
e-mail: hong.tang@yale.edu

© Springer International Publishing Switzerland 2016

R.H. Hadfield and G. Johansson (eds.), *Superconducting Devices in Quantum Optics*, Quantum Science and Technology,
DOI 10.1007/978-3-319-24091-6_4

solution based on superconducting nanowires fabricated on top of nanophotonic waveguides. Such waveguide integrated superconducting nanowire single photon detectors are particularly attractive because they can be realized with scalable fabrication procedures and thus fulfil one of the key requirements for linear optical quantum computing [3].

Scalability by microfabrication has been the underlying principle for the success of integrated electrical circuits (ICs) that are nowadays found in virtually all aspects of our daily life. Through controllable and highly developed fabrication processes complex systems can be realized by arranging individually designed and optimized components in circuit form. Interconnection between the individual building blocks is achieved with printed electrical wires which allow for electrical signal exchange while being amenable to convenient layout using computer aided design (CAD). As the employed micro- and nanofabrication routines are inherently designed for mass production, such circuits can be replicated with high yield for up-scaling towards complex yet low-cost applications. In direct analogy to the design of ICs, a similar route has been followed for building multi-component optical systems out of microfabricated individual photonic elements [4, 5]. In this case nanophotonic waveguides take on the role of electrical wires to connect functional elements into nanophotonic integrated circuits. Such circuits find a multitude of applications in the telecommunications world, for optical signal processing and optical sensing applications. In particular nanophotonic circuits made from silicon and CMOS compatible materials are attractive because they can be readily combined with ICs to replace electrical connections with optical interconnects [6, 7]. In addition, already established fabrication routines for the realization of ICs can be used to fabricate the optical components and thus form a self-consistent production platform for opto-electronic components.

In recent years it was realized that such nanophotonic circuits hold promise for other applications [8, 9]. By solely relying on waveguide based photonic devices the on-chip equivalents of free-space optical components such as beam-splitters [10, 11] and phase-shifters [12] can be implemented, offering a new route towards scalable linear optical quantum computing [1]. In this case nanophotonic circuits are operated in the single photon regime in contrast to the relatively high optical intensities that are used for classical optical applications. However, for fully integrated non-classical circuits not only passive devices but also active components such as single photon sources and integrated single photon detectors are required [2]. A monolithic implementation of all building blocks of a quantum photonic circuit would then be able to overcome the stability and scalability limitations of bulk optic realizations that are overwhelmingly used to date. In the case of linear optical quantum computing this can be achieved with nanophotonic waveguides on silicon chips [13–15]. Low-loss interfacing with single-photon sources and detectors is then optimally achieved by embedding the sources and detectors directly into the optical circuitry on-chip.

In order to design optimal detectors for integrated photonic circuits, several performance metrics have to be addressed. In particular, a low dark count rate (DCR), high detection efficiency (DE) and accurate timing resolution are the three most desired features of a single photon detector [16]. These characteristics can be combined into one figure-of-merit for single photon detectors as for example the noise equivalent power:

$$NEP = \frac{h\nu \cdot \sqrt{2 \cdot DCR}}{DE} \tag{4.1}$$

where $h\nu$ is the energy of a photon of wavelength $\lambda = c/\nu$. Alternatively, a figure of merit, H, taking into account the timing performance (timing jitter δt) of the single photon detector in addition can be used as suggested in [2] with the expression

$$H = \frac{DE}{DCR \cdot \delta t} \tag{4.2}$$

Detectors with low NEP are increasingly sought after for applications in both quantum and classical technology because they enable extended measurement runs with high signal to noise ratio. To highlight a prominent example (covered in Chap. 5 of this volume), quantum key distribution (QKD) implementations are currently limited in rate and range by imperfect detector characteristics [17]. Other applications which require high performance detectors include the characterization of quantum emitters [18, 19], optical time domain reflectometry [20] as well as picosecond imaging circuit analysis [21]. All these applications will greatly benefit from improved single-photon detection systems which not only achieve high H-figure (low NEP) but are also directly embedded within integrated waveguide designs.

One of the most promising approaches to achieving low noise detector characteristics consists in using superconducting nanowire single photon detectors (SNSPDs) [22]. These detectors are made from ultrathin superconducting films typically with a thickness below 5 nm that are patterned lithographically into narrow nanowires on the order of 100 nm width [23]. Usually the nanowires are arranged in a suitable meander pattern (as described in Chap. 1) and then illuminated with single photons by coupling to optical fibers. The detection mechanism relies on single-photon induced loss of superconductivity in a nanowire which is current biased closed to its critical current. The detection process is characterized by a fast recovery time and relatively high quantum efficiency both for visible and infrared wavelength photons [24], thus providing broadband detection capability covering in particular the telecommunication bands [25]. Most of the state-of-the-art SNSPDs however, are stand-alone units absorbing fiber-coupled photons under normal incidence. Because the nanowire meanders are prepared from ultrathin films, much thinner than the incident photons' wavelength, the optical absorption in a single pass is limited which restricts the usefulness of such detectors. More advanced optical architectures need to be employed. Optical cavities are discussed in Chap. 2 (Sect. 2.2.2). Here we focus on the alternative of integrating superconducting nanowires with optical waveguides suitable for large scale photon counting applications.

Such difficulties can be overcome by realizing superconducting nanowire detectors with varying geometries fabricated directly on top of nanophotonic waveguides [26–29]. Depending on the material system from which the waveguides are prepared different wavelength regimes can be targeted. In the case of silicon, a relatively small bandgap restricts the operation window to wavelengths above 1100 nm. The high surface quality of silicon substrates, on the other hand, provides a good template for the realization of high quality superconducting thin films and thus good detector performance. When broader wavelength access is required, silicon related materials such as silicon nitride (Si_3N_4) can be used to enhance the detection window for visible and near-infrared wavelengths [30, 31].

In terms of superconducting materials suitable for SNSPD manufacture niobium nitride (NbN) is a common choice [32]. NbN offers high critical temperatures above the boiling point of liquid helium and thus does not require dilution refrigeration for the operation of a corresponding single photon detector [23]. Other superconducting materials, which have recently been investigated for single-photon detection include tungsten silicide (WSi [33]), tantalum nitride (TaN [34]), niobium (Nb [35]) and niobium titanium nitride (NbTiN, [36]). The latter has an even higher critical temperature than NbN, which makes it particularly attractive for SNSPD applications [37]. NbTiN is a superior material choice for many detector performance parameters, especially for achieving low noise and low dark count rates. In this chapter we will provide an introduction to waveguide based single photon detectors on silicon photonic platforms using both NbN and NbTiN thin films for the preparation of superconducting nanowires. By engineering the detector and waveguide dimensions at the nanoscale we simultaneously achieve highly efficient single photon detection with high timing accuracy and low dark count rate. Most notably, such waveguide integrated detectors feature very low noise equivalent power (NEP) on a fully scalable platform, thus addressing the most urgent needs of integrated optical quantum information processing and time correlated single photon counting applications.

4.2 Single Photon Detection in Superconducting Nanowires

Superconducting nanowire single-photon detectors are generally prepared from ultrathin superconducting films which feature a film thickness of 5 nm or less. The first SNSPD was realized using NbN nanowires and demonstrated originally by Gol'tsman and coworkers [22]. Using electron beam lithography (EBL) and reactive ion etching (RIE), an initially continuous superconducting thin film is patterned into a narrow line with lateral dimensions below 100 nm. The core detector structure is called a nanowire because its cross-section in both dimensions lies in the nanometer regime. To enable the detection of single photons the nanowire is cooled down to temperatures well below the superconducting transition temperature [38] and operated under current bias. In the low-temperature regime the superconducting nanowire exhibits a clear discontinuity in its current-voltage characteristic once the bias current exceeds a critical current value I_C. When the bias current is kept below the critical current at first, the voltage measured across the nanowire remains zero. When the kinetic energy of the Cooper pairs reaches the binding energy, the supercurrent breaks down at a characteristic value which determines I_C.

For single photon measurements the nanowire is illuminated with low intensity light guided to the detector with an optical fiber. Visible and near-infrared wavelength photons reaching the nanowire deposit energies exceeding the superconducting energy gap Δ, which for typical superconductors used for SNSPDs is around $\Delta \approx 5\,\text{meV}$ [39]. Upon photon absorption Cooper pairs will thus break up into quasiparticles, which possess high energy and are able to start a cascade to generate further quasiparticles. The resulting reduction of the density of Cooper pairs creates a region of suppressed superconductivity around the absorption site. Since the nanowire bias current was set near I_C, the supercurrent is exceeding I_C in the whole cross-sectional area of the nanowire, which transitions to the normal conducting state. The normal conducting zone gives rise to an abrupt voltage pulse which is registered as a count event with the electronic readout circuitry. Once the energy provided by the absorbed photon is dissipated from of the normal zone, the nanowire will eventually reset to its initial superconducting state. The reset time scale defines the time window during which no further photon event can be registered and thus is the dead time of the detector, which for typical SNSPDs is in the range of nanoseconds.

A crucial characteristic of a SNSPD is the probability by which a photon reaching the detector leads to a counting event, which is called the detection efficiency (DE). The exact definition of DE depends on the details of the detector implementation and measurement apparatus, as well as on how the optical losses are taken into account. Thus each detector is characterized by a system detection efficiency (SDE), which is the number of count events registered by the electronic readout unit divided by the number of photons that were sent into the apparatus in a given time interval. The intrinsic detection efficiency or internal device quantum efficiency (QE) describes the number of output pulses divided by the number of photons that have been absorbed inside the nanowire. For waveguide integrated SNSPDs we furthermore define the on-chip detection efficiency (OCDE), which is the product of the internal device quantum efficiency, QE, and the probability for photons inside the waveguide to be absorbed by the nanowire. All of the contributions to the detection efficiency have a spectral dependence. This overall dependence defines the spectral bandwidth of the system which can cover a wide wavelength range.

In addition to detection events due to photon absorption, there is a certain probability that counting events occur in the absence of any illumination. These false counts contribute the dark count rate (DCR) and can be considered as noise. The dark counts provide a lower limit for the minimum photon flux that can be resolved by the detector. Depending on the application, an additional detector characteristic is the temporal evolution of the voltage pulse upon photon absorption. Within the duration of the energy relaxation processes which resets the detector back to the superconducting initial state, the SNSPD is insensitive to additional photons which may be absorbed. The resulting dead time limits the maximum count rates which can be achieved and thus the maximum photon flux that can be measured. For time-correlated single-photon counting (TCSPC) applications, the time variance (jitter) between photon absorption and triggering a counting event is an additional important property. The timing jitter is directly related to the minimum time intervals that can be measured with SNSPDs. In general, these properties scale favourably with smaller nanowire lengths, under the constraint that the probability for photon absorption is not affected. While this constraint is hard to fulfil with traditional SNSPDs, the move towards waveguide-integrated detectors drastically relaxes this condition and allows for the realization of ultra-short SNSPDs [29].

4.3 Silicon Photonic Circuits for SNSPD Integration

This chapter is concerned primarily with single photon detectors compatible with integrated optical devices. In this section we first discuss two material platforms which allow for the realization of high quality optical devices and high quality detector devices. Among the available options, a prominent material used in modern integrated optics is silicon, which has led to the establishment of the rapidly expanding research field of silicon photonics [40]. After the invention of reliable manufacturing procedures for silicon on insulator (SOI) substrates, research on silicon photonics rapidly evolved into one of the most fruitful areas of integrated photonics to date [41]. The relatively high refractive index of silicon enables tight confinement of light in nanoscale waveguides and therefore very compact optical structures can be realized. In addition, extensive fabrication procedures for the manufacture of silicon devices have been developed for the electronics industry and can therefore be relied on for the fabrication of sophisticated nanophotonic devices. Silicon has excellent material properties that are important in photonic devices [42]. These include high thermal conductivity (10 times higher than GaAs), high optical damage threshold (10 times higher than GaAs), and high third-order optical nonlinearities. A major boost for research on silicon based integrated photonic devices was the development of high quality SOI templates. The additional oxide layer underneath a thin silicon top layer allows for single-mode optical waveguiding in the near-infrared wavelength regime, which is the preferred wavelength range for telecommunication applications. Several methods for the fabrication of SOI substrates with a buried oxide layer exist to date [43]. The most recent fabrication routine for SOI is the wafer splitting or SmartCut process [44]. Hydrogen implantation into oxidized silicon wafers is used to create a damage layer beneath the oxide surface. After the implantation a second silicon wafer is bonded onto the oxide surface, to form the buried oxide layer. Through thermal processing the compound wafer is split along the hydrogen damaged layer. In order to reduce surface roughness a chemical-mechanical polishing step (CMP) is applied. If thicker silicon layers are required epitaxial silicon growth can be applied to increase the thickness of the top silicon layer.

The silicon top layer can be structured into nanophotonic components using high resolution electron beam lithography in combination with reactive ion etching. To etch silicon, chlorine based etching chemistry is often employed, leading to near vertical sidewalls and low surface roughness [45]. Owing to the high refractive index of silicon across the telecommunication bands, single mode waveguiding requires sub-wavelength photonic components, typically a few 100 nm in width. This way light can be guided around sharp bends with low loss which is a prerequisite for

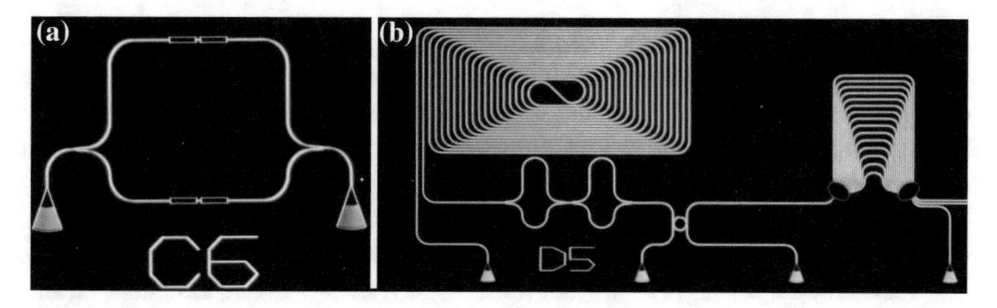

Fig. 4.1 Silicon photonic circuits: **a** On-chip Mach-Zehnder interferometer made from silicon with high extinction ratio. **b** SOI based photonic circuit comprising several cascaded elements and multiple input/output grating coupler ports

photonic devices with small footprint. Using such fabrication approaches a rich library of photonic components has been realized, including the devices shown in Fig. 4.1. This includes high quality waveguides with low propagation loss below 1 dB/cm, high quality microresonators in the form of microrings [46] or photonic crystal cavities, as well as on-chip interferometers (Fig. 4.1a) and diffractive elements such as on-chip spectrometers. All these components can be assembled into sophisticated circuits for routing, splitting and filtering optical signals as shown in Fig. 4.1b. For coupling light into on-chip photonic structures several approaches have been developed, including inverse tapers for butt-coupling or out-of plane access through grating couplers [47]. While the latter coupling structures provide slightly higher coupling losses, they are convenient devices for connecting integrated optical components to multiple optical fibers.

A major advantage of a silicon platform with respect to single photon detector manufacture is the high surface quality of SOI templates. The silicon top layer provides an atomically flat surface and therefore a good starting layer for the deposition of superconducting thin films with high surface uniformity [36]. From such ultra-thin films highly uniform nanowires can be fabricated as the basis of high performance single photon detectors. Furthermore, the possibility to realize such material templates on a wafer scale makes it possible to fabricated large numbers of identical detectors and also allows for combining many detectors in complex systems [48]. The combination with CMOS electronics is another intriguing opportunity for joined detection and readout in a single device.

4.4 Silicon Nitride Photonic Circuits for Broadband Single Photon Applications

While silicon is a superior photonic platform for guiding photons in the near infrared wavelength regime, the relatively small bandgap of 1.1 eV prohibits applications in the visible wavelength range. Therefore alternative material options with a larger bandgap are highly sought after, among them silicon nitride as a CMOS compatible material. Silicon nitride has been used for integrated photonic devices such as waveguides for decades [49], similar to silicon. Being well established in electronic ICs it can be fully integrated in CMOS processes. Compared to silicon, Si_3N_4 has a much larger band gap of around 5 eV and therefore exhibits transparency for wavelengths from the visible to the far infrared [50], thus allowing guiding modes for a wide range of frequencies. Due to its relatively high refractive index ($n \sim 2$) it is also very suitable for highly compact devices [51]. Low pressure chemical vapour deposition (LPCVD) is used to produce Si_3N_4 films, with precisely controlled thickness, uniform refractive index and low surface roughness; the latter being important to minimise scattering losses at the top and bottom interface of the waveguide [52]. Therefore, high quality waveguides with low propagation losses can be fabricated using Si_3N_4 films.

For the fabrication of photonic circuitry the nitride layer needs to be surrounded by optical materials with a lower refractive index to achieve good optical mode confinement. In analogy to SOI, buried oxides with lower refractive index (1.45) compared to silicon nitride are thus frequently employed to realize silicon nitride-on-insulator substrates. Photonic devices are then patterned using EBL with organic resists which provide sufficient protection for subsequent dry etching. Since naturally written EBL resist contains residual roughness, reflow procedures can be applied, which greatly increases the quality of the fabricated devices [53].

In order to measure small changes in the group index or attenuation coefficient of a waveguide mode, it is necessary to realize nanophotonic circuits with very low propagation loss and sensitivity to small changes in the phase of a propagating optical mode. Mach-Zehnder interferometers (MZIs) especially are very useful for measuring the attenuation of waveguide-coupled devices [54], while both rings [55] and MZIs can be used to enhance the modulation depth of optical modulators [56]. In addition, ring resonators can not only be used as filters and many other applications but are also excellent tools to quantify the scattering loss of fabricated waveguides [57]. This is because the optical quality factor of a ring resonator is directly related to the propagation loss in the waveguide and therefore can be used to find optimal parameters for the fabrication process and the waveguide geometry. Ring resonators with very high quality factors are of much interest themselves [58–60]. The average Q-factor of such resonators exceeds 10^6, corresponding to propagation losses of 26 and 21 dB/m respectively. The propagation losses are comparatively low, especially because no sophisticated fabrication steps or special waveguide geometries with large bending radii have to be used. In the C-band at a wavelength of 1550 nm Si_3N_4 waveguides with a propagation loss of 3 dB/m at 2 mm bend radius and 8 dB/m at 0.5 mm bend radius have been demonstrated [61]. In [55] ring resonators with quality factors of 7×10^6 at a much larger bend radius of 2 mm have been presented, corresponding to propagation losses of 2.9 dB/m. This way high quality nanophotonic circuits can be realized that are operational over a wide wavelength range from the near-ultraviolet region to infrared wavelengths. Furthermore, silicon nitride thin films provide very low surface roughness after deposition and thus are suitable substrates for further thin-film deposition. This is of particular interest with respect to superconducting nanowire devices as elucidated in the next section. However, when considering operation of devices at longer wavelengths silicon will still be the preferred material platform. In addition, on a high refractive index platform such as SOI, the absorption length of superconducting nanowires in the near-field is typically shorter, leading to reduced device dimensions.

4.5 Absorption Engineering of Superconducting Nanowire Devices

The material platforms outlined above can be readily prepared for single photon detector applications through hybrid integration with superconducting nanowires as described above. In an integrated photonic circuit architecture the nanowire detectors are placed directly on top of a waveguide in order to realize a travelling wave geometry. In this way propagating photons are absorbed through evanescent coupling from the waveguide to the nanowire along their direction of propagation [26–29]. In order to realize a high performance waveguide integrated single photon detector it is crucial to maximize the absorption efficiency while maintaining a minimal cross-section of the nanowire, which reduces reflections [62]. Using numerical simulations optimal geometries can be found which are then experimentally confirmed by measuring the photon absorption rate per detector unit length. In order to determine the absorption coefficient for a given nanowire geometry it is convenient to analyse transmission through on-chip waveguides equipped with corresponding superconducting nanowires fabricated on top. In this configuration the propagating optical mode is evanescently coupled to the superconducting wire and the interaction length with the detector is thus given by the length of the wire. A schematic view of photonic circuits suitable for balanced detection is shown in Fig. 4.2a.

In order to allow for balanced measurements each nanophotonic circuit includes two equal waveguide arms with the same length. One arm is used as the reference path for light propagating in a waveguide without any nanowire; the other arm is equipped with a U-shaped nanowire on top. The waveguides are each terminated with focusing grating couplers [63] connected via a 50:50 Y-splitter. The focusing grating couplers are used to couple light from optical single mode fibers into the chip. The geometric parameters of the 50:50 Y-splitter are optimized experimentally to obtain even splitting ratio, allowing for division of the incoming mode with an uncertainty below 0.5 %. A schematic view of the nanowire on top of the waveguide is shown in Fig. 4.2b, including the parameters which are varied to optimize the absorption properties, i.e. the length (l_{nw}), width (w_{nw}), and gap (g_{nw}) between the nanowires. The nanowire gap is defined as the clear distance between two wires (red line in Fig. 4.2b.

Numerical analysis of mode propagation and nanowire absorption can be conveniently carried out through finite-element simulations with COMSOL MultiPhysics, using material parameters as given in [64]. In the presence of the NbN nanowire the

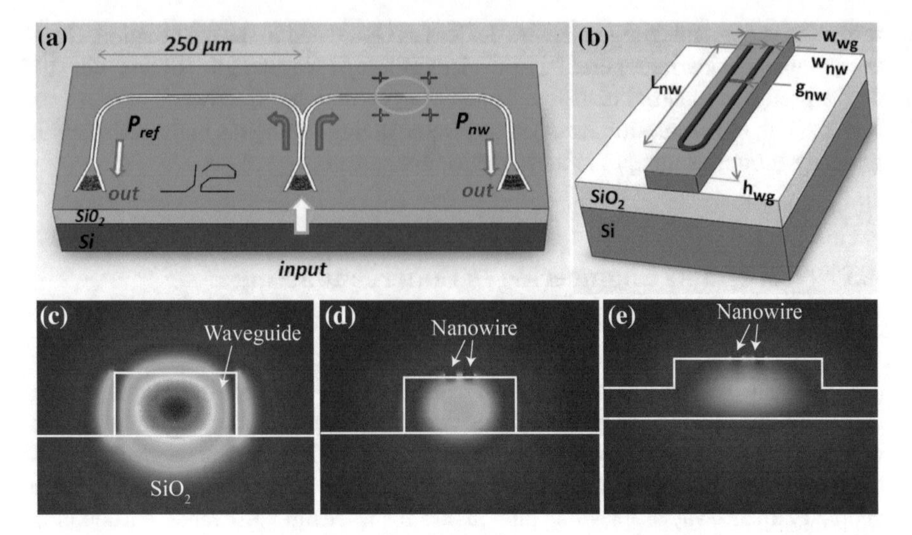

Fig. 4.2 Nanophotonic circuit design: **a** Schematic view of the nanophotonic circuit used to optimize the nanowire absorption. **b** Cross-section of a nanophotonic waveguide covered with a *U-shaped* NbN nanowire. **c** The simulated distribution of the electric field in the x-direction (TE-like mode) for 1550 nm wavelength. Light intensities are shown with a linear color scale. **d** Simulated distribution of the TE-like mode for 1550 nm wavelength in the NbN nanowire covered region. **e** Simulated distribution of the TE-like mode for 1550 nm wavelength for a half-etched waveguide in the NbN nanowire covered region. Adapted from [62]

evanescent tail of the guided mode is strongly coupled to the NbN wire. From the numerical simulations the complex refractive index of the propagating mode can be extracted which is directly related to the propagation loss. From the imaginary part of the refractive index n_i the absorption coefficient α can be calculated using the expression

$$\alpha = 4.34\frac{4\pi n_i}{\lambda} \tag{4.3}$$

in units of dB/μm. The numerical results from modal analysis can also be compared to time domain simulations based on the finite difference time domain method (FDTD) yielding good agreement [29].

While the absorption properties of superconducting nanowires deposited on silicon nitride waveguides are on the order of 0.2 dB/μm, stronger absorption can be achieved by moving to material systems with strong optical confinement because of a larger refractive index. In silicon waveguide devices for example, typical absorption values are on the order of 1 dB/μm [29]. Stronger absorption allows shorter detector devices to achieve a fixed attenuation; ultra-short detector devices realized on this high index material system are described in the next section. Using silicon based devices the absorption efficiency can reach 90 % for a detector length of 10 μm, with direct implications on detector reset time and thus maximum detection rate.

4.6 Waveguide Integrated Single Photon Detectors

Using optimized devices as outlined in the previous section efficient single photon detectors can be realized which are fully embedded in nanophotonic circuits. This is because the travelling wave geometry is directly compatible with the photonic wiring philosophy applied in integrated optics [65]. In the case of single photon circuits traditional low-loss waveguides are terminated with waveguide single photon detectors which absorb incoming photons with ultra-high probability as outlined above. The detector devices are compatible with the typical dimensions of integrated optical components and therefore seamlessly fit into the optical circuit architecture.

The detector design concept can be applied in principle to any material system that supports guidance of optical modes, including silica, III–V semiconductors such as gallium arsenide (GaAs) or lithium niobate [26–28, 66]. Integration of SNSPDs with GaAs waveguide circuits is discussed in Chap. 3. Here we restrict ourselves to the CMOS compatible silicon based materials described in the previous sections. Both on silicon and silicon nitride, high quality superconducting thin films can be deposited such that the same design approach is used to realize high performance single photon detectors [29–31]. These devices are analysed by providing both optical access as well as electrical access to on-chip circuits. To be able to readout many devices simultaneously, multi-optical access is realized via fiber arrays [67], while several electrical contacts are achieved with radio frequency (RF) probes. In our case we evaluate the detector performance in head-to-head configuration. This approach can be conveniently integrated into compact cryogenic measurement setups. We use liquid helium refrigeration to cool the detector devices below the superconducting transition temperature to a base temperature below 2 K (^{4}He has a boiling point of 4.2 K at atmospheric pressure; the temperature can be reduced by pumping on the ^{4}He bath). The detectors are then current biased with a low-noise current source and connected to low noise readout electronics as described elsewhere [30, 31].

The photonic circuit architecture provides multi-port optical access, such that the detector performance can be determined with high accuracy using calibrated optical measurements. In a waveguide framework the number of photons propagating towards a detector can be assessed through suitably added reference ports. Since the optical circuits are fabricated with established fabrication recipes, they feature high reproducibility which means that the performance of each circuit component can be calibrated accurately. This is particularly important for the input ports in the form of grating couplers with a typical insertion loss below 10 dB.

Typical detector performance is shown in Fig. 4.3a for NbN nanowires on top of silicon waveguides. Shown is the measured detection efficiency in dependence of normalized bias current. In our detectors we achieve a maximum on-chip detection efficiency of 91 % for photons at 1550 nm input wavelength. Silicon nitride waveguides are also transparent in the visible wavelength regime enabling us to measure detection efficiency at 780 nm wavelength. For equivalent detectors made from NbTiN deposited onto silicon nitride waveguides, we find a maximum on-chip detection efficiency of 82 % as shown in Fig. 4.3b. By decreasing the nanowire

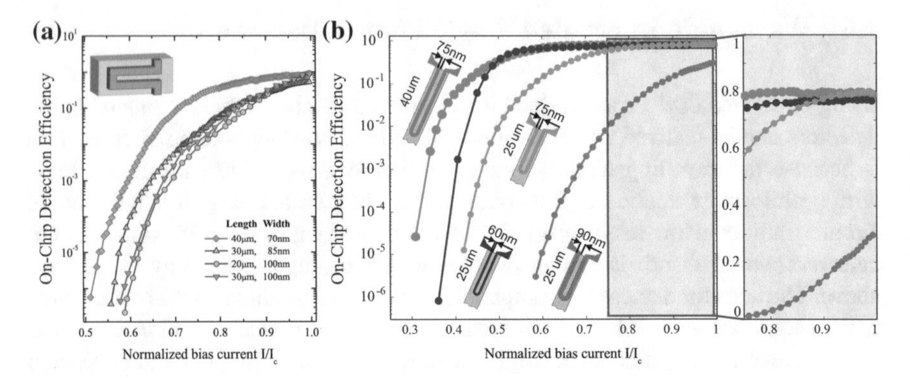

Fig. 4.3 Characterization of waveguide integrated SNSPDs: **a** Measured on-chip detection efficiency as a function of normalized bias current for an NbN nanowire on silicon detector. Traces are shown for varying nanowire geometries measured at 1550 nm input wavelength [29]. **b** Measured on-chip detection efficiency for an NbTiN on silicon nitride detector. Several device geometries measured at 780 nm input wavelength are shown [31]. *Inset* zoom into the high bias current region in a linear plot, showing the plateau behavior for narrow nanowires. Reproduced with permission of Nature Publishing Group from [29] and [31]

width to 60 nm we find clear plateau behavior at higher bias current, which implies ultra-high internal quantum efficiency. Besides high on-chip detection efficiency, our detectors exhibit low DCRs. For the silicon detector devices we obtain minimal dark count rates of 50 Hz when approaching I_C. Using NbTiN this DCR can be further reduced to measured rates in the mHz range [31], which allows us to reach measured NEP below $10^{-19}W/\sqrt{Hz}$ at 1550 nm input wavelength and at 780 nm wavelength.

Since the waveguide-integrated detectors described above are fabricated with a small device footprint, the overall meander length is significantly shorter than in traditional fiber-coupled SNSPDs. This results in an equivalent reduction of the kinetic inductance of the detector and thus reduced reset time [68]. For the silicon based detector devices we find a minimum relaxation time of 500 ps, which allows for single photon detection rates above 1 GHz. Furthermore, the devices provide very low timing jitter below 20 ps FWHM, which is attractive for on-chip time-resolved measurements [29]. Thus the hybrid integration with a nanophotonic waveguide architecture allows for combining essentially all desired detector metrics advantageously in a single device.

4.7 Applications

Waveguide integrated single photon detectors are promising devices because they combine many desirable detector properties all in one device. Of particular importance in this respect are the high timing resolution, as well as the very low dark count rate and hence low noise-equivalent power. In the following sections we illustrate two applications which showcase each of these properties, i.e. ballistic photon transport on chip to underline the high timing resolution and optical time domain reflectometry to exploit the low noise-equivalent power.

4.7.1 Ballistic Photon Transport in Silicon Microring Resonators

The high quantum efficiency and fast response of the waveguide integrated single photon detectors enable time-domain analysis and optical multiplexing in integrated photonic circuits. In order to demonstrate the applicability for fast on-chip single-photon measurements, variable photon-delay from a micro-ring resonator provides a convenient example. Microring resonators allow for the generation of pulse trains by evanescent coupling to a feeding waveguide as illustrated in the schematic in Fig. 4.4a, when exciting the circuit with a pulsed laser source. By adjusting the diameter of the ring the roundtrip time and thus the separation between individual pulses can be selected. The corresponding device used in our experiments consists of a micro-ring resonator, which is coupled to two waveguides providing measurement capabilities in both through and drop port configuration [69]. By coupling the waveguides to optical grating couplers, the ring resonator can be studied both in the time domain with an on-chip SNSPD, as well as in the frequency domain by sweeping the input wavelength. The optical output from one of the drop lines is split with an on-chip 50/50 splitter as discussed above and fed into a grating output port and an integrated SNSPD [29].

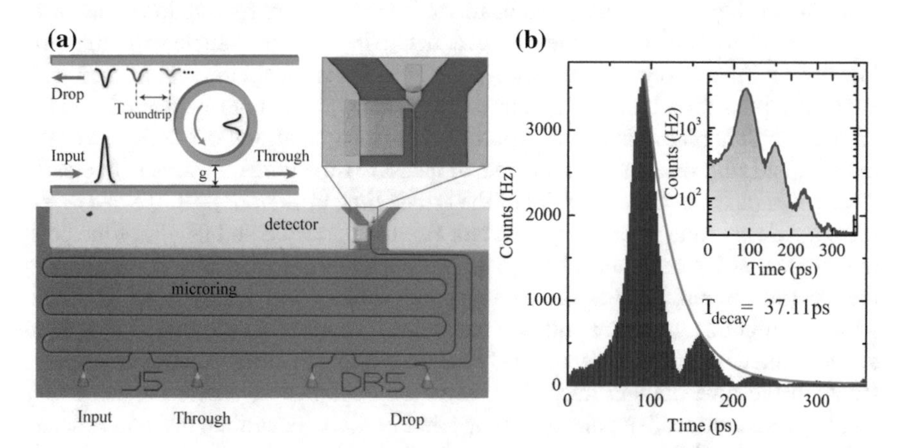

Fig. 4.4 Ring resonator characterization, reproduced with permission of Nature Publishing Group from [29]: **a** Schematic of a ring resonator photonic circuit coupled to two waveguides for the generation of pulse trains on chip. The optical micrograph below shows the physical realization on a silicon chip. *Inset* a higher magnification image of the detector region. **b** Arrival time histogram of single photon detection events as recorded with a SNSPD of < 20 ps timing accuracy. *Inset* logarithmic representation, showing 4 consecutive pulses emerging from the ring resonator

The ring resonator used in our experiments features a total circumference of 5.8 mm, laid out in a meander form to preserve chip real estate and to fit into a single write field during EBL to avoid stitching errors, as shown in Fig. 4.4a. Depending on the strength of the evanescent coupling of the input waveguide to the ring, different optical regimes can be studied, ranging from the weak to the strong coupling regime [70]. When the gap is small between input waveguide and resonators, the resonator is strongly coupled to the feeding waveguide and thus a significant portion of the input light is transferred into the ring. However, the light circulating inside the ring is also coupled out efficiently to the second waveguide (the drop port). Thus the circulating intensity drops quickly over time and hence the pulses within the outgoing pulse train rapidly decay in intensity.

The strong coupling reduces the optical quality factor of the microring because the cavity is overloaded. This can be seen directly in the frequency domain by scanning the resonant spectrum of the ring [29]. We measure the transmission spectrum in the through port with a tunable continuous wave (CW) laser source in order to assess the quality factor of the ring. Due to the large circumference of the ring resonator the free spectral range (FSR) is small, leading to dense transmission dips at the optical resonances in the spectrum. Fitting the dips with a Lorentzian function yields a quality factor of 14,000 in the overcoupled case. When the coupling gap, g, is increased, less light is coupled into and out of the ring resonator. Therefore, light circulating inside the ring decays slower and produces elongated pulse trains when the ring is excited with a pulsed laser source. In this weakly coupled case we measure the expected improved optical quality factors around 24,000.

The above characterization is in accordance with the traditional spectral analysis of photonic resonators. In addition, however the high timing resolution of our detectors enables us to deduce the quality of the resonators by ringdown measurements. We therefore analyse the ring parameters in the time-domain exploiting the low timing jitter of our single photon detectors. The optical circuit is excited with attenuated picosecond laser pulses and photon detection events are registered with the on-chip SNSPD placed in the drop port of the resonator, after the Y-splitter. For the overcoupled ring resonator we are able to measure time-domain traces with clearly discernible peaks in the linear plot of the arrival time histogram [29]. We can determine the decay time of the ring resonator from the position and height of the peak amplitudes, which amounts to 19.3 ps in an exponential fit to the data. Converted into the spectral domain, the decay time corresponds to an optical quality factor of 11,900 which is in good agreement with the measured spectral value (14,000 as described above). Likewise, in the undercoupled case we are able to observe consecutive pulse fronts in the time-domain trace. For the weakly coupled resonator we can easily identify four consecutive pulses, as shown in Fig. 4.4b, illustrating the slower decay out of the ring. Fitting peak positions with an exponential function reveals a decay time of 37.1 ps corresponding to a spectral width of 67 pm or equivalently an optical quality factor of 22,900. The positions of the pulse peaks are separated by a delay time of 72.7 ps, which is determined by the length of the ring resonator and the group index of the waveguide profile.

The results illustrate that such waveguide-coupled detectors can be directly employed for ultra-fast measurements in the time domain and provide new tools to study photonic devices without the need for external probes. This is of interest for ultra-fast measurements and also for the analysis of time-domain switching processes in integrated photonic devices.

4.7.2 Optical Time Domain Reflectometry

While the ballistic transport measurements make use of the high timing resolution of the on-chip single photon detectors, the low dark count rate enables further practical applications in fiber optic networks, such as optical time domain reflectometry (OTDR). OTDR is an efficient, non-destructive technique to diagnose the physical condition of an optical fiber in situ [20, 71, 72]. By launching laser pulses into the fiber and detecting the returning light from reflecting and scattering sites it is possible to get information about attenuation properties, loss and refractive index changes in the fiber-under-test (FUT) [73]. This allows for localizing defects in a given fiber-link with high spatial resolution over distances of more than a hundred kilometers by analysing the returning optical signal in the time-domain, in analogy to the approach described in the previous section. For best results the detector used to record the back-scattered signal should provide high timing resolution and high signal to noise ratio to be sensitive to low light levels. With increasing distance the light scattered or reflected along the fiber suffers from stronger attenuation and eventually reaches the detector noise level. Hence, the sensitivity of an OTDR system is determined by the noise equivalent power (NEP) of the detector which ultimately limits the measurement range. Given a detector of sufficient timing accuracy the achievable two-point resolution is then determined by the pulse length of the interrogation light source used for the measurements. Using high laser power results in a larger number of photons scattered back towards the detector and thus reduces the data acquisition time needed to achieve a given OTDR measurement range. To unambiguously identify defects in the fiber it is furthermore necessary to adjust the pulse repetition rate to the total length of the FUT, in order to avoid overlapping echoes from two consecutive pulses.

To investigate the use of waveguide coupled SNSPDs for OTDR applications we perform measurements with a customized pulsed laser system with variable repetition rate as shown in Fig. 4.5a [74]. The laser is launched via an optical circulator into a fiber link consisting of several connected spools with a total length of 263 km. To maximize the OTDR measurement sensitivity we adjust the clock rate to 300 Hz, exceeding the length of the FUT such that photons reflected from the open fiber end are able to travel back to the circulator before the next pulse is launched. The backscattered photons from the FUT are coupled out at circulator port 3 and guided to a travelling wave NbTiN superconducting nanowire single-photon detector. Coupling of light from the optical fiber into the on-chip photonic waveguide is achieved with an optical grating coupler. For the NbTiN SNSPD we measure a low dark count

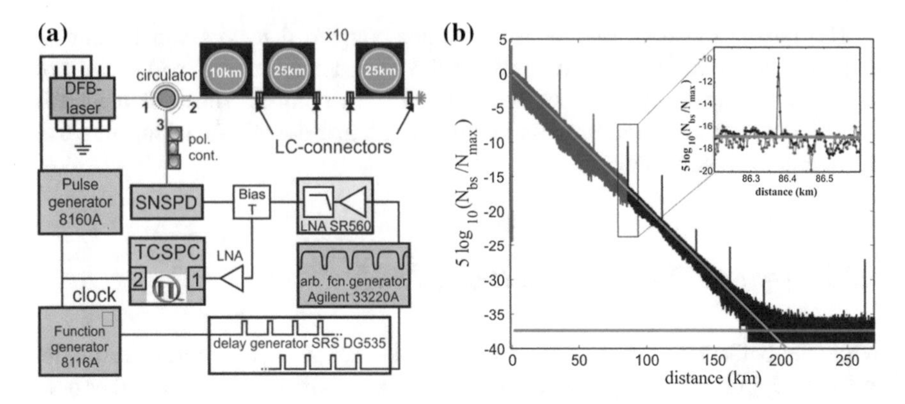

Fig. 4.5 Optical time domain reflectometry (OTDR): **a** The measurement setup used for OTDR with waveguide integrated single photon detectors. **b** Measured OTDR trace for a 263 km long fiber spool. The slope of the trace results from fiber scattering loss of roughly 0.2 dB/km, while the sharp spikes denote reflection from the LC-connectors. *Inset* zoom into one reflection peak to illustrate the spatial resolution of the measurement. Reproduced with permission from [74]. Copyright 2013, AIP Publishing LLC

rate of less than 10 Hz over the entire bias current range, mainly limited by stray light [74]. The resulting system NEP $\approx 10^{-17} - 10^{-18} W/\sqrt{Hz}$ system close to the critical current and is dependent on ambient light conditions.

The SNSPD output signal is amplified with high-bandwidth low noise amplifiers and fed into a time correlated single-photon counting system. We then create list-files of all arrival times for signals from the detector channel (1) and a clock channel (2). We then calculate the time delay Δt between photon detection events in channel 1 with respect to the clock signal recorded in channel 2. This time delay translates to the distance

$$\Delta s = \frac{\Delta t}{2c_f} \tag{4.4}$$

which the photon travels before being scattered or reflected back. Since the SNSPD is operated in free running mode while a pulse is propagating in the FUT, the entire OTDR trace is reconstructed by calculating the waiting time distribution recorded with the TCSPC unit as shown in Fig. 4.5b.

The recorded OTDR traces exhibit several peaks caused by Fresnel reflections from the refractive index change at the tiny air gap between two fibers connecting adjacent spools. Despite the noise at the end of the measurement range the strong reflection from the glass to air transition at the open fiber end after 263 km (eleven fiber spools) is still visible in our OTDR measurement. The OTDR data also allow us to extract the round-trip attenuation due to Rayleigh scattering in the FUT from the fit to the linear slope of the trace for each fiber spool. We find attenuation of 0.2 dB/km, as expected for typical low-loss optical fibers which are available commercially.

In our measurements the ultimate measurement range is limited both by the residual dark count rate of the detectors, as well as the power of the input laser. By providing improved shielding against stray light into the detector, as wells as by avoiding coupling of black-body radiation present in the cryostat cooling system the dark count rate can be significantly reduced. At the same time, high power fiber lasers could be used as input source thus extending the measurement range of SNSPD-based OTDR systems beyond current technical solutions. Thus OTDR can be used as a prime application to demonstrate the high signal to noise ratio achievable with a waveguide detector architecture.

4.7.3 Outlook on Applications of Waveguide Integrated SNSPDs

The two examples outlined above illustrate the tremendous potential of waveguide integrated single photon detectors. Since the use of a travelling wave geometry allows for drastically reducing the length of traditional meander SNSPDs, waveguide coupled devices offer attractive performance metrics in almost every respect. Particularly noteworthy is the scalability of this approach, i.e. the possibility to realize large numbers of detector devices in a single fabrication run. This is extremely important for emerging applications in linear optical quantum computation and simulations. The high on-chip detection efficiency, allowing detection of photons propagating on chip with near-perfect probability, is a prerequisite for scalable optical quantum technologies and therefore overcome key challenges that have hindered progress so far.

4.8 Conclusions

In this chapter we have reviewed progress on integrating superconducting single photon detectors with silicon photonic circuits. We have shown that superconducting nanowire based devices (SNSPDs) offer a highly promising solution for on-chip single-photon detection at telecom wavelengths. Unlike competing semiconductor detector technologies, SNSPD devices consist of a single electron-beam patterned superconducting layer, and the fabrication is compatible with standard CMOS processing. Waveguide integrated single-photon detectors can be fabricated with high yield and uniform performance across a large area, making these devices prime candidates for multi-detector architectures. For realizing scalable quantum technology solutions it is further desirable to embed SNSPDs with non-classical light sources and nanophotonic circuits on a single silicon chip. The core components of such a chip-scale quantum information processing unit have been shown with the advent of silicon integrated single photon sources [75–77], quantum photonic circuits [78] and the waveguide integrated single photon detectors discussed in this chapter. Considerable engineering challenges remain in integrating state-of-the-art components together with high yield on a single platform, and operating these devices in concert. For example, optical pump power from single-photon or photon pair generation must be rejected, and development of low power dissipation high-speed optical switches compatible with low temperature detectors [12] must continue. Also bespoke high speed low temperature control electronics will be required for control and readout of such advanced quantum photonic circuits. Generating, processing and detecting quantum information on a CMOS compatible platform will thus eventually allow for realizing scalable linear optic quantum computing [79, 80]. Through co-integration of nanophotonic circuits with waveguide coupled detectors all required elements for future quantum photonic networks can be monolithically prepared as a robust platform for emerging applications in optical quantum technology.

References

1. J.L. O'Brien, A. Furusawa, J. Vuckovic, Photonic quantum technologies. Nat. Photon. **3**, 687 (2009)
2. R.H. Hadfield, Single-photon detectors for optical quantum information applications. Nat. Photon. **3**, 696 (2009)
3. J.L. O'Brien, Optical quantum computing. Science **318**, 1567 (2007)
4. A. Alduino, M. Paniccia, Interconnects: wiring electronics with light. Nat. Photonics **1**, 153 (2007)
5. R. Kirchain, L. Kimerling, A roadmap for nanophotonics. Nat. Photonics **1**, 303 (2007)
6. D.A.B. Miller, Rationale and challenges for optical interconnects to electronic chips. Proc. IEEE **88**, 728 (2000)
7. G. Guillot, L. Pavesi, *Optical Interconnects* (Springer, Berlin, 2006)
8. D. Dai, J. Bauters, J.E. Bowers, Passive technologies for future large-scale photonic integrated circuits on silicon: polarization handling, light non-reciprocity and loss reduction. Light: Sci. Appl. 1, e1 (2012)
9. D. Bonneau et al., Quantum interference and manipulation of entanglement in silicon wire waveguide quantum circuits. New J. Phys. **14**, 045003 (2012)
10. E.A.J. Marcatili, Dielectric rectangular waveguide and directional coupler for integrated optics. Bell Syst. Tech. J. **48**, 2071 (1969)
11. X. Xu et al., Near-infrared Hong-Ou-Mandel interference on a silicon quantum photonic chip. Opt. Express **21**, 5014 (2013)
12. M. Poot, H.X. Tang, Broadband nanoelectromechanical phase shifting of light on a chip. Appl. Phys. Lett. **104**, 061101 (2014)
13. A. Politi, M.J. Cryan, J.G. Rarity, S. Yu, J.L. O'Brien, Silica on silicon waveguide quantum circuits. Science **320**, 646 (2008)
14. J.C.F. Matthews et al., Manipulation of multiphoton entanglement in waveguide quantum circuits. Nat. Photon. **3**, 346 (2009)
15. P.J. Shadbolt et al., Generating, manipulating and measuring entanglement and mixture with a reconfigurable photonic circuit. Nat. Photon. **6**, 45 (2012)
16. M.D. Eisaman et al., Invited review article: Single-photon sources and detectors. Rev. Sci. Instrum. **82**, 071101 (2011)
17. C. Gobby, Z.L. Yuan, A.J. Shields, Quantum key distribution over 122 km of standard telecom fiber. Appl. Phys. Lett. **84**, 3762 (2004)

18. A.J. Shields, Semiconductor quantum light sources. Nat. Photon. **1**, 215 (2007)
19. B. Lounis, M. Orrit, Single-photon sources. Rep. Prog. Phys. **68**, 1129 (2005)
20. P. Eraerds et al., Photon counting OTDR: advantages and limitations. J. Lightwave Technol. **28**, 952 (2010)
21. W. Becker et al., Fluorescence lifetime imaging by time-correlated single-photon counting. Microsc. Res. Tech. **63**, 58 (2004)
22. G.N. Gol'tsman et al., Picosecond superconducting single-photon optical detector. Appl. Phys. Lett. **79**, 705 (2001)
23. C.N. Natarajan, M.G. Tanner, R.H. Hadfield, Superconducting nanowire single-photon detectors: physics and applications. Supercond. Sci. Technol. **25**, 063001 (2012)
24. F. Marsili et al., Efficient single photon detection from 500 nm to 5 μm wavelength. Nano Lett. **12**, 4799 (2012)
25. C. Zinoni et al., Single-photon experiments at telecommunication wavelengths using nanowire superconducting detectors. Appl. Phys. Lett. **91**, 031106 (2007)
26. X.L. Hu, C.W. Holzwarth, D. Masciarelli, E.A. Dauler, K.K. Berggren, Efficiently coupling light to superconducting nanowire single-photon detectors. IEEE Trans. Appl. Supercond. **19**, 336 (2009)
27. J.P. Sprengers et al., Waveguide superconducting single-photon detectors for integrated quantum photonic circuits. Appl. Phys. Lett. **99**, 181110 (2011)
28. T. Gerrits et al., On-chip, photon-number-resolving, telecommunication-band detectors for scalable photonic information processing. Phys. Rev. A **84**, 060301 (2011)
29. W.H.P. Pernice et al., High-speed and high-efficiency travelling wave single-photon detectors embedded in nanophotonic circuits. Nat. Commun. **3**, 1325 (2012)
30. C. Schuck, W.H.P. Pernice, H.X. Tang, NbTiN superconducting nanowire detectors for visible and telecom wavelengths single photon counting on Si_3N_4 photonic circuits. Appl. Phys. Lett. **102**, 051101 (2013)
31. C. Schuck, W.H.P. Pernice, H.X. Tang, Waveguide integrated low noise NbTiN nanowire single-photon detectors with milli-Hz dark count rate. Sci. Rep. **3**, 1893 (2013)
32. A. Verevkin et al., Detection efficiency of large-active-area NbN single-photon superconducting detectors in the ultraviolet to near-infrared range. Appl. Phys. Lett. **80**, 4687 (2002)
33. F. Marsili et al., Detecting single infrared photons with 93% system efficiency. Nat. Photonics **7**, 210 (2013)
34. A. Engel et al., Tantalum nitride superconducting single-photon detectors with low cut-off energy. Appl. Phys. Lett. **100**, 062061 (2012)
35. A. Annunziata et al., Niobium superconducting nanowire single-photon detectors. IEEE Trans. Appl. Supercond. **19**, 327 (2009)
36. S.N. Dorenbos et al., Low noise superconducting single photon detectors on silicon. Appl. Phys. Lett. **93**, 131101 (2008)
37. M.G. Tanner et al., Enhanced telecom wavelength single-photon detection with NbTiN superconducting nanowires on oxidized silicon. Appl. Phys. Lett. **96**, 221109 (2010)
38. R. Sobolewski et al., Ultrafast superconducting single-photon optical detectors and their applications. IEEE Trans. Appl. Supercond. **13**, 1151 (2003)
39. V.B. Verma et al., Superconducting nanowire single photon detectors fabricated from amorphous $Mo_{0.75}Ge_{0.25}$ thin film. Appl. Phys. Lett. **102**, 022602 (2014)
40. L. Pavesi, D.J. Lockwood, *Silicon Photonics* (Springer, Berlin, 2004)
41. M. Bruel, Silicon on insulator material technology. Electron. Lett. **31**, 1201 (1995)
42. M.A. Green, M.J. Keevers, Optical properties of intrinsic silicon at 300 K. Prog. Photovolt.: Res. Appl. **3**, 189 (1995)
43. G.K. Celler, S. Cristoloveanu, Frontiers of silicon-on-insulator. J. Appl. Phys. **93**, 4955 (2003)
44. M. Bruel, B. Aspar, A.-J. Auberton-Herve, Smart-Cut: a new silicon on insulator material technology based on hydrogen implantation and wafer bonding. Jpn. J. Appl. Phys. **36**, 1636 (1997)
45. T. Baehr-Jones et al., High-Q optical resonators in silicon-on-insulator-based slot waveguides. Appl. Phys. Lett. **86**, 081101 (2005)

46. M. Soltani, S. Yegnanarayanan, A. Adibi, Ultra-high Q planar silicon microdisk resonators for chip-scale silicon photonics. Opt. Express **15**, 4694 (2007)
47. D. Taillaert et al., An out-of-plane grating coupler for efficient butt-coupling between compact planar waveguides and single-mode fibers. IEEE J. Quantum Electron. **38**, 949 (2002)
48. C. Schuck et al., Matrix of integrated superconducting single-photon detectors with high timing resolution. IEEE Trans. Appl. Supercond. **23**, 2201007 (2013)
49. W. Stutius, W. Streifer, Silicon nitride films on silicon for optical waveguides. Appl. Opt. **16**, 3218 (1977)
50. S.V. Deshpande, E. Gulari, S.W. Brown, S.C. Rand, Optical properties of silicon nitride films deposited by hot lament chemical vapor deposition. J. Appl. Phys. **77**, 6534 (1995)
51. S. Zheng, H. Chen, A. Poon, Microring-resonator cross-connect filters in silicon nitride: rib waveguide dimensions dependence. IEEE J. Sel. Top. Quantum Electron. **12**, 1380 (2006)
52. F. Morichetti, A. Melloni, M. Martinelli, R.G. Heideman, A. Leinse, D.H. Geuzebroek, A. Borreman, Box-shaped dielectric waveguides: a new concept in integrated optics. J. Lightwave Technol. **25**, 2579 (2007)
53. A. Gondarenko, J.S. Levy, M. Lipson, High confinement micron-scale silicon nitride high Q ring resonator. Opt. Express **17**, 11366 (2009)
54. N. Gruhler et al., High-quality Si_3N_4 circuits as a platform for graphene-based nanophotonic devices. Opt. Express **21**, 31678 (2013)
55. M.-C. Tien et al., Ultra-high quality factor planar Si3N4 ring resonators on Si substrates. Opt. Express **19**, 13551 (2011)
56. G.T. Reed et al., Silicon optical modulators. Nat. Photonics **4**, 518 (2010)
57. C. Xiong, W.H.P. Pernice, H.X. Tang, Low-loss, silicon integrated, aluminum nitride photonic circuits and their use for electro-optic signal processing. Nano Lett. **12**, 3562 (2012)
58. A.B. Matsko, *Practical Applications of Microresonators in Optics and Photonics* (CRC Press, Florida, 2009)
59. T.J. Kippenberg, K.J. Vahala, Cavity opto-mechanics. Opt. Express **15**, 17172 (2007)
60. K. Hennessy et al., Quantum nature of a strongly coupled single quantum dot-cavity system. Nature **445**, 896 (2007)
61. J. Bauters et al., Ultra-low-loss high-aspect-ratio Si_3N_4 wavequides. Opt. Express **19**, 3163 (2011)
62. V. Kovalyuk et al., Absorption engineering of NbN nanowires deposited on silicon nitride nanophotonic circuits. Opt. Express **21**, 22683 (2013)
63. D. Taillaert, P. Bienstman, R. Baets, Compact efficient broadband grating coupler for silicon-on-insulator waveguides. Opt. Lett. **29**, 2749 (2004)
64. X. Hu, Ph.D.-thesis (MIT, Cambridge, 2011)
65. S.E. Miller, Integrated optics: an introduction. Bell Syst. Tech. J. **48**, 2059 (1969)
66. M.G. Tanner et al., A superconducting nanowire single photon detector on lithium niobate. Nanotechnology **23**, 505201 (2012)
67. P. Dumon et al., Compact wavelength router based on a silicon-on-insulator arrayed waveguide grating pigtailed to a fiber array. Opt. Express **14**, 664 (2006)
68. A.J. Kerman et al., Kinetic-inductance-limited reset time of superconducting nanowire photon counters. Appl. Phys. Lett. **88**, 111116 (2006)
69. B.F. Little et al., Microring resonator channel dropping filters. J. Lightwave Technol. **15**, 998 (1997)
70. A. Yariv, Critical coupling and its control in optical waveguide-ring resonator systems. IEEE Photonics Technol. Lett. **14**, 483 (2002)
71. M.K. Barnoski, S.M. Jensen, Fiber waveguides: a novel technique for investigating attenuation characteristics. Appl. Opt. **15**, 2112 (1976)
72. M.K. Barnoski, M.D. Rourke, S.M. Jensen, R.T. Melville, Optical time domain reflectometer. Appl. Opt. **16**, 2375 (1977)
73. D. Derickson (ed.), *Fiber Optic Test and Measurement* (Prentice Hall, New Jersey, 1998)
74. C. Schuck et al., Optical time domain reflectometry with low noise waveguide-coupled superconducting nanowire single-photon detectors. Appl. Phys. Lett. **102**, 191104 (2013)

75. J.W. Silverstone et al., On-chip quantum interference between silicon photon-pair sources. Nat. Photon. **8**, 104 (2014)
76. M.J. Collins et al., Integrated spatial multiplexing of heralded single-photon sources. Nat. Commun. **4**, 2582 (2013)
77. G. Reithmaier et al., On-chip time resolved detection of quantum dot emission using integrated superconducting single photon detectors. Sci. Rep. **3**, 1901 (2013)
78. J.W. Silverstone et al., Qubit entanglement on a silicon photonic chip (2014). arXiv:1410.8332
79. E. Knill, R. Laflamme, G.J. Milburn, A scheme for efficient quantum computation with linear optics. Nature **409**, 4652 (2001)
80. R. Raussendorf, H.J. Briegel, A one-way quantum computer. Phys. Rev. Lett. **86**, 5188 (2001)

Chapter 5
Quantum Information Networks with Superconducting Nanowire Single-Photon Detectors

Shigehito Miki, Mikio Fujiwara, Rui-Bo Jin, Takashi Yamamoto and Masahide Sasaki

Abstract The advent of high performance practical superconducting nanowire single photon detectors (SNSPDs) has enabled rapid progress in a range of quantum information technologies, including quantum key distribution, characterization of single photon sources and quantum interface technologies. This chapter gives an overview of these recent advances.

5.1 Introduction

Quantum information (QI) technology, which can be constructed by controlling the wave and quantum nature of photons based on the principles of superposition and quantum uncertainty, promises to someday provide the ultimate transmission efficiency and unconditional levels of security in communication networks. For the realization of a future QI network, technology for control over the quantum state of

S. Miki (✉)
Advanced ICT Research Institute, National Institute of Information and Communications Technology, 588-2, Iwaoka, Nishi-ku, Kobe, Hyogo 651-2492, Japan
e-mail: s-miki@nict.go.jp

M. Fujiwara · M. Sasaki
National Institute of Information and Communication Technology, 4-2-1 Nukui-kitamachi, Koganei, Tokyo 184-8795, Japan
e-mail: fujiwara@nict.go.jp

M. Sasaki
e-mail: psasaki@nict.go.jp

R.-B. Jin
Advanced ICT Research Institute, National Institute of Information and Communication Technology, 4-2-1 Nukui-kitamachi, Koganei, Tokyo 184-8795, Japan
e-mail: ruibo@nict.go.jp

T. Yamamoto
Graduate School of Engineering Science, Osaka University, Toyonaka, Osaka 560-8531, Japan
e-mail: yamamoto@osaka-u.ac.jp

© Springer International Publishing Switzerland 2016

R.H. Hadfield and G. Johansson (eds.), *Superconducting Devices in Quantum Optics*, Quantum Science and Technology,
DOI 10.1007/978-3-319-24091-6_5

photons must be extensively exploited, and in this regard, single-photon detection technology is indispensable. For example, the performance of the detector has a direct impact on the distance, speed, and security level possible for quantum key distribution (QKD). Therefore, single-photon detectors with high detection efficiency, a low dark count rate (DCR), low timing jitter, and a high speed are a pressing requirement.

Superconducting nanowire single photon detectors (SNSPDs [1]), which are described in detail in Chap. 1 of this volume, have recently shown promise owing to a high sensitivity, ranging from ultraviolet to infrared wavelengths, an extremely low dark rate, and a precise timing resolution [2]. In addition, SNSPDs do not require any gating, making them extremely useful in the construction of simple systems for the utilization of QI technologies. Because of several technical advances, such as a low-loss optical coupling [3] and the introduction of liquid cryogen-free refrigerators [4, 5], SNSPD systems can be used as an accessible detector for potential users, and have to date been utilized in many QI experiments [6–24]. Even in the first implementations (2006–2008), an SNSPD system with only a low system detection efficiency (SDE) at a 1550 nm wavelength proved its validity in QI technologies thanks to its extremely low DCR [4, 6–11]. For example, a successful QKD experiment conducted over a 200 km distance was the longest QKD at the time [8]. In addition, efforts to improve the detector system performance have been continuously made, and an SNSPD system with a SDE of $\sim$20 % at 1550 nm was realized in 2010 by adding an optical cavity structure [3, 25], which enabled more ambitious QI demonstrations, such as the field demonstration of a QKD network in Japan [16]. Furthermore, recent progress pushed this SDE up to near unity (>80 %) [26–29], which promises to bring about new innovations in QI technologies in the near future and accelerate progress to widespread real-world applications.

This chapter describes recent accomplishments in QI technologies achieved through utilization of SNSPD systems. In Sect. 5.2, we introduce a practical QKD network demonstrated in a metropolitan area in Japan and the characterization of the field fiber network used in this QKD demonstration. In Sect. 5.3, a recent study of the characterization of single-photon sources using an SNSPD is described. In Sect. 5.4, quantum interface technologies for wavelength conversion of a single photon without destroying its quantum state are introduced.

5.2 Quantum Key Distribution Using SNSPDs

Data theft is on the increase and is set to rise dramatically in the coming years; in addition, technologies for information security have become a critical issue for advanced information and communications networks. QKD [30] is a method enabling two legitimate parties to share a secret key without leaking information regarding the final secure key to any third party. Provably secure cipher communication can be achieved when the key obtained is used with a one-time pad (OTP) [31]. The first proof-of-principle experimental demonstrations of QKD were carried out in the 1980s, with many institutes worldwide entering the field in the 1990s. These developments were transferred from a controlled laboratory environment into a real-world environment for practical use during the 2000s [30]. Although many commercial and experimental demonstrations have been carried out in field network environments around the world for examining the practical utility of QKD [32–44], the typical QKD link performance achieved in field networks thus far has been represented by a secure key rate of a few kbps at a distance of a few tens of km, which is only sufficient to encrypt voice data through a real-time OTP, or to feed the primary session key to a classical encryptor. The secure key rate needs to be significantly improved for other applications. To expand the QKD distance, we need to rely on a key relay through trusted nodes. The Tokyo QKD Network, which was started in the metropolitan area of Tokyo, Japan in 2010, has enabled the world's first demonstration of OTP encryption of movie data over 45 km of field fiber [13]. In this network an SNSPD has had a significant role in achieving a high secure key rate. In this section, we outline this practical network and the related QKD systems in which SNSPDs have been utilized. SNSPDs are also a powerful tool for characterizing a fiber network thanks to their low DCR. This section also describes the characterization of practical field fibers that are used in the Tokyo QKD Network.

5.2.1 Outline of the Tokyo QKD Network

The Tokyo QKD Network consists of parts of the National Institute of Communications Technology (NICT) open testbed network, called Japan Gigabit Network 2 plus (JGN2plus), which has four access points, Koganei, Otemachi, Hakusan, and Hongo. The access points are connected by a bundle of commercial fibers. They include many splicing points and connectors, and even run partly through the air over utility poles. The percentage of aerial fibers is about 50 %, causing the links to be quite lossy and susceptible to environmental fluctuations. The loss rate is roughly 0.3 dB/km on average for the Koganei-Otemachi link, and as high as 0.5 dB/km on average for the other two links. The fibers are also noisy, i.e., photon leakage from neighboring fibers causing inter-fiber crosstalk in the same cable is frequently observed, which is described in next section [19]. In 2010, nine organizations from Japan and the EU collaborated in the operation of the Tokyo QKD Network. Using their own QKD devices, a total of six interleaved QKD links were formed, as shown in Fig. 5.1a.

The Tokyo QKD Network adopted a three-layer architecture based on a key relay through trusted nodes, as shown in Fig. 5.1b, which is similar to that of the SECOQC network realized in Vienna in 2009 [36]. The quantum layer consists of point-to-point quantum links, forming a quantum backbone (QBB). Each link generates a secure key in its own way. The QKD protocols and the format and size of the key materials can be used arbitrarily. QKD devices push the key materials to the middle layer, called the key management layer. In this layer, a key management agent (KMA) located at each site receives the key material. KMAs have a variety of tasks for key management such as the resizing of key materials, storage of key materials and statistical data, and the forwarding of all such data to a key management server (KMS). The KMS coordinates and oversees all links in the network. Networking

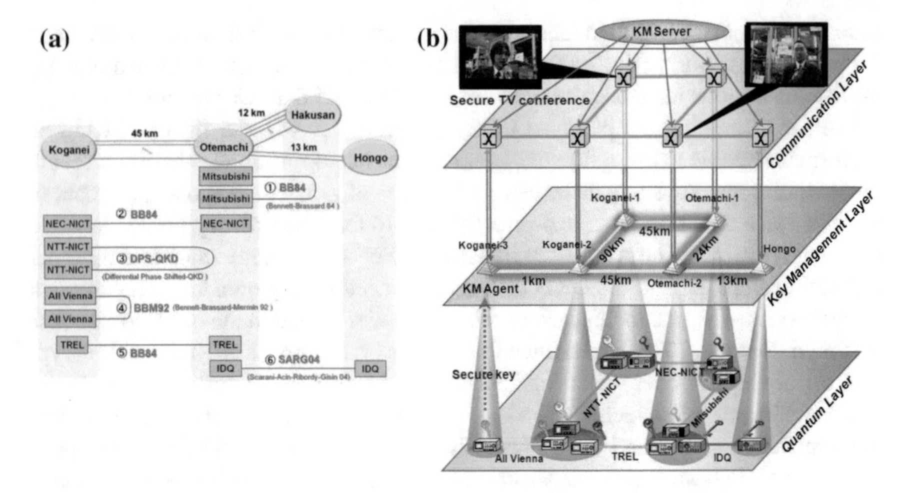

Fig. 5.1 **a** Physical link configuration of the Tokyo QKD Network, which is a mesh-type network consisting of four access points, Koganei, Otemachi, Hakusan, and Hongo. In total, six kinds of QKD systems are installed at these access points. Some QKD links are connected in a loop-back configuration with parallel fibers. **b** Three-layer architecture of the Tokyo QKD Network consisting of the quantum, key management, and communication layers. Figure reproduced with permission of the Optical Society of America, after [13]

functions are conducted entirely in the KM layer by software under the supervision of the KMS. A KMA can relay a secure key shared with one node to a second node by OTP-encrypting the key using another key shared with the second node. Thus, a secure key can be shared between nodes that are not directly connected to each other through a quantum link. In the communication layer, secure communication is ensured using distributed keys for the encryption and decryption of text, audio, or video data produced by various applications. User data are sent to the KMAs, and encrypted and decrypted by the OTP in stored key mode.

Among the QKD links, a SNSPD system was used in link no. 2 through a decoy-state BB84 protocol [45–47] in a 45 km link (NEC-NICT system), and in link no. 3 through a DPS-QKD [48] (NTT system) for the challenge of achieving a long-distance QKD of over 90 km in a loop-back configuration. A decoy-state BB84 system in link no. 2 was designed for a multi-channel QKD scheme using wavelength division multiplexing (WDM) to realize high-speed QKD for metropolitan-scale distances. The overall details of this QKD system are described in [13].

5.2.2 Decoy State BB84 Protocol System with SNSPDs

Figure 5.2 shows the photon transmission setup of the NEC-NICT QKD system in link no. 2. In the transmitter, a laser diode produces 1550 nm photon pulses with a 100 ps width at a repetition rate of 1.25 GHz. A 2-by-2 asymmetric Mach-Zehnder interferometer (AMZI) made of a polarization-free planar-lightwave-circuit (PLC)

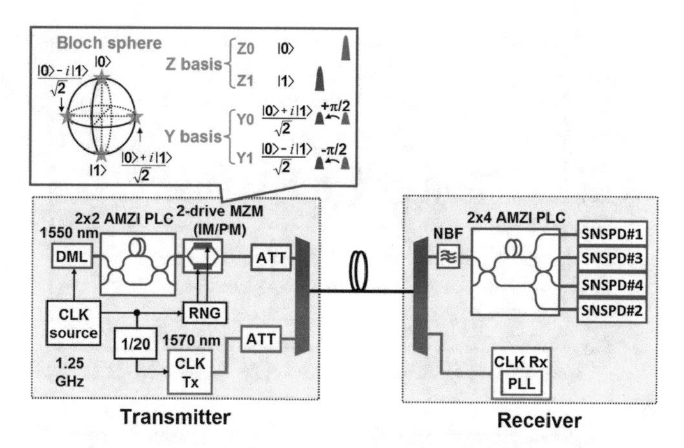

Fig. 5.2 Photon transmission schematic of the NEC-NICT system. *AMZI* Asymmetrical Mach-Zehnder interferometer; *PLC* planar lightwave circuit; *DML* directly modulated laser; *IM/PM* intensity modulation/phase modulation; *CLK* clock; *RNG* random number generator; *MZM* Mach-Zehnder modulator; *IM* intensity modulation; *PM* phase modulation; *ATT* attenuator; *NBF* narrow bandpass filter; *PLL* phase locked loop; *SNSPD* superconducting nanowire single photon detector. Figure reproduced with permission of the Optical Society of America, after [13]

splits these pulses into pairs of double pulses with a 400 ps delay. A dual-drive Mach-Zehnder modulator produces four quantum states via time-bin encoding, based on the pseudorandom numbers provided by the controller. The quantum signal is combined with the clock and frame synchronization signals using a WDM coupler, all of which are transmitted through the same fiber for precise and automatic synchronization. In the receiver, the quantum and synchronization signals are divided by a WDM filter. The quantum signal is discriminated using a 2-by-4 asymmetric and totally passive PLC-MZI, and is then detected by four-channel SNSPDs, which are free from afterpulse effects and are embedded with complex gate timing control. The SDE and DCR of the SNSPDs are about 15 % and 100 cps, respectively [3]. When combined with the QKD system, however, the total detection efficiency is reduced to about 7 %, and the noise count rate increases to 500 cps owing to the presence of stray light. The reduction in the total detection efficiency is due to the fact that the active window imposed on the time-bin signal cannot cover the whole temporal spread of the pulse after the fiber transmission.

The key distillation engine conducts frame synchronization, sifting, random permutations (RPs), error correction (EC), and privacy amplification (PA). The sifted key is processed in block units of 1 Mbits, and the RPs, EC, and PA are executed in real time, i.e., within 200 ms, for this particular block size. A low-density parity check code with a 1 Mbit code length is implemented for the EC. The coding rate can be adjusted at an appropriate value depending on the quantum bit error rate (QBER), such as 0.75, 0.65, or 0.55 for a QBER of <3.5, 5.5, or 7.5 %, respectively. This means that a PA after an EC is conducted using a modified Toeplitz matrix [49] with a block size of 750, 650, or 550 kbits.

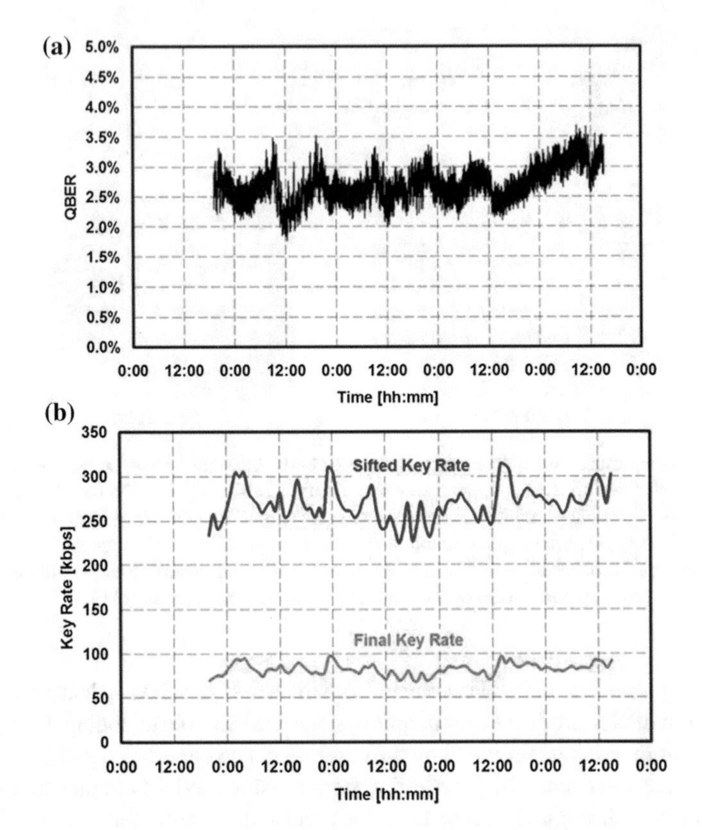

Fig. 5.3 **a** Measured quantum bit error rate (QBER) and **b** sifted and secure (final) key rates. Figure reproduced with permission of the Optical Society of America, after [13]

Figure 5.3 shows the temporal fluctuation of the measured QBER and the sifted and final key rates after a transmission distance of 45 km (14.5 dB channel loss), where one out of eight channels was coupled to the SNSPD. The decoy method [45–47] was realized using three kinds of pulses: signal, decoy, and vacuum pulses [47, 50, 51]. The averaged photon numbers of the signal and decoy pulses were 0.5 and 0.2 photons/pulse, respectively. The averaged QBER was 2.7 %, and the averaged sifted key rate was 268.9 kbps. According to the estimation of the leaked information when applying a decoy state method analysis, the averaged final secure key rate was estimated to be 81.7 kbps. It should be noted that the sifted key generation rate of this decoy state BB84 protocol system was improved after a live demonstration in 2010 by employing a wavelength-division multiplexing (WDM) scheme to over 1 Mbps, as described in [14, 15].

5.2.3 DPS-QKD System with SNSPDs

The differential phase shift QKD (DPS-QKD) scheme is especially suitable for fiber-optic transmission [8, 48], and is known to be secure against general individual attacks [52]. Figure 5.4 shows the experimental setup of the DPS-QKD system utilized in link no. 3. In the transmitter, a 1551-nm continuous light wave from a semiconductor laser is changed into a 70 ps width pulse stream with a 1 GHz repetition rate using a $LiNbO_3$ intensity modulator. Each pulse is randomly phase-modulated using $\{0, \pi\}$ with a $LiNbO_3$ phase modulator driven by a random bit signal from an FPGA board [53]. The optical pulse is attenuated at 0.2 photons/pulse and then transmitted to Bob through an optical fiber acting as a quantum channel with a total loss of 27 dB. The 100 kHz synchronization signal, which makes up the head of a 10 kbit block of random bits, is also generated by the FPGA board, converted into a 1560 nm optical pulse by a distributed feedback laser with an electro-absorption modulator (EA-DFB), and then sent over another optical fiber with a total loss of 30.0 dB.

At the receiver, the 1 GHz pulse stream is input into a PLC-MZI. The output ports of the MZI are connected to SNSPDs, which have detection efficiencies and DCRs of about 15 % and 100 cps, respectively. The detected signals are input into a time interval analyzer (TIA) through a logic gate to record the photon detection events. The optical synchronization pulses are first amplified by an erbium-doped fiber amplifier (EDFA), then received by a photo detector (PD), and finally used as a reference time in the TIA.

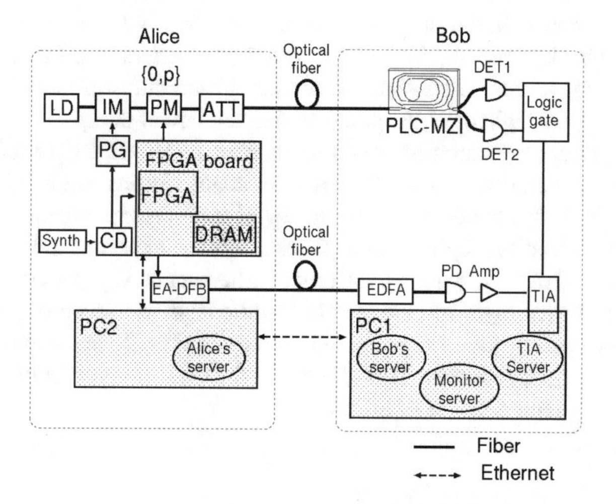

Fig. 5.4 Experimental setup of the DPS-QKD system. *LD* Laser diode; *IM* intensity modulator; *PG* pulse generator; *PM* phase modulator; *ATT* attenuator; *PLC-MZI* planar lightwave circuit Mach-Zehnder interferometer; *Synth* synthesizer; *CD* clock divider; *FPGA* field programmable gate arrays; *PD* photodiode; *TIA* time interval analyzer; *PC* personal computer. Figure reproduced with permission of the Optical Society of America, after [13]

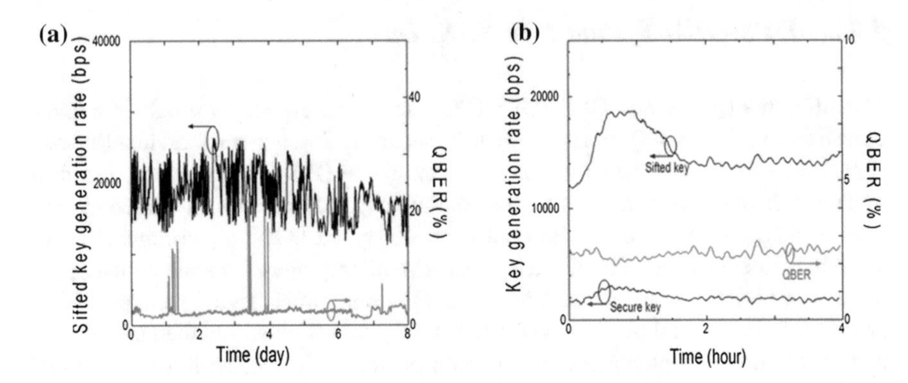

Fig. 5.5 Experimental results of **a** sifted key generation and **b** secure key generation rates. Figure reproduced with permission of the Optical Society of America, after [13]

The detection events from the TIA are sent to Bob's server through the TIA server. Bob's server generates his sifted key and sends the time information to Alice's server through an ethernet connection. Alice's server generates her key based on the phase modulation information obtained from the DRAM on the FPGA board through the gigabit ethernet interface, along with the time information from Bob's server. Both servers send 10 % of their keys to a monitor server, which estimates the key generation rate and QBERs. The remaining 90 % of the keys are sent to a key distillation engine developed by NEC, which conducts error correction and privacy amplification to distill the secure keys in the same manner as in an NEC-NICT system.

First, we checked the stability of the sifted key generation. Figure 5.5a shows the experimental results. Ultra-stable sifted key generation was demonstrated for a period of more than eight days. A spike-like degradation of the QBER was caused by the eavesdropping demonstration conducted during the UQCC2010 conference [54]. The sifted key generation rate and QBER were about 18 kbps and 2.2 % on average, respectively. Next, we conducted a secure key generation experiment by combining with the key distillation engine. Figure 5.5b shows the experimental results. Stable operation over about 4 h was demonstrated. The secure key generation rate was about 2.1 kbps on average, and the distilled secure keys were secure against general individual attacks. The sifted key generation rate and QBER were about 15 kbps and 2.3 % on average, respectively. This allows for the OTP encryption of voice data in real time, even over a distance of 90 km.

5.2.4 *Demonstration of Secure Network Operation*

A live demonstration of secure TV conferencing, eavesdropping detection, and the rerouting of QKD links on the Tokyo QKD Network was conducted and the results made public in October, 2010 [55]. The configuration of the secure TV conferencing

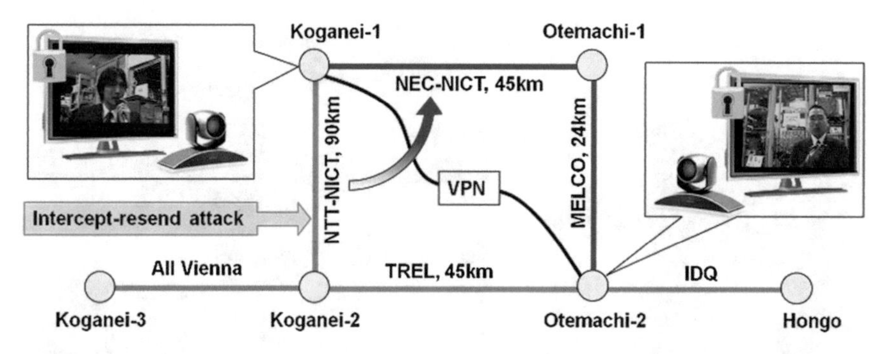

Fig. 5.6 The network configuration for secure TV conferencing between Koganei and Otemachi. For the VPN used for video transmission, two relaying QKD routes, i.e., the *red line* through Koganei-2, and the *blue line* through Otemachi-1, can be chosen. Figure reproduced with permission of the Optical Society of America, after [13]

is shown in Fig. 5.6. In the secure TV conferencing, Polycom Video Conference Systems were set up in Koganei-1 and Otemachi-2, which were directly connected through a JGN2plus L2-VPN. A live video stream for this VPN was encrypted by OTP at the KMAs in stored key mode. The encryption rate was 128 kbps. Secure keys were provided by one of two QKD relay routes, i.e., through Koganei-2 with a total distance of 135 km, or through Otemachi-1 with a total distance of 69 km, as shown by the red and blue lines, respectively. The former was used as the primary route. The 90 km QKD link worked flawlessly, and secure TV conferencing was also successfully demonstrated against an intercept resend attack using attack detection and a switching operation to a secondary route [13, 55]. A key relay was also tested and operated successfully, not only for the above secure video streaming but also in testing various relay routes, including nodes Koganei-3 and Hongo, with the teams of All Vienna and ID Quantique.

5.2.5 *Characterization of Field Practical Fibers Using a SNSPD*

In a practical QKD system, low loss and "dark" fibers are indispensable because QKD is particularly vulnerable to losses and noise. Practical fibers installed in the field, however, are both lossy and noisy, and are built by splicing many short-length fibers from different companies at intervals of a few kilometers within an urban area. The use of many spliced fiber junctions causes significant losses. Such junctions are installed not only underground but also through aerial cables, which are frequently used owing to their ease in routing. Thus, the transmission characteristics are sensitive to weather conditions. Fiber characterization measurements are indispensable for designing practical QKD systems. In this section, a characterization technique using an SNSPD is described [19].

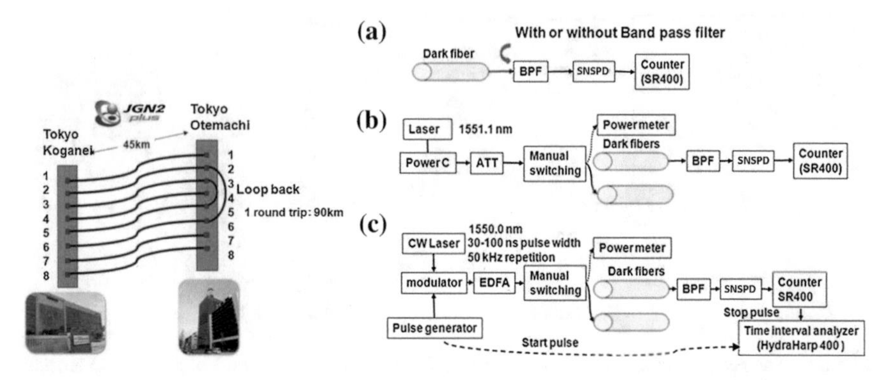

Fig. 5.7 **a** Conceptual diagram of dark fibers at the Koganei-Otemachi link in JGN2plus. Eight dark fibers in the cable are allotted to JGN2plus. The proportion of fiber on overhead lines is 50 %. **b**, **c** Conceptual image of fiber characteristic test for crosstalk estimation between neighboring fibers, and identification of the leakage points, respectively. *Power C* Power controller; *ATT* attenuator; *BPF* band pass filter; *EDFA* erbium-doped fiber amplifier. Reproduced with permission of the Optical Society of America, after [19]

The measured fibers are for the JGN2plus network, as already described, and used in the Tokyo QKD Network [56]. Dark fibers (#1 through #8) are laid between Koganei and Otemachi (located in central Tokyo), as shown in Fig. 5.7a. The fiber length totals 45 km, the total attenuation amounts to 14 dB on average, and ~50 % of the fibers are aerial lines. An NbN SNSPD is used for the characterization of the dark fibers. The SNSPD has high sensitivity for a wide wavelength region, including visible light, as discussed in Chap. 1. The bias current of the SNSPD is set such that the SDE is around 15–17 %, and the DCR is around 100 cps.

Figure 5.8a shows DCRs in daylight and at night in a loopback line of over 90 km of connecting fibers #2 and #6, when the dark fiber is connected directly to an SNSPD. Both show almost the same level of around 2000 cps. This result implies that dark counts owing to daylight are not dominant in this fiber, and a high DCR may be attributable to stray light at the telecom wavelengths. To specify the wavelength, we measure the spectra using a variable narrow band-pass filter, as shown in Fig. 5.7b, which has attenuation rate and bandwidth of 2 dB and 1 nm respectively. Figure 5.8b shows the DCRs as functions of the wavelength in the #2–6 loopback lines, and in lines #7 and #8, shown in Fig. 5.7a. A few clear peaks can be recognized in all fibers around the wavelengths that are widely used in telecom systems. Note that no light sources are input into the measured fibers or other backup lines. Thus, crosstalk takes place in the cable, and photons in the telecom wavelengths used in other commercial fibers leak into the JGN2plus fibers. This kind of crosstalk, which even occurs between covered fibers in a cable, should be taken into account in designing a QKD system, even if a dark fiber is dedicated for a quantum channel. Actually, parallel fibers are adopted to realize a QKD system in many cases. Of the so-called "classical channels," a synchronous signal, base-matching, and/or information for error correction are sent using bright light signals, which may contaminate the

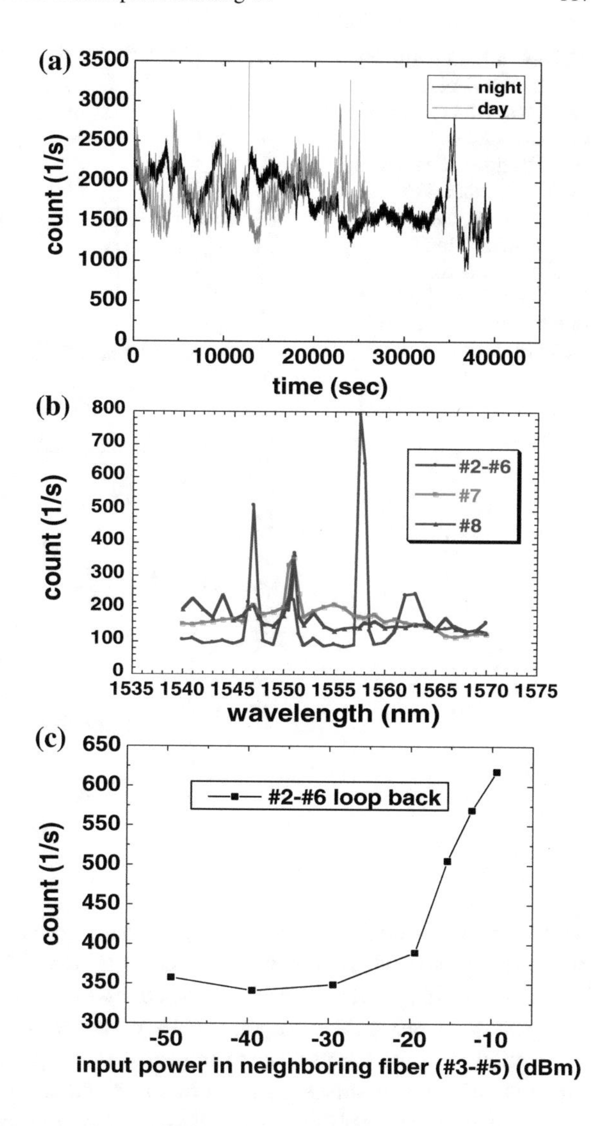

Fig. 5.8 **a** DCR of JGN2plus fibers without a band pass filter, **b** DCR dependency on the wavelength, and **c** DCR as a function of input power in the neighboring fiber. Reproduced with permission of the Optical Society of America, after [19]

neighboring quantum channel through crosstalk. To affirm the crosstalk phenomenon between parallel fibers in a quantitative manner, the CW laser is input into the #3–5 loopback lines, and measures the DCR in the #2–6 loopback lines, which are simply neighboring fiber lines, as shown in Fig. 5.7c. A significant increase in the dark count can be recognized when the input power exceeds −20 dBm, as shown in Fig. 5.8c.

The leakage points can be estimated using a method resembling an optical time-domain reflectometer, as shown in Fig. 5.7c. Light of 1550 nm wavelength from a CW laser is pulse-shaped by a modulator. The pulse width is 30–100 ns, and the repetition rate is 50 kHz. This pulse is amplified to 4–7 dBm by an erbium-doped fiber amplifier (EDFA). When optical pulses are input into the fiber, start pulses are

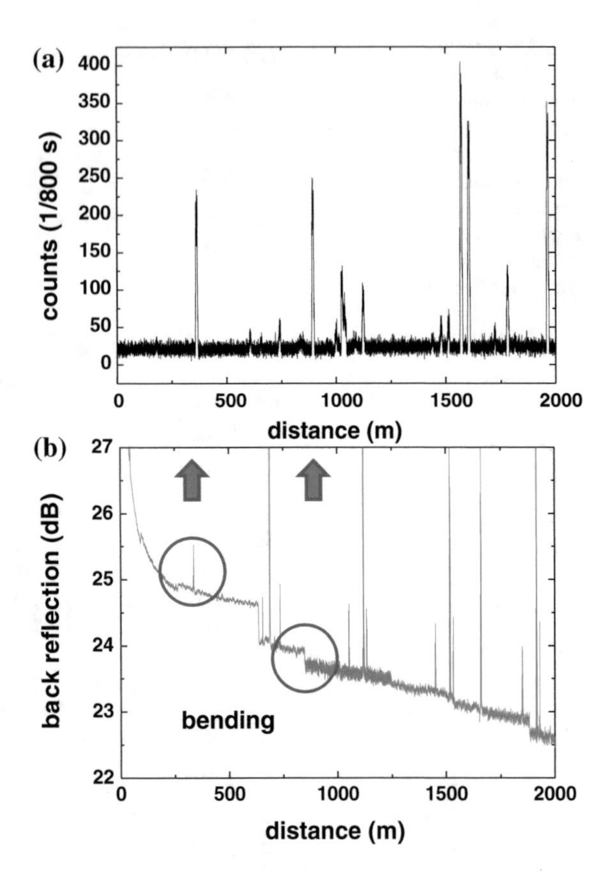

Fig. 5.9 **a** Back scattering counts of the leakage photons as a function of the transmission length, and **b** OTDR trace of an optical pulse input fiber. The large reflection and plunge step in (**b**) correspond to the bending positions. Reproduced with permission of the Optical Society of America, after [19]

sent to the time-interval analyzer. Stop pulses are given by the SNSPD-connected neighboring fiber when photons are detected by the SNSPD. If the photon leakage occurs at specific points, the peaks of the counts at the corresponding back reflecting term should be recognized. Figure 5.9a shows the leakage photon count distribution as a function of the corresponding transmission length in the #2–6 loopback lines (see Fig. 5.7a) when pulses are input into a neighboring line of the #3–5 loopback. The pulse width is 100 ns with a 50-kHz repetition rate and an input power of 7 dBm. To compare the fiber characteristics, an optical time-domain reflectrometer (OTDR) trace of the #3–5 loopback lines is shown in Fig. 5.9b. The peaks in Fig. 5.9a do not necessarily match those in Fig. 5.9b. Namely, splicing points and fiber joints that are recognized in the OTDR trace in Fig. 5.9b owing to large reflections, do not match the photon leakage points. Meanwhile, leakage points match the cable bending positions (ascertained directly), at least within 1000 m.

To confirm the bending positions more carefully, the measurement results of the short-distance fibers on the premises of an NICT site is shown. The repetition rate of the optical pulse train was 50 kHz when the data shown in Fig. 5.9a were

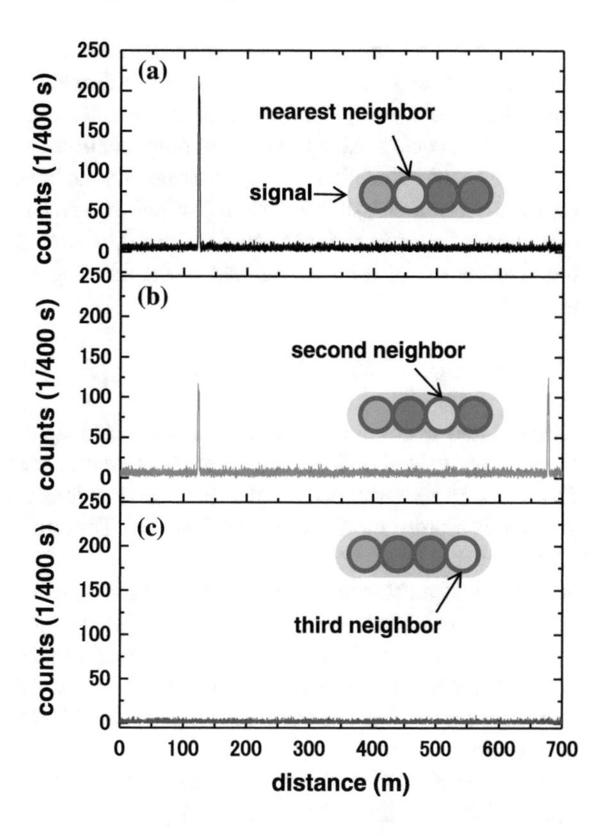

Fig. 5.10 Extent of photon leakage in parallel fibers: **a** nearest neighbor, **b** second-nearest neighbor, and **c** third-nearest neighbor. Reproduced with permission of the Optical Society of America, after [19]

acquired. We cannot deny that ghost signals may have been measured in the data in Fig. 5.9a under a 50 kHz repetition rate because this rate is too fast to estimate the back reflection in a fiber of as long as 90 km. The fiber length at this time is only 700 m, but we can explicitly and directly determine the conditions of the field-installed fiber. Figure 5.10 shows the photon leakage to the nearest-, second-, and third-nearest neighboring fibers within the 700 m of field-installed fibers. Half a dozen four-core tapes, including the fibers used in this experiment, are bundled into a single optical cable [57]. The pulse width is set to 30 ns with the same repetition rate of 50 kHz and an input power of 4 dBm. In the nearest- and second-nearest neighboring fibers, reflections of the leaking photons are observed at the 130 m mark. At around this point, the optical cable is bent into a hand-hole, and there are no splicing points or joint connections. In Fig. 5.10a–c, a significant amount of photon leakage can be seen in the second-nearest neighboring fiber at the 680 m mark. At this point, photon leakage cannot be observed in the nearest-neighboring fiber, however. The optical cable is also rolled at around this point. Photon leakage also cannot be observed in the third-nearest neighboring fiber. Such leakage photons cannot be seen in Fig. 5.9a. The data shown in Fig. 5.9a were obtained at a temperature of around 12 °C, and those in Fig. 5.10 were obtained at a temperature of around 22 °C. These results imply that the fiber cable bending is the cause of crosstalk between the parallel fibers, and that the photon leakage reaches as far as the second-nearest neighboring fiber. Moreover, the leakage points vary depending on the weather conditions.

Crosstalk occurs mainly at the damping locations, such as at the splicing or bending points. The results in Fig. 5.10 indicate that the bending of the optical cable is the cause of photon crosstalk between parallel fibers. The bending of an optical cable is a known factor in the transmission loss. The loss photons are radiated owing to the degradation of the confinement effect. Homogeneous bending loss α_B [58] is given by

$$\alpha_B = \sqrt{\pi} u^2 \exp\left[-4\Delta w^3 R / \left(3 a v^2\right)\right] / \left[2 w^{3/2} V^2 (Ra)^{1/2} K_1^2(w)\right], \qquad (5.1)$$

where u and w are normalization factors of propagation constants, R is the bending radius, a is the fiber core radius, Δ is the relative index difference, $V = 2\pi n_0 a(2\Delta)^{1/2}/\lambda$ (where n_0 is the clad glass index, and λ is the wavelength), and K_1 is a first-order modified Bessel function. Therefore, α_B depends strongly on R. The allowable bending radius of our optical cable is 100 mm, and the cable is laid at a radius of more than 200 mm. However, crosstalk is observed around the rolled sections. The dependence of the crosstalk on the weather conditions can be explained by the expansion or shrinking of the fiber coating material from the changes in the ambient temperature affecting the photon coupling efficiency. A detailed study on this phenomenon must be conducted to provide a clear understanding of the crosstalk probability.

Let us evaluate the influence of stray light in the JGN2plus fibers on the QKD system. Based on the DCR shown in Fig. 5.8a, if an InGaAs APD with a DE of 15 % at a 1 ns time-width in Geiger mode is applied to a QKD system, the dark count probability can be estimated as 2×10^{-6} (2000 cps $\times 10^{-9}$ s), even if the APD has no dark counts. This value is 20 % of the widely used APD dark count probability [59] and is sufficiently large to cause deterioration of the QKD system because the decoy-state QKD is susceptible to dark counts. Therefore, stray light in the JGN2plus fibers is a significant barrier to a QKD system. In addition, the fact that photons leak from the parallel fiber indicate that a classical channel parallel to the quantum channel sending synchronous signal and base-matching data, and/or information for error correction, may be a potential source of dark count noise. Moreover, data theft through the bending of the fibers should be understood as a genuine threat. Data theft using an SNSPD is demonstrated in Fig. 5.11. An optical intensity-modulated signal is input into one of the pair fibers, and the other fiber is connected to the SNSPD. When the pair of fibers is bent at a right angle, the optical signal is detected by the SNSPD. This result implies that data theft can be accomplished without the need for a special tapping device. It is possible to eavesdrop on the data by using a highly sensitive photon detector from the neighboring fiber.

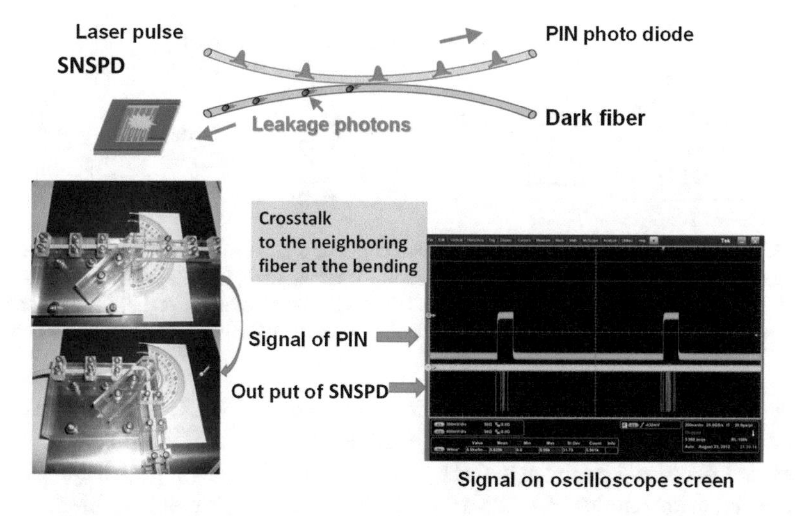

Fig. 5.11 Eavesdropping demonstration using a SNSPD with fiber bending. The signals can be detected when the fiber is bent at a sharp angle

5.3 Characterization of Single Photon Sources with SNSPDs

Over the last two decades, QI technology has become an area of rapid advancement in physics, and considerable experimental efforts have been devoted to demonstrations of complicated QI protocols. For further development of this field, more sophisticated photon engineering and detection technologies are indispensable, but such development has been limited by the performance of conventional photon sources and detectors. In this part of the chapter, we report recent progress in telecom wavelength single- and entangled photon sources, enabled by advanced SNSPD technology. This part of the chapter is organized as follows. Section 5.3.1 describes the detection of heralded single photon emission using SNSPDs. Section 5.3.2 characterizes an entangled photon source using twin SNSPDs [22], and in Sect. 5.3.3, non-classical interference between two independent sources detected by SNSPDs is demonstrated. Finally, some concluding remarks are given in Sect. 5.3.4.

5.3.1 Detection of Heralded Single Photon Emission Using Twin SNSPDs

In this section, we demonstrate the application of an SNSPD for the detection of a single photon source [20] using the setup shown in Fig. 5.12. Picosecond laser pulses (76 MHz, wavelength of 792 nm, temporal duration of approximately 2 ps, and horizontal polarization) from a mode-locked titanium sapphire laser (Mira900, Coherent, Inc.) were focused using an f $=$ 200 mm lens and pumped a 30 mm long periodically poled $KTiOPO_4$ (PPKTP) crystal with a poling period of 46.1 mm

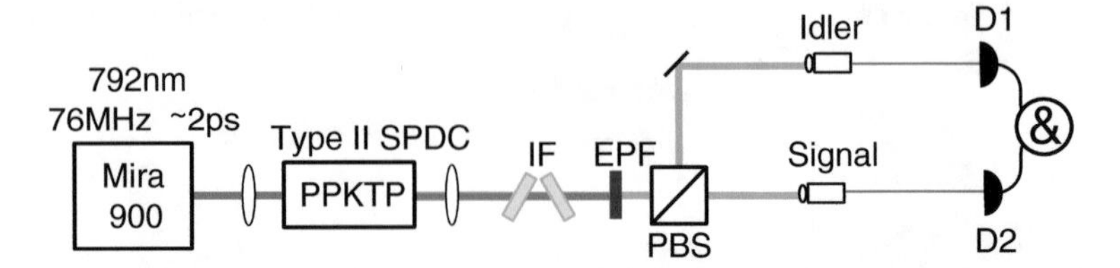

Fig. 5.12 The experimental setup for characterization of a heralded single-photon source. Mira 900, mode-locked titanium sapphire laser (Mira900, Coherent, Inc.); *SPDC* spontaneous parametric down-conversion; *PPKTP* periodically poled $KTiOPO_4$; *IF* interference filter; *EPF* edge pass filter; *PBS* polarizing beam splitter; D1, D2, single photon detectors (SNSPDs); & coincidence counter

for a type-II spontaneous parametric down-conversion (SPDC). The PPKTP was maintained at 32.5 °C to achieve a degenerate wavelength at 1584 nm. The down-converted photons, i.e., the signal (H polarization) and idler (V polarization), were collimated using another $f = 200$ mm lens, filtered through two interference filters (IF) and an edge pass filter (EPF), separated by a polarizing beam splitter (PBS), and then collected into two single-mode fibers (SMFs). The signal and idler were connected to two SNSPDs and a coincidence counter (&) for testing the coincidence counts. Owing to the group-velocity matching condition [62], the photon source from a PPKTP crystal has a high spectral purity of 0.82 and wide tunability at telecom wavelengths [21]. SNSPDs are fabricated using 5–9 nm thick and 80–100 nm wide niobium nitride (NbN) or niobium titanium nitride (NbTiN) meander nanowires on thermally oxidized silicon substrates [28]. A nanowire covers an area of 15 μm × 15 μm. The SNSPDs used here were installed in a Gifford-McMahon cryocooler system and cooled to 2.1 K. The maximum SDE is 79 % with a DCR of 2000 cps. The measured timing jitter and dead time (recovery time) were 68 ps [28] and 40 ns [29]. The measured spectral range can cover at least 1470–1630 nm [20]. In this experiment, the SDEs of the two SNSPDs were set to 70 and 68 %, corresponding to DCRs of less than 1000 cps.

We measured the single counts and coincidence counts as a function of the pump power, as shown in Fig. 5.13a. The single (coincidence) counts reached 2.14×10^5 cps (4.5×10^4 cps), 1.91×10^6 cps (4.06×10^5 cps), and 5.23×10^6 cps (1.17×10^6 cps) at pump powers of 10, 100 and 400 mW respectively. To the best of our knowledge, these are the highest coincidence counts ever reported at telecom wavelengths. We also calculated the generated photon pairs and average photons per pulse, as shown in Fig. 5.13b. Compared with the previous PPKTP source at telecom wavelengths by Evans et al. [60], the calculated generation rates of the photon pairs per second are comparable, whereas our detected coincidence counts are about 90 times higher. It is also noteworthy to compare the brightness of our source with the previous best results at near infrared (NIR) wavelengths [61, 64, 65]. We noticed that the maximum coincidence counts in our system are comparable to the results of these experiments, which means that using an SNSPD and PPKTP crystal at telecom wavelengths can make the multi-photon coincidence counts similar to or even higher than those within the NIR range.

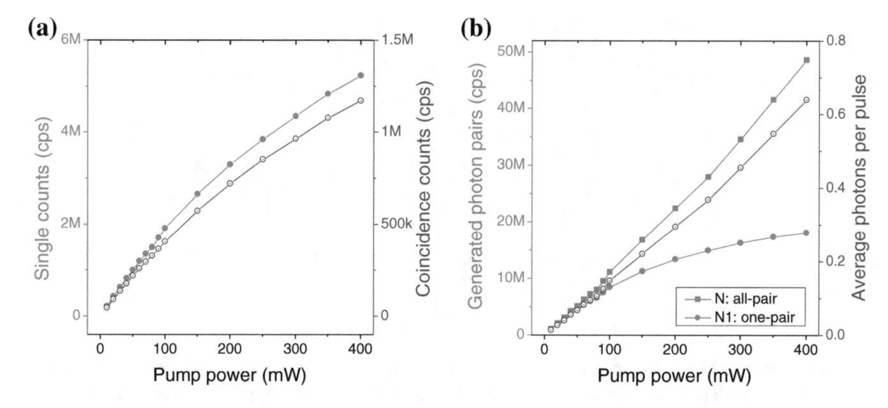

Fig. 5.13 **a** Measured single and coincidence counts as functions of pump power, and **b** calculated all-pair and one-pair component generation rates and the average photon numbers per pulse as functions of the pump power. After [20], copyright permission courtesy of Elsevier

5.3.2 Demonstration of an Entangled Photon Source with SNSPDs

After the detection of a single photon emission from the source, we updated the source configuration to demonstrate entangled photon generation, which was realized by inserting a PPKTP crystal into a Sagnac interferometer [22]. The experimental setup is shown in Fig. 5.14. Picosecond laser pulses from a mode-locked Titanium sapphire laser passed through an optical isolator (OI), a half-wave plate (HWP), and a quarter-wave plate (QWP). The pulses were then focused by an $f = 200\,\text{mm}$ lens (beam

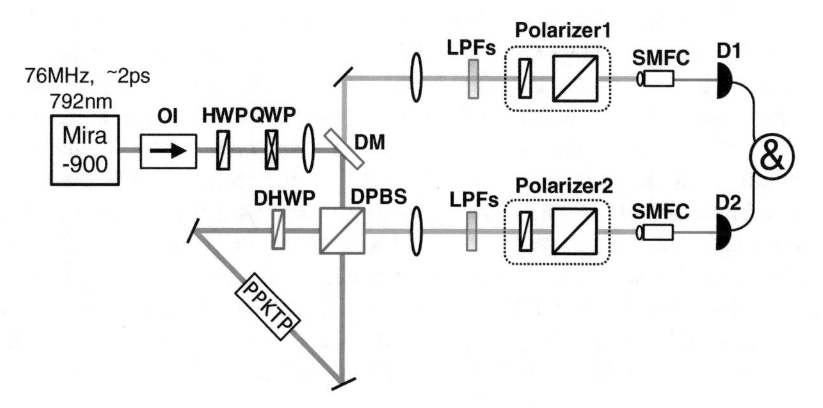

Fig. 5.14 The experimental setup for demonstration of an entangled photon source. *OI* Optical isolator; *HWP* half-wave plate; *QWP* quarter-wave plate; *DM* dichroic mirror; *DHWP, DHWP* dual-wavelength HWP; *DPBS* dual-wavelength polarization beam splitter; *LPFs*, long-pass filters; *SMFC*, single-mode fibers using two couplers; & coincidence counter. After [22], reproduced with permission of Optical Society of America

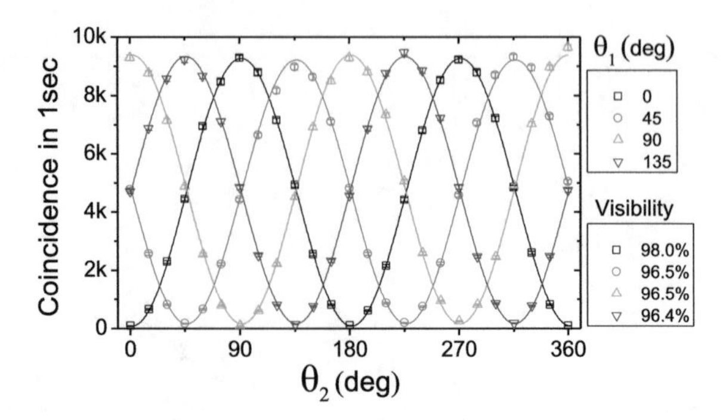

Fig. 5.15 Two-fold coincidence counts in one second as a function of two polarizers with a pump power of 10 mW. The background counts were subtracted. The error bars were added by assuming the Poisson statistics of these coincidence counts. After [22], reproduced with permission of the Optical Society of America

waist of $\phi 45\,\mu$m), reflected by a dichroic mirror (DM), and guided into a Sagnac-loop. The Sagnac-loop consisted of a dual-wavelength polarization beam splitter (DPBS), a dual-wavelength HWP (DHWP), and a 30 mm long PPKTP crystal with a polling period of 46.1 mm for a type-II collinear group-velocity matched SPDC. The temperature of the PPKTP was maintained at 32.5 °C to achieve wavelength degeneracy at 1584 nm. The PPKTP crystal was pumped using clockwise (CW) and counterclockwise (CCW) laser pulses concurrently. The DHWP was set at 45° to make the CCW pump horizontally polarized. The down-converted photons, i.e., the signal and idler, were collimated using two other $f = 200$ mm lenses, filtered by long-pass filters (LPFs), and then coupled into single-mode fibers using two couplers (SMFC). Finally, all of the collected photons were sent to two SNSPDs, which were connected to a coincidence counter (&). We set the pump power to 10 mW and carried out a polarization correlation measurement by recording the coincidence counts while changing angles q1 and q2 of Polarizer 1 and Polarizer 2, respectively. All of the resultant fringe visibilities shown in Fig. 5.15 were higher than 96 %. We also achieved an S value of 2.76 ± 0.001 using Bell's inequality measurement, and fidelities of 0.98 ± 0.0002 using quantum state tomography [22]. The measured maximum coincidence count in Fig. 5.15 was 1×10^4 cps, which corresponds to a coincidence of 2×10^5 cps without polarizers. Comparing our scheme with the previous entangled photon source with a PPKTP crystal in a calcite beam displacer configuration at telecom wavelengths [60], our count rates were more than 100 times higher, mainly thanks to our highly efficient SNSPDs and fine alignment in the Sagnac-loop. Obtaining entangled photon pairs with a high count rate and a low pump power at telecom wavelengths is an important feature of our scheme.

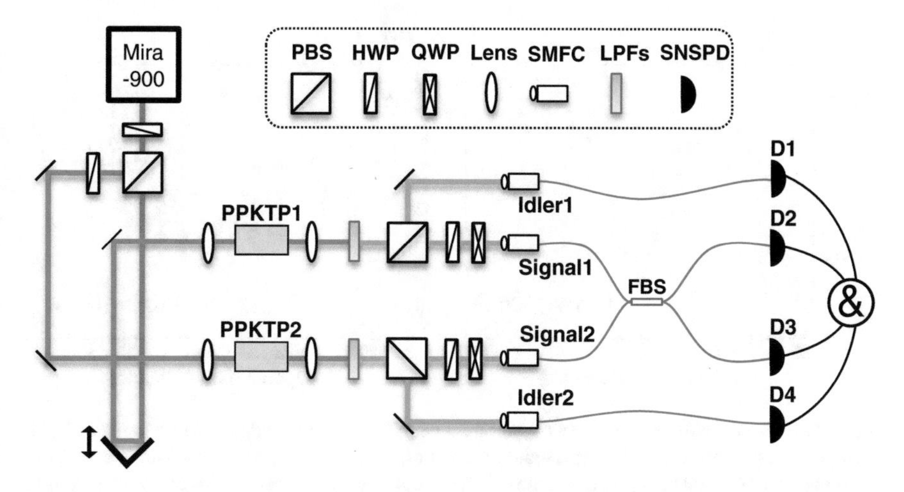

Fig. 5.16 The experimental setup for the demonstration of interference between two independent heralded single photon sources. *PPKTP* Periodically poled $KTiOPO_4$; *FBS* fiber beam splitter; D1-4, detectors 1-4; &, coincidence counter; *HWP* half-wave plate; *PBS* polarizing beam splitter; *QWP* quarter-wave plate; *LPFs* long-pass filters; *SMFC* single-mode fibers using two couplers; *SNSPD* superconducting nanowire single photon detector. After [23], reproduced with permission of the American Physical Society

5.3.3 Demonstration of Interference Between Two Independent Single Photon Sources Using SNSPDs

After the detection of a single photon source and an entangled photon source with only one PPKTP crystal, we next expanded the system to two PPKTP crystals. We demonstrated Hong-Ou-Mandel interference between two independent, intrinsically pure, heralded single photons from an SPDC source at a telecom wavelength [23]. The experimental setup is shown in Fig. 5.16. The components are similar to those shown in Figs. 5.12 and 5.14. Owing to the group-velocity matched SPDC and highly efficient SNSPDs, the four-fold coincidence counts shown in Fig. 5.17a are one-order higher than in previous experiments. A visibility of $85{:}5 \pm 8{:}3\,\%$ was achieved without the use of a bandpass filter, as shown in Fig. 5.17b. Because this work was carried out before the SNSPDs with high SDE used in Figs. 5.12 and 5.14 had been developed, the efficiency of the four SNSPDs in this experiment was around 20 %.

5.3.4 Photon Source Characterization: Conclusions

To summarize, we characterized the performance of a heralded single photon source, an entangled source, and the interference between two independent heralded single photon sources. Owing to the use of highly efficient SNSPDs and an ultra-bright single-photon source from a group-velocity-matching PPKTP crystal, the coinci-

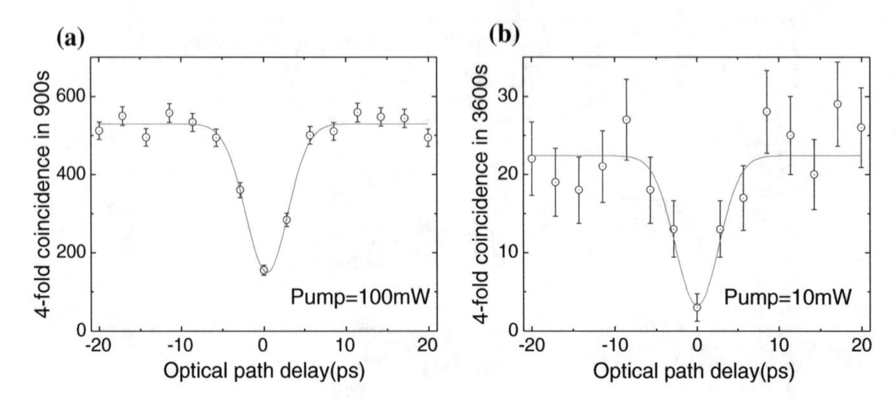

Fig. 5.17 Experimental results of two-photon interference between independent heralded single-photon sources. The *solid lines* represent Gaussian fits of the data points. Error bars are equal to the square-root of each data point, assuming Poisson counting statistics. Here, **a** and **b** are four-fold coincidence counts of D1, D2, D3, and D4 with a pump power of 100 and 10 mW for each PPKTP crystal, respectively. After correcting the background counts, the visibilities of **(a)** and **(b)** were 71.8 ± 2.7, and $85.5 \pm 8.3\,\%$, respectively. After [23], reproduced with permission of the American Physical Society

dence count rate obtained is much higher than that of previous schemes. The combination of highly efficient SNSPDs and the high brightness photon source in our scheme has opened up opportunities for various future applications at telecom wavelengths, e.g., multi-photon interference in fiber networks, entanglement swapping, and free-space quantum key distribution.

5.4 Quantum Interface Technology Enabled by SNSPDs

In this section, we introduce a quantum interface for the wavelength conversion of a single photon without destroying its quantum state. Such a quantum interface is useful for linking diverse solid state or atomic quantum systems through the quantum states of light. In particular, when we look at long-distance quantum communication, entangling the distantly located solid state quantum memories through photons is an elemental building block for quantum repeaters [66–68]. Many experimental demonstrations related to this issue have been conducted over the past decade. However, the wavelengths of the photons do not fit a telecom wavelength of around 1550 nm for optical fiber communication in many of the experiments conducted owing to the limitations of the resonant wavelengths of the quantum systems [69, 70]. Therefore, such a quantum interface enabling wavelength conversion has attracted significant interest. Clearly, SNSPDs play a significant role in achieving the faithful operation of the quantum interfaces and quantum repeaters.

5.4.1 *Quantum Interface for the Coherent Wavelength Conversion of a Single Photon*

Our approach is to employ a second-order nonlinear optical effect for the wavelength conversion, which is referred to as wave mixing. Under a wavelength conversion, the energy conservation law $\omega_a = \omega_b + \omega_c$ is satisfied, where $\omega_{a,b,c}$ represents the angular frequency of the light in mode a, b, c. The interaction Hamiltonian of this process is described as

$$\hat{H} = ig\hat{a}\hat{b}^\dagger\hat{c}^\dagger + h.c., \tag{5.2}$$

where $\hat{a}$, $\hat{b}$ and $\hat{c}$ represent the annihilation operator in modes a, b, and c, respectively. ($h.c.$ denotes the Hermitian conjugate.) The wavelength conversion we employ here is the difference frequency generation (DFG) with a strong pump laser in mode b, as shown in Fig. 5.18a. In this case, the above Hamiltonian is approximated by

$$\hat{H} = \beta\hat{a}\hat{c}^\dagger + h.c., \tag{5.3}$$

where β is a complex amplitude proportional to the complex amplitude of the pump laser. Under the Hamiltonian, the state transformation is represented as

$$|0\rangle_a|1\rangle_c = e^{i\varphi}\sin\theta|1\rangle_a|0\rangle_c + \cos\theta|0\rangle_a|1\rangle_c, \tag{5.4}$$

where LHS and RHS represent the output and input states, respectively. Angle θ is determined by the amplitude of the pump laser, the nonlinear coefficient of the material, and the interaction time. When $\theta = \pi/2$, the input single photon in mode a

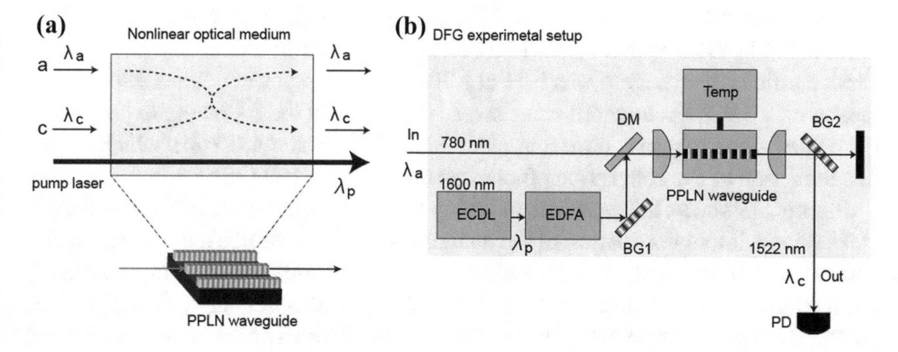

Fig. 5.18 Quantum interface for a wavelength conversion (as employed in [75–77], modified figure). **a** Concept of a coherent wavelength conversion through a second-order nonlinear optical medium. Difference (sum) frequency generation with a properly adjusted pump light converts the input light of wavelength $\lambda_a(\lambda_c)$ into wavelength $\lambda_c(\lambda_a)$. **b** Experimental setup for the difference frequency generation using a PPLN waveguide. A 1600 nm ECDL pump light amplified by an EDFA and a 780 nm signal light are mixed using a dichroic mirror (DM) and coupled into a PPLN waveguide. The converted 1522 nm light is extracted using a Bragg grating (BG2)

is converted into a single photon in mode c with the probability of unity. The coherent property is ensured by the stability of phase φ, which is determined by the coherence time of the pump laser. The wavelength conversion employed here was first proposed by P. Kumar [71] and has been demonstrated experimentally in various systems [72–74]. Below, we introduce the quantum interface for the wavelength conversion from visible to telecom wavelengths [16, 17, 75], which plays a significant role in quantum repeater systems.

5.4.2 Experiments with a Waveguide PPLN Crystal

A periodically-poled lithium niobate (PPLN) crystal is one of the most suitable choices for such a wavelength conversion because a large variety of wavelength selections is possible by tuning the poling period, and a large interaction length is achieved from the waveguide structure. In our experiment, we used a PPLN waveguide with a core of Zn-doped lithium niobate and a cladding layer of lithium tantalite [76]. The poling period is 18 μm for the type-0 quasi-phase matching condition. The waveguide structure is a ridged type with a length of 20 mm and a square cross section width of 8 μm. Although along PPLN crystal has a small acceptable phase-matching bandwidth, it is calculated to be about 0.3 nm for a 780 nm signal light, which is sufficient for many applications. The temperature of the PPLN waveguide is adjusted to 50 °C for maximum conversion efficiency. The pump laser for the DFG is based on a 1600 nm external cavity diode laser (ECDL) with a bandwidth of $\delta f < 150\,\text{kHz}$, which is amplified by an EDFA. The amplified spontaneous emission (ASE) of the EDFA is eliminated by the Bragg grating (BG1) with a bandwidth of 1 nm. The 780 nm signal light is converted to 1522 nm owing to the energy conservation law of the DFG. After the conversion at the PPLN waveguide, the 1522 nm converted light is extracted by BG3 with a bandwidth of 1 nm. The maximum internal photon conversion efficiency observed was 0.71 at a 700 mW pump power when a 780 nm CW laser with a 3 MHz bandwidth was used as the input. In the following experiments, the pulsed signal photons have a broader bandwidth of about 0.2 nm, which results in a reduction of the conversion efficiency from 0.71 to 0.62 [75].

Figure 5.19 shows the experimental setup for the demonstration of the wavelength conversion without destruction of the quantum states. The source emits a maximally entangled photon pair at the polarization degree of freedom, which is prepared using a spontaneous parametric down-conversion (SPDC) with a frequency-doubled 1.2 ps pulsed Ti:sapphire laser. One of the photon pairs at 780 nm was sent to the DFG-based wavelength converter. The wavelength converter we demonstrated works only for the vertical polarization state of the photons owing to the type-0 quasi-phase matching condition. Therefore, asymmetric interferometers composed of a polarization beam splitter (PBS), a half wave plate (HWP), and a non-polarizing beam splitter (BS) interchange the physical system between the polarization degree of freedom and the temporal one. After transmitting the first interferometer, the photon is in two temporary separated time bins with vertical polarization. Therefore, it is crucial that

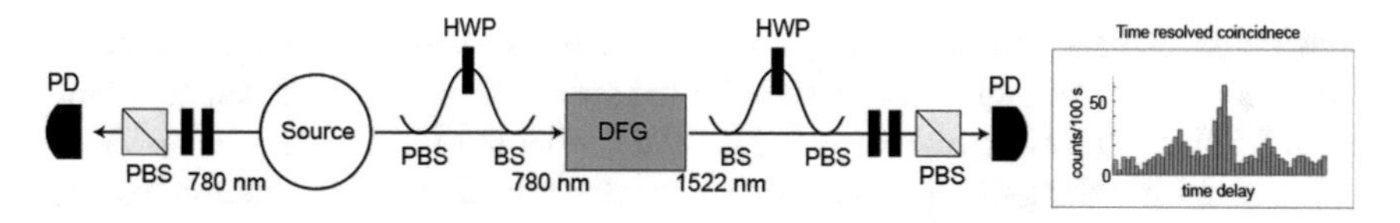

Fig. 5.19 Experimental setup for the DFG with the entangled photon pair (after [75], modified figure). A polarization entangled photon pair is generated at the source. The wavelength of one of the photons is converted using the DFG module shown in Fig. 5.18. Two unbalanced interferometers convert between the polarization degree and time-bin degree using the same polarization. An analysis of this process was conducted through quantum state tomography using photon detectors (PD)

the DFS process preserve the phase information of the single photon states encoded in the two time bins. In our experiment, this is ensured by the large coherence time of the narrow bandwidth CW pump laser. When the photon passes the long and short arms of two interferometers after both interferometers are transmitted, the photon is adjusted to be in the initial polarization state. Note that this dual interferometry scheme reduces the overall efficiency by 1/4 at least. The half wave plates (HWP) and the quarter wave plates (QWP) and PBSs just before the detectors are used for the polarization state analysis, which is used for the reconstruction of the density matrix of the photon pair states. The inset in Fig. 5.19 shows the histogram of the time-resolved coincidence photon detection based on APDs [75]. The central peak corresponds to the events in which the state is preserved. However, in the histogram, we see continuous background photon counting events that also contribute to the central peak. The origin of the background is considered to be the dark counting of the photon detectors and the Raman scattering through the strong pump laser.

Photon detectors with a low DCR and a low timing jitter with high detection efficiency were required to reduce the effects of the background and achieve a high-fidelity quantum interface. In our experiment [16], to fulfill such requirements of the photon detectors, we used two types of SNSPD for detecting the 780 nm photons and 1522 nm photons [3]. The SNSPDs are composed of a 100 nm thick Ag mirror, a $\lambda/4$ SiO cavity, and a 4 mm thick niobium nitride meander nanowire on a 0.4 mm thick MgO substrate from the top. The nanowire is 80 nm wide, and covers an area of 15 μm $\times$ 15 μm. The SDE for 780 and 1522 nm wavelengths is 32 and 12.5 %, respectively. Each SNSPD chip coupled with a single-mode optical fiber through a small-gradient index lens is in a copper holder cooled at 2.28 ± 0.02 K by the Gifford-McMahon cryocooler system. Note that the SDE was recently improved to up to 80 % [28]. When the APDs were used for this experiment, the measured time distribution was $\approx$350 ps. On the other hand, when the SNSPDs were used, the distribution reduced to $\approx$150 ps. As mentioned previously, the wavelength converter continuously generates noise photons owing to the Raman scattering from the strong pump laser. Therefore, a high timing-resolution measurement using the SNSPDs reveals the intrinsic performance of our quantum interface, and achieves the required performance for the next step of our experiments.

A well-known method to determine how well the quantum interface operates is to quantify the output-entangled states from the tomographic reconstruction of the density matrix based on the polarization analysis of two photons. As a reference, we observed the two-photon state prepared at the source. The observed density matrix had a fidelity value of $\langle\Phi|\rho_{\mathrm{in}}|\Phi\rangle = 0.97 \pm 0.01$, where $|\Phi\rangle \equiv (|\mathrm{H}\rangle|\mathrm{H}\rangle + |\mathrm{V}\rangle|\mathrm{V}\rangle)\sqrt{2}$ is a maximally entangled state. This shows that the prepared 780 nm photon pair was highly entangled. Even after the wavelength conversion, we observed a fidelity value of 0.93 ± 0.04, which is very close to the initial fidelity value [11]. Note that when APDs were used for this measurement, the fidelity was 0.75 ± 0.06 [10]. Here, we can clearly see the advantages of using the SNSPDs.

5.4.3 HOM Interference Over the Quantum Interface with SNSPD

As illustrated in Fig. 5.20a, establishing an entanglement between the distantly located quantum memories (QM) through two-photon interference [17] is the next step toward the achievement of long-distance optical fiber quantum communication based on quantum repeaters. For such a realization, two down-converted telecom photons from the separated QMs need to have the ability to complete a Bell measurement (BM) at the intermediate node. This implies that the two telecom photons must be indistinguishable, resulting in high-visibility Hong-Ou-Mandel (HOM) interference. To conduct our experiment, we used the same quantum interface as used for the wavelength conversion. The conceptual setup of the experiment is shown in Fig. 5.20b. To shorten the measurement time, we used a weak coherent pulse at 780 nm instead of a single photon as an input. In contrast, the other input is a heralded single photon prepared from the SPDC. We used an SNSPD in the visible range for the heralding, and two SNSPDs in the telecom range to determine the HOM interference. The results of the delayed coincidence measurement show a dip in the HOM interference. The visibility observed was 0.76 ± 0.12, which clearly surpasses the classical limit of 0.5. Our theoretical analysis shows that the main reason for the degradation of the visibility from unity is the generation of multiple photons from the weak coherent light pulse in one input mode, which suggests that the demonstrated quantum interface has the ability to achieve a visibility of 0.91, which is very close to unity [17].

5.4.4 Quantum Interface Technology: Summary

In this section, we introduced a quantum interface for wavelength conversion, which aims at long-distance optical fiber quantum communication. Although we have focused here on a specific wavelength translation choice, i.e., 780–1522 nm, which is suitable for Rb-based quantum memories (QM), quantum interfaces for other QMs have also been actively studied. Remarkable studies have been performed on quantum dots [77–79] and nitrogen vacancy (NV) centers in diamond [18]. The

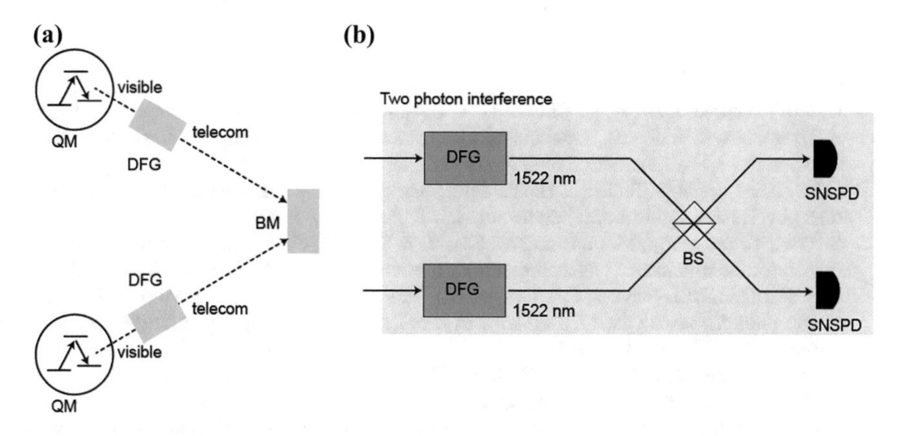

Fig. 5.20 Two-photon interference toward a quantum repeater including the wavelength converter (as employed in [17]): **a** concept of a fundamental quantum repeater node with a wavelength converter, and **b** experimental setup for two-photon interference required for the BM after the wavelength conversion

quantum interface for a wavelength conversion has also attracted significant interest in the context of manipulating the quantum states encoded within the wavelength (frequency) degree of freedom. Further work in this direction has demonstrated coherence between both inputs of a wavelength converter [24].

5.5 Conclusions

As described in this chapter, SNSPDs have been successfully employed and have shown their superiority in enabling the realization of various kinds of QI technologies. In particular, several important QI demonstrations already utilized recently-developed SNSPDs with high practical efficiency. SNSPDs enabled the realization of the Tokyo QKD network. Moreover high performance SNSPDs have enabled the advancement of a variety of emerging QI technologies such as telecom wavelength entangled photon sources and quantum interfaces, which hitherto were out of reach due to the limitations of available photon counting technology. We anticipate that the various other types of next generation SNSPD device described in this volume (such as waveguide SNSPD described in Chaps. 3 and 4, the photon number resolving SNSPDs introduced in Chap. 1, and multi pixel SNSPD discussed in Chap. 1) will also be employed in QI demonstrations in the near future, spurring the development of QI networks as a real-world technology.

References

1. G. Gol'tsman, O. Okunev, G. Chulkova, A. Lipatov, A. Semenov, K. Smirnov, B. Voronov, A. Dzardanov, C. Williams, R. Sobolewski, Picosecond superconducting single-photon optical detector. Appl. Phys. Lett. **79**, 705–707 (2001)
2. C.M. Natarajan, M.G. Tanner, R.H. Hadfield, Superconducting nanowire single-photon detectors: physics and applications. Supercond. Sci. Technol. **25**, 063001 (2012)
3. S. Miki, T. Yamashita, M. Fujiwara, M. Sasaki, Z. Wang, Multichannel SNSPD system with high detection efficiency at telecommunication wavelength. Opt. Lett. **35**, 2133–2135 (2010)
4. R.H. Hadfield, M.J. Stevens, S.S. Gruber, A.J. Miller, R.E. Schwall, R.P. Mirin, S.W. Nam, Single photon source characterization with a superconducting single photon detector. Opt. Express **13**, 10846–10853 (2005)
5. S. Miki, M. Fujiwara, M. Sasaki, Z. Wang, Development of SNSPD System with Gifford-McMahon Cryocooler. IEEE Trans. Appl. Supercond. **19**, 332–335 (2009)
6. R.H. Hadfield, J.L. Habif, J. Schlafer, R.E. Schwall, S.W. Nam, Quantum key distribution at 1550 nm with twin superconducting single-photon detectors. Appl. Phys. Lett. **89**, 241129 (2006)
7. R.J. Collins, R.H. Hadfield, V. Fernandez, S.W. Nam, G.S. Buller, Low timing jitter detector for gigahertz quantum key distribution. Electron. Lett. **43**, 180–182 (2007)
8. H. Takesue, S.W. Nam, Q. Zhang, R.H. Hadfield, T. Honjo, K. Tamaki, Y. Yamamoto, Quantum key distribution over a 40-dB channel loss using superconducting single-photon detectors. Nat. Photon. **1**, 343–348 (2007)
9. D. Stucki, N. Walenta, F. Vannel, R.T. Thew, N. Gisin, H. Zbinden, S. Gray, C.R. Towery, S. Ten, High rate, long-distance quantum key distribution over 250 km of ultra low loss fibres. New J. Phys. **11**, 075003 (2009)
10. A. Tanaka, M. Fujiwara, S.W. Nam, Y. Nambu, S. Takahashi, W. Maeda, K. Yoshino, S. Miki, B. Baek, Z. Wang, A. Tajima, M. Sasaki, A. Tomita, Ultra fast quantum key distribution over a 97 km installed telecom fiber with wavelength division multiplexing clock synchronization. Opt. Express **16**, 11354–11360 (2008)
11. T. Honjo, S.W. Nam, H. Takesue, Q. Zhang, H. Kamada1, Y. Nishida, O. Tadanaga, M. Asobe, B. Baek, R. Hadfield, S. Miki, M. Fujiwara, M. Sasaki, Z. Wang, K. Inoue, Y. Yamamoto, Long-distance entanglement-based quantum key distribution over optical fiber. Opt. Express **16**, 19118–19126 (2008)
12. C.M. Natarajan, A. Peruzzo, S. Miki, M. Sasaki, Z. Wang, B. Baek, S. Nam, R.H. Hadfield, J.L. O'Brien, Operating quantum waveguide circuits with superconducting single-photon detectors. Appl. Phys. Lett. **96**, 211101 (2010)
13. M. Sasaki, M. Fujiwara, H. Ishizuka, W. Klaus, K. Wakui, M. Takeoka, S. Miki, T. Yamashita, Z. Wang, A. Tanaka, K. Yoshino, Y. Nambu, S. Takahashi, A. Tajima, A. Tomita, T. Domeki, T. Hasegawa, Y. Sasaki, H. Kobayashi, T. Asai, K. Shimizu, T. Tokura, T. Tsurumaru, M. Matsui, T. Honjo, K. Tamaki, H. Takesue, Y. Tokura, J.F. Dynes, A.R. Dixon, A.W. Sharpe, Z.L. Yuan, A.J. Shields, S. Uchikoga, M. Legre, S. Robyr, P. Trinkler, L. Monat, J.-B. Page, G. Ribordy, A. Poppe, A. Allacher, O. Maurhart, T. Langer, M. Peev, A. Zeilinger, Field test of quantum key distribution in the Tokyo QKD Network. Opt. Express **19**, 10387–10409 (2011)
14. K. Yoshino, M. Fujiwara, A. Tanaka, S. Takahashi, Y. Nambu, A. Tomita, S. Miki, T. Yamashita, Z. Wang, M. Sasaki, A. Tajima, High-speed wavelength-division multiplexing quantum key distribution system. Opt. Lett. **37**, 223–225 (2012)
15. A. Tanaka, M. Fujiwara, K. Yoshino, S. Takahashi, Y. Nambu, A. Tomita, S. Miki, T. Yamashita, Z. Wang, M. Sasaki, A. Tajima, High-speed quantum key distribution system for 1-Mbps real-time key generation. IEEE J. Quantum Electron. **48**, 542–550 (2012)
16. R. Ikuta, H. Kato, Y. Kusaka, S. Miki, T. Yamashita, H. Terai, M. Fujiwara, T. Yamamoto, M. Koashi, M. Sasaki, Z. Wang, N. Imoto, High-fidelity conversion of photonic quantum information to telecommunication wavelength with superconducting single-photon detectors. Phys. Rev. A **87**, 010301(R) (2013)

17. R. Ikuta, T. Kobayashi, H. Kato, S. Miki, T. Yamashita, H. Terai, M. Fujiwara, T. Yamamoto, M. Koashi, M. Sasaki, Z. Wang, N. Imoto, Nonclassical two-photon interference between independent telecommunication light pulses converted by difference-frequency generation. Phys. Rev. A **88**, 042317 (2013)
18. R. Ikuta, T. Kobayashi, H. Kato, S. Miki, T. Yamashita, H. Terai, M. Fujiwara, T. Yamamoto, M. Koashi, M. Sasaki, Z. Wang, N. Imoto, Observation of two output light pulses from a partial wavelength converter preserving phase of an input light at a single-photon level. Opt. Express **21**, 27865 (2013)
19. M. Fujiwara, S. Miki, T. Yamashita, Z. Wang, M. Sasaki, Photon level crosstalk between parallel fibers installed in urban area. Opt. Express **18**, 22199–22207 (2010)
20. R.B. Jin, M. Fujiwara, T. Yamashita, S. Miki, H. Terai, Z. Wang, K. Wakui, R. Shimizu, M. Sasaki, Efficient detection of a highly bright photon source using superconducting nanowire single photon detectors. Opt. Commun. **336**, 47 (2015)
21. R.B. Jin, R. Shimizu, K. Wakui, H. Benichi, M. Sasaki, Widely tunable single photon source with high purity at telecom wavelength. Opt. Express **21**, 10659–10666 (2013)
22. R.B. Jin, R. Shimizu, K. Wakui, M. Fujiwara, T. Yamashita, S. Miki, H. Terai, Z. Wang, M. Sasaki, Pulsed Sagnac polarization-entangled photon source with a PPKTP crystal at telecom wavelength. Opt. Express **22**(10), 11498–11507 (2014)
23. R.B. Jin, K. Wakui, R. Shimizu, H. Benichi, S. Miki, T. Yamashita, H. Terai, Z. Wang, M. Fujiwara, M. Sasaki, Nonclassical interference between independent intrinsically pure single photons at telecommunication wavelength. Phys. Rev. A **87**, 063801 (2013)
24. R. Ikuta, T. Kobayashi, S. Yasui, S. Miki, T. Yamashita, H. Terai, M. Fujiwara, T. Yamamoto, M. Koashi, M. Sasaki, Z. Wang, N. Imoto, Frequency down-conversion of 637 nm light to the telecommunication band for non-classical light emitted from NV centers in diamond. Opt. Express **22**, 11205 (2014)
25. X. Hu, T. Zhong, J.E. White, E.A. Dauler, F. Najafi, C.H. Herder, F.N.C. Wong, K. Berggren, Fiber-coupled nanowire photon counter at 1550 nm with 24 % system detection efficiency. Opt. Lett. **34**, 3607–3609 (2009)
26. F. Marsili, V.B. Verma, J.A. Stern, S. Harrington, A.E. Lita, T. Gerrits, I. Vayshenker, B. Baek, M.D. Shaw, R.P. Mirin, S.W. Nam, Detecting single infrared photons with 93 % system efficiency. Nat. Photon. **7**, 210–214 (2013)
27. D. Rosenberg, A.J. Kerman, R.J. Molnar, E.A. Dauler, High-speed and high-efficiency superconducting nanowire single photon detector array. Opt. Express **21**, 1440–1447 (2013)
28. S. Miki, T. Yamashita, H. Terai, Z. Wang, High performance fiber-coupled NbTiN superconducting nanowire single photon detectors with Gifford-McMahon cryocooler. Opt. Express **21**, 10208 (2013)
29. E.A. Dauler, M.E. Grein, A.J. Kerman, F. Marsili, S. Miki, S.W. Nam, M.D. Shaw, H. Terai, V.B. Verma, T. Yamashita, Review of superconducting nanowire single-photon detector system design options and demonstrated performance. Opt. Eng. **53**, 081907 (2014)
30. N. Gisin, G. Ribordy, W. Tittel, H. Zbinden, Quantum cryptography. Rev. Mod. Phys. **74**, 145–195 (2002)
31. G.S. Vernam, Secret signaling system, U.S. Patent 1 310 719, Jul 1919
32. ID Quantique. http://www.idquantique.com/
33. MagiQ Technologies, Inc. http://www.magiqtech.com/MagiQ/Home.html
34. QuintessenceLabs Pty Ltd. http://www.quintessencelabs.com/
35. C. Elliott, A. Colvin, D. Pearson, O. Pikalo, J. Schlafer, H. Yeh, Current status of the DARPA quantum network, in *Quantum Information and Computation III, Proceedings of SPIE*, ed. by E.J. Donkor, A.R. Pirich, H.E. Brandt, vol. 5815, pp. 138-149 (2005). arXiv:quant-ph/0503058v2
36. M. Peev, C. Pacher, R. Alléaume, C. Barreiro, J. Bouda, W. Boxleitner, T. Debuisschert, E. Diamanti, M. Dianati, J.F. Dynes, S. Fasel, S. Fossier, M. Fürst, J.-D. Gautier, O. Gay, N. Gisin, P. Grangier, A. Happe, Y. Hasani, M. Hentschel, H. Hübel, G. Humer, T. Länger, M. Legré, R. Lieger, J. Lodewyck, T. Lorünser, N. Lütkenhaus, A. Marhold, T. Matyus, O. Maurhart, L. Monat, S. Nauerth, J.-B. Page, A. Poppe, E. Querasser, G. Ribordy, S. Robyr, L. Salvail,

A.W. Sharpe, A.J. Shields, D. Stucki, M. Suda, C. Tamas, T. Themel, R.T. Thew, Y. Thoma, A. Treiber, P. Trinkler, R. Tualle-Brouri, F. Vannel, N. Walenta, H. Weier, H. Weinfurter, I. Wimberger, Z.L. Yuan, H. Zbinden, A. Zeilinger, The SECOQC quantum key distribution network in Vienna. New J. Phys. **11**, 075001 (2009)
37. T. Länger, G. Lenhart, Standardization of quantum key distribution and the ETSI standardization initiative ISG-QKD. New J. Phys. **11**, 055051 (2009)
38. Swiss quantum. http://www.swissquantum.com/
39. A. Mirza, F. Petruccione, Realizing long-term quantum cryptography. J. Opt. Soc. Am. B **27**, A185–A188 (2010)
40. Z.L. Yuan, A.J. Shields, Continuous operation of a one-way quantum key distribution system over installed telecom fibre. Opt. Express **13**, 660–665 (2005)
41. T.E. Chapuran, P. Toliver, N.A. Peters, J. Jackel, M.S. Goodman, R.J. Runser, S.R. McNown, N. Dallmann, R.J. Hughes, K.P. McCabe, J.E. Nordholt, C.G. Peterson, K.T. Tyagi, L. Mercer, H. Dardy, Optical networking for quantum key distribution and quantum communications. New J. Phys. **11**, 105001 (2009)
42. D. Lancho, J. Martinez-Mateo, D. Elkouss, M. Soto, V. Martin, QKD in standard optical telecommunications networks, *Quantum Communication and Quantum Networking*, vol. 36, Lecture Notes of the Institute for Computer Sciences, Social Informatics and Telecommunications Engineering (Springer, Heidelberg, 2010), pp. 142–149
43. S. Wang, W. Chen, Z.-Q. Yin, Y. Zhang, T. Zhang, H.-W. Li, F.-X. Xu, Z. Zhou, Y. Yang, D.-J. Huang, L.-J. Zhang, F.-Y. Li, D. Liu, Y.-G. Wang, G.-C. Guo, Z.-F. Han, Field test of wavelength-saving quantum key distribution network. Opt. Lett. **35**(14), 2454–2456 (2010)
44. T.-Y. Chen, J. Wang, H. Liang, W.-Y. Liu, Y. Liu, X. Jiang, Y. Wang, X. Wan, W.-Q. Cai, L. Ju, L.-K. Chen, L.-J. Wang, Y. Gao, K. Chen, C.-Z. Peng, Z.-B. Chen, J.-W. Pan, Metropolitan all-pass and inter-city quantum communication network. Opt. Express **18**, 27217–27225 (2010)
45. W.-Y. Hwang, Quantum key distribution with high loss: toward global secure communication. Phys. Rev. Lett. **91**, 057901 (2003)
46. H.-K. Lo, X. Ma, K. Chen, Decoy state quantum key distribution. Phys. Rev. Lett. **94**, 230504 (2005)
47. X.-B. Wang, Beating the photon-number-splitting attack in practical quantum cryptography. Phys. Rev. Lett. **94**, 230503 (2005)
48. K. Inoue, E. Waks, Y. Yamamoto, Differential-phase-shift quantum key distribution using coherent light. Phys. Rev. A **68**, 022317 (2003)
49. S. Obana, A. Tanaka, General purpose hash function family computer and shared key creating system. Patent WO/2007/034685 (29 March 2007)
50. X. Ma, B. Qi, Y. Zhao, H.-K. Lo, Practical decoy state for quantum key distribution. Phys. Rev. A. **72**, 012326 (2005)
51. Y. Zhao, B. Qi, X. Ma, H.-K. Lo, L. Qian, Experimental quantum key distribution with decoy states. Phys. Rev. Lett. **96**, 230503 (2006)
52. E. Waks, H. Takesue, Y. Yamamoto, Security of differential-phase-shift quantum key distribution against individual attacks. Phys. Rev. A **73**, 012344 (2006)
53. T. Honjo, A. Uchida, K. Amano, K. Hirano, H. Someya, H. Okumura, K. Yoshimura, P. Davis, Y. Tokura, Differential-phase-shift quantum key distribution experiment using fast physical random bit generator with chaotic semiconductor lasers. Opt. Express **17**, 9053–9061 (2009)
54. The Third International Conference on Updating Quantum Cryptography and Communications (UQCC2010). http://www.uqcc2010.org/
55. The "Tokyo QKD Network video" of the network operation demonstrated during the UQCC2010 conference is available at http://www.uqcc2010.org/
56. Japan Gigabit Network 2 plus. http://www.jgn.nict.go.jp/jgn2plus/english/index.html
57. http://www.Optigate.jp/products/cable/szcable.html
58. D.C. Chang, E.F. Kuester, Radiation and propagation of a surface-wave mode on a curved open waveguide of arbitrary cross section. Radio Sci. **11**, 449–457 (1976)
59. A.R. Dixon, Z.L. Yuan, J.F. Dynes, A.W. Sharpe, A.J. Shields, Continuous operation of high bit rate quantum key distribution. Appl. Phys. Lett. **96**, 161102 (2010)

60. P.G. Evans, R.S. Bennink, W.P. Grice, T.S. Humble, J. Schaake, Bright source of spectrally uncorrelated polarization-entangled photons with nearly single-mode emission. Phys. Rev. Lett. **105**, 253–601 (2010)
61. Y.F. Huang, B.H. Liu, L. Peng, Y.H. Li, L. Li, C.F. Li, G.C. Guo, Experimental generation of an eight-photon Greenberger-Horne-Zeilinger state, Nat. Commun. **2** 546(1–6) (2011)
62. F. König, F.N.C. Wong, Extended phase matching of second-harmonic generation in periodically poled $KTiOPO_4$ with zero group-velocity mismatch. Appl. Phys. Lett. **84**, 1644 (2004)
63. S. Miki, M. Fujiwara, M. Sasaki, Z. Wang, NbN superconducting single-photon detectors prepared on single-crystal MgO substrates. IEEE Trans. Appl. Superconduct. **17**, 285–288 (2007)
64. X.C. Yao, T.X. Wang, P. Xu, H. Lu, G.S. Pan, X.H. Bao, C.Z. Peng, C.Y. Lu, Y.A. Chen, J.W. Pan, Observation of eight-photon entanglement. Nat. Photon. **6**, 225–228 (2012)
65. J. Yin, J.G. Ren, H. Lu, Y. Cao, H.L. Yong, T.P. Wu, C. Liu, S.K. Liao, F. Zhou, Y. Jiang, X.D. Cai, P. Xu, G.S. Pan, J.J. Jia, Y.M. Huang, H. Yin, J.Y. Wang, Y.A. Chen, C.Z. Peng, J.W. Pan, Quantum teleportation and entanglement distribution over 100-kilometre freespace channels. Nature **488**, 185–188 (2012)
66. H.-J. Briegel, W. Dür, J.I. Cirac, P. Zoller, Quantum repeaters: the role of imperfect local operations in quantum communication. Phys. Rev. Lett. **81**, 5932–5935 (1998)
67. L.-M. Duan, M.D. Lukin, J.I. Cirac, P. Zoller, Long-distance quantum communication with atomic ensembles and linear optics. Nature **414**, 413–418 (2001)
68. N. Gisin, R. Thew, Quantum communication. Nat. Photonics **1**, 165–171 (2007)
69. J. Hofmann, M. Krug, N. Ortegel, L. Gérard, M. Weber, W. Rosenfeld, H. Weinfurter, Heralded entanglement between widely separated atoms. Science **337**, 72 (2012)
70. S. Ritter, C. Nölleke, C. Hahn, A. Reiserer, A. Neuzner, M. Uphoff, M. Mücke, E. Figueroa, J. Bochmann, G. Rempe, An elementary quantum network of single atoms in optical cavities. Nature **484**, 195–200 (2012)
71. P. Kumar, Quantum frequency conversion. Opt. Lett. **15**, 1476–1478 (1990)
72. S. Tanzilli, W. Tittel, M. Halder, O. Alibart, P. Baldi, N. Gisin, H. Zbinden, A photonic quantum information interface. Nature **437**, 116–120 (2005)
73. C. Langrock, E. Diamanti, R.V. Roussev, Y. Yamamoto, M.M. Fejer, H. Takesue, Highly efficient single-photon detection at communication wavelengths by use of upconversion in reverse-proton-exchanged periodically poled LiNbO3 waveguides. Opt. Lett. **30**, 1725–1727 (2005)
74. A.G. Radnaev, Y.O. Dudin, R. Zhao, H.H. Jen, S.D. Jenkins, A. Kuzmich, T.A.B. Kennedy, A quantum memory with telecom-wavelength conversion. Nat. Phys. **6**, 894–899 (2010)
75. R. Ikuta, Y. Kusaka, T. Kitano, H. Kato, T. Yamamoto, M. Koashi, N. Imoto, Wide-band quantum interface for visible-to-telecommunication wavelength conversion. Nat. Commun. **2**, 537 (2011)
76. T. Nishikawa, A. Ozawa, Y. Nishida, M. Asobe, F.-L. Hong, T.W. Hänsch, Efficient 494 mW sum-frequency generation of sodium resonance radiation at 589 nm by using a periodically poled Zn:LiNbo3 ridge waveguide. Opt. Express **17**, 17792–17800 (2009)
77. S. Zaske, A. Lenhard, C.A. Keßler, J. Kettler, C. Hepp, C. Arend, R. Albrecht, W.-M. Schulz, M. Jetter, P. Michler, C. Becher, Visible-to-telecom quantum frequency conversion of light from a single quantum emitter. Phys. Rev. Lett. **109**, 147404 (2012)
78. S. Ates, I. Agha, A. Gulinatti, I. Rech, M.T. Rakher, A. Badolato, K. Srinivasan, Two-photon interference using background-free quantum frequency conversion of single photons emitted by an InAs quantum dot. Phys. Rev. Lett. **109**, 147405 (2012)
79. J.S. Pelc, L. Yu, K. De Greve, P.L. McMahon, C.M. Natarajan, V. Esfandyarpour, S. Maier, C. Schneider, M. Kamp, S. Höfling, R.H. Hadfield, A. Forchel, Y. Yamamoto, M.M. Fejer, Downconversion quantum interface for a single quantum dot spin and 1550-nm single-photon channel. Opt. Express **20**, 27510–27519 (2012)

Part II
Superconducting Quantum Circuits: Microwave Photon Detection, Feedback and Quantum Acoustics

Chapter 6
Microwave Quantum Photonics

Bixuan Fan, Gerard J. Milburn and Thomas M. Stace

Abstract The past decade has witnessed rapid and remarkable development of electronic circuits that can probe the quantum regime. Very strong light-matter interactions and the improvement in manipulating artificial atoms and microwave resonators have brought the field to the single-photon regime, enabling the study of dynamics involving a single microwave photon. This chapter focusses on the development and status of microwave electronics in the few-quanta regime, and specifically on the twin aspects of measurement and generation of single microwave photons.

6.1 Introduction

A traditional view of bulk solid-state systems is that their many degrees of freedom provide many channels by which quantum coherence may be lost. As a consequence, it may be expected that coherence times will be short, and discrete transitions will be smeared into a continuum of modes. Over the last two decades, with the development of nano-fabrication techniques and advances in quantum measurement and control, a new class of engineered quantum systems, such as superconducting circuits, micro/nano-mechanical systems and quantum-dot systems have demonstrated excel-

B. Fan (✉)
College of Physics, Communication and Electronics, Jiangxi Normal University,
Nanchang 330022, China
e-mail: bixuan.fan@uqconnect.edu.au

G.J. Milburn · T.M. Stace
ARC Centre for Engineered Quantum Systems, University of Queensland,
Brisbane 4072, Australia
e-mail: milburn@physics.uq.edu.au

T.M. Stace
e-mail: stace@physics.uq.edu.au

© Springer International Publishing Switzerland 2016

R.H. Hadfield and G. Johansson (eds.), *Superconducting Devices in Quantum Optics*, Quantum Science and Technology,
DOI 10.1007/978-3-319-24091-6_6

lent quantum coherence, and offer many advantages over 'natural' atomic system. In particular, many of the features of atomic systems can be replicated in solid-state devices. These include discrete quantum systems with well resolved individual energy levels with low degeneracy, and long lifetimes.

By virtue of their atom-like properties, these systems are often described as artificial atoms, in the sense that their properties are engineered during fabrication. When artificial atomic systems are coupled to microwave waveguides or resonators, the atom-resonator coupling can be significant compared to coupling to the environment. This strong coupling that has been achieved has enabled the demonstration of cavity quantum electrodynamics (QED), in the deep quantum regime [1, 2]. Compared to its counterpart in natural atomic systems, circuit QED has advantages in terms of stability and flexibility in control and manipulation. As a result, it has become a leading platform for demonstrating fundamentally quantum phenomena, as well as finding applications in quantum information processing.

We start this chapter with a brief introduction of a few key elements in superconducting circuit systems—the microwave coplanar waveguide, microwave resonators and superconducting artificial atoms. This naturally leads into the discussion of an artificial atom, e.g. a superconducting qubit or quantum dot, interacting with a single-mode of the electromagnetic field in a microwave resonator. In the final part we will discuss two applications of circuit QED: single microwave photon detection and single microwave photon generation.

6.2 Key Ingredients of Superconducting Circuit

Since much of the progress in this field has been centred on engineered superconducting systems, as a prelude, we give a brief overview of the quantisation of simple superconducting circuits. More detailed discussions can be found in [3–6].

6.2.1 Superconducting Artificial Atoms

Circuits consisting of linear elements such as capacitors and inductors act as coupled harmonic oscillators, whose (uncoupled) energy levels are evenly-spaced. Conversely, natural atoms are strongly anharmonic, due to the $1/r^2$ Coloumb potential, leading to unevenly spaced energy levels. To replicate the anharmonicity of natural atoms in electronic circuits, coherent, nonlinear elements are necessary. One very important example of such an element is the Josephson junction, which is a junction with two superconducting electrodes sandwiching a thin oxide layer, behaving as a dissipationless nonlinear inductor. It provides the requisite nonlinearity for a variety of quantum electronic devices, including superconducing qubits [7], Josephson parametric amplifiers [8], and superconducting-quantum-interference-devices (SQUID) [9].

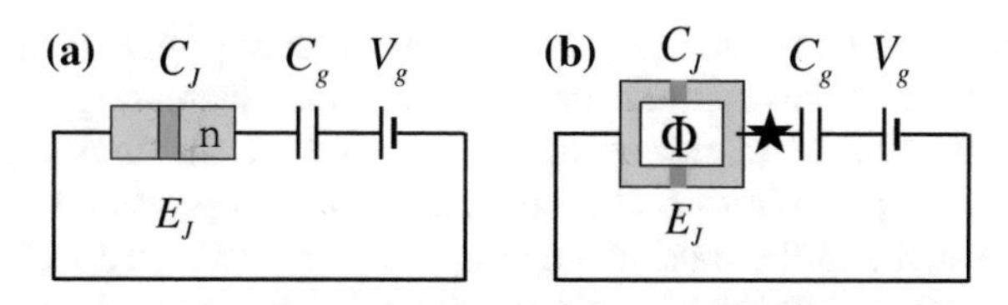

Fig. 6.1 **a** The basic model of a CPB **b** A CPB with the adjustable Josephson tunneling energy $E_J(\Phi)$. The single Josephson junction is replaced by a loop of two junctions (SQUID) and so that E_J can be tuned by varying the flux of the loop. The star indicates the circuit node used for quantisation

There are a variety of artificial superconducting qubits, including the charge qubit [10, 11], the flux qubit [12, 13] and the phase qubit [14]. These devices are named according to the different ratio of the charge energy over the Josephson tunneling energy, and are characterised by different sensitivities to environmental noise. There are also superconducting qubits in the intermediate regimes of the three categories, such as the transmon [15] and fluxonium [16]. Although these are mesoscopic devices, they can be engineered to have rather simple energy structures.

The first experimentally demonstrated superconducting two-level "atom" was the Cooper-Pair Box (CPB), with coherence times of nanoseconds [11], and theoretically described in [17]. As shown in Fig. 6.1a, in the simplest CPB setup, a superconducting island is connected to a reservoir through a Josephson junction and capacitively coupled to a gate voltage V_g through a capacitor C_g. In the superconducting phase, electrons in superconducting material form a BCS fluid which can coherently tunnel from the island to the reservoir, consisting of the rest of the circuit. There are two important quantities that determine the dynamics of the system: the charging energy E_C, defined as

$$E_C = e^2/(2C_\Sigma), \tag{6.1}$$

which depends on the total capacitance $C_\Sigma = C_J + C_g$, in which C_J is the capacitance of the junction and C_g is the gate capacitance; and the Josephson energy E_J, which is proportional to the critical current of the Josephson junction I_c,

$$E_J = \hbar I_c/(2e). \tag{6.2}$$

A CPB with a tunable E_J can be made by replacing the single junction with a two-junction loop (i.e. a SQUID) with a controllable flux passing through the loop, shown in Fig. 6.1b. Energy level spacings of 5–20 GHz are typical of CPB devices.

The CPB is very sensitive to noise arising from charge fluctuations in the vicinity of the island, which limits the coherence time of charge qubits. To overcome this, the 'transmission-line shunted plasma oscillation' or *transmon* was developed in 2007 as a charge insensitive superconducting qubit [6, 15]. A transmon consists of a tunable CPB shunted by a capacitance, to increase the effective junction capacitance. This means that C_J can be varied without significantly affecting E_J. This capacitor acts as a high-pass filter for noise on the CPB, and reduces the ratio E_C/E_J making the transmon relatively insensitive to charge fluctuations, increasing coherence times to ~1 μs. On the other hand, this capacitor decreases the ahamonicity of the qubit. Nevertheless, it is possible to find an optimal point of operation, at which noise sensitivity is small and the transmon energy levels are still well resolved and addressable. Typical anharmonicities of a few hundred MHz (corresponding to ~10 % of the energy level spacing), together with coherence times of a few microseconds are achievable. Many important milestones of quantum circuits have been achieved with transmon devices, such as the observation of vacuum Rabi splitting [18] and giant cross-Kerr effect [19], and the implementation of a Toffoli gate [20] in quantum circuits. Recently, '3D transmon' qubits have been demonstrated in bulk microwave resonators, exhibiting long coherence times, up to ~100 μs [21].

The transmon is similar in form to a CPB with tunable E_J, as shown in Fig. 6.1b. In the following the Hamiltonian for a CPB/transmon will be derived (we neglect the shunt capacitance, which can be absorbed into the junction capacitance). By analogy with a mechanical system, the charging energy plays the role of kinetic energy. We choose to describe the circuit dynamics with respect to the degree of freedom at the node marked ⋆ (this is effectively a choice of gauge). The 'kinetic' term is then

$$T = \frac{C_g}{2}(\dot{\Phi}_J - V_g)^2 + \frac{C_J}{2}\dot{\Phi}_J^2 \tag{6.3}$$

where Φ_J is the flux threading the circuit in Fig. 6.1a, and the Josephson coupling energy gives the 'potential' term:

$$U = -E_J \cos(2\pi\Phi/\Phi_0) \tag{6.4}$$

where $\Phi_0 = h/(2e)$ is the flux quantum, and $\phi = 2\pi\Phi/\Phi_0$ is the phase across the junction. As usual, the Lagrangian is $\mathcal{L} = T - U$, so the canonical momentum (here the charge variable) is $Q_J = \frac{\partial\mathcal{L}}{\partial\dot{\Phi}_J} = C_\Sigma\dot{\Phi}_J - C_g V_g$. The Hamiltonian is then

$$\begin{aligned} \mathcal{H} &= Q_J\dot{\Phi}_J - \mathcal{L} \\ &= \frac{(Q_J + C_g V_g)^2}{2C_\Sigma} - E_J\cos(\phi), \\ &= 4E_C(n - n_g)^2 - E_J\cos(\phi). \end{aligned} \tag{6.5}$$

where $n = Q_J/(2e)$ and $n_g = -C_g V_g/(2e)$ is the gate induced charge offset.

We quantise the junction charge, $\hat{n}$, and its conjugate degree of freedom is the phase operator $\hat{\phi}$. In the charge basis,

$$\hat{n} = \sum_n n\,|n\rangle\,\langle n| \tag{6.6}$$

while the conjugate phase operator is best defined through the exponential form (though we note this should formally be defined in a limiting sense [22])

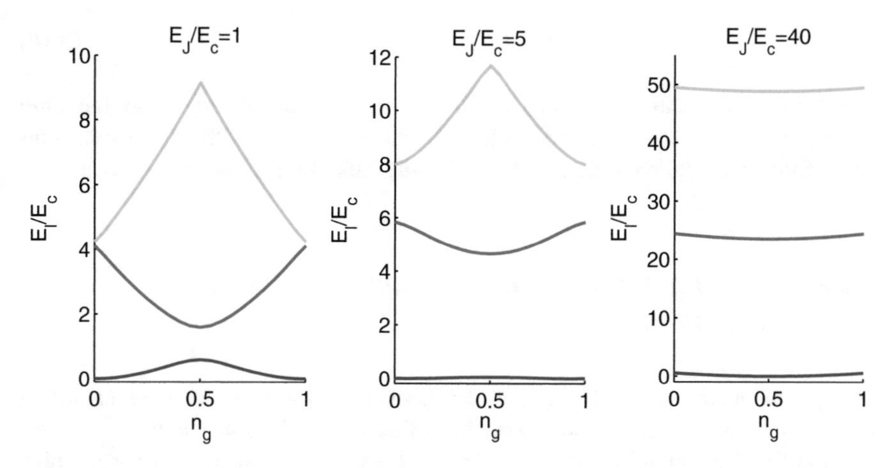

Fig. 6.2 The lowest three energy levels of a CPB/transmon versus the gate charge n_g at three different ratios of E_J/E_C

$$e^{i\hat{\phi}} = \sum_{-\infty}^{+\infty} |n\rangle \langle n+1| , \tag{6.7}$$

which satisfies $e^{i\hat{\phi}} |n+1\rangle = |n\rangle$ and $[\hat{\phi}, \hat{n}] = i$. Then, we have the quantized Hamiltonian of a CPB/transmon:

$$\hat{H} = 4E_C \sum_{n=0}^{\infty} (n - n_g)^2 |n\rangle \langle n| - \frac{E_J}{2} \sum_{n=0}^{\infty} (|n+1\rangle \langle n| + |n\rangle \langle n+1|). \tag{6.8}$$

For charge qubits, the $E_C \gg E_J$, whilst for a transmon $E_J/E_C \gg 1$, typically between 20 and 100, making the transmon insensitive to charge fluctuations. The influence of the ratio E_J/E_C on the energy structure of a CPB is presented in Fig. 6.2. The eigenenergies are calculated by diagonalising the Hamiltonian Eq. (6.8) in a truncated number basis. It is seen from Fig. 6.2 that the larger the ratio E_J/E_C, the better the immunity from the charge noise. However, this is at the price of the reduction of anharmonicity, though the anharmonicity of a transmon is sufficient for individual addressing different transitions.

Now we turn to evaluate the anharmonicity of a transmon. In the regime $E_J/E_C \gg 1$, there is an approximate solution for the eigenenergies of a transmon [15]:

$$E_l \approx E_J + \sqrt{8E_J E_C}(l + \frac{1}{2}) - E_C(6l^2 + 6l + 3)/12. \tag{6.9}$$

It is easy to know that the first transition energy $E_{1,0} = \sqrt{8E_J E_C} - E_C \approx \sqrt{8E_J E_C}$ and the n_{th} transition energy is $\sqrt{8E_J E_C} - nE_C$. Therefore, the anharmonicity α is

$$\alpha \approx E_C. \tag{6.10}$$

The relative anharmonicity $\alpha/E_{1,0} \approx \sqrt{E_C/(8E_J)}$ which decreases as the ratio E_J/E_C increases, consistent with Fig. 6.2. For a transmon with 7 GHz first transition frequency and the ratio $E_J/E_C = 50$, the anharmonicity $\alpha \approx 364$ MHz.

6.2.2 *Coplanar Transmission Lines and Microwave Resonators*

In superconducting circuits, the role of a transmission line is like a waveguide in optical systems, supporting the propagation of traveling microwave fields. The transmission line we would like to discuss here is the coplanar transmission line, which is consisted of a central conductor and two lateral ground plane in the same plane, see Fig. 6.3a. The central conductor is used for propagating microwave signals and the lateral ground planes are used as the reference potential. Figure 6.3c is the circuit representation of a coplanar transmission line and it can be taken as an infinitely-long chain of L-C oscillators for small conductance (G_0) and small resistance (R_0).

The Lagrangian is

$$\mathcal{L}(\phi_1, \dot{\phi}_1, \ldots \phi_N, \dot{\phi}_N) = \sum_{n=1}^{N} \left(\frac{C_0 \dot{\phi}_n^2}{2} - \frac{(\phi_n - \phi_{n-1})^2}{2L_0} \right), \tag{6.11}$$

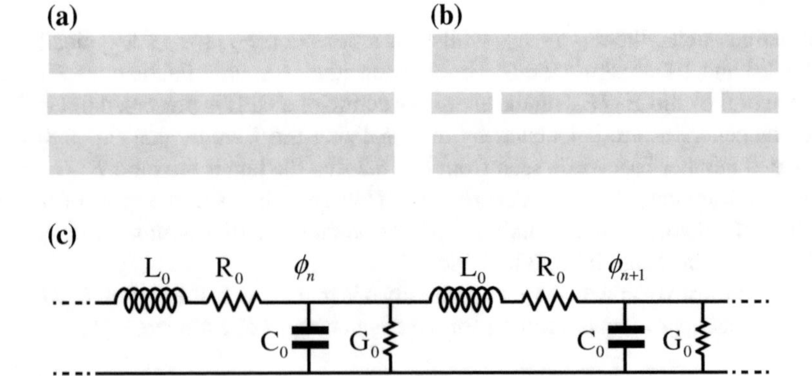

Fig. 6.3 **a** Sketch of a coplanar transmission line, in which 2D metallic gates (*shaded*) are laid out in the surface of a wafer. **b** Sketch of a coplanar transmission line resonator, in which the central conductor of the resonator is broken to create scattering centres. **c** Circuit representation of a coplanar transmission line. L_0, C_0, R_0 and G_0 are the inductance, capacitance, resistance and conductance per unit length of the circuit

Equation (6.11) can be transformed to the frequency domain as

$$\mathcal{L}(\Phi_1, \dot{\Phi}_1, \ldots) = \sum_k^{\infty} \left(\frac{C_k \dot{\Phi}_k^2}{2} - \frac{\Phi_k^2}{2L_k} \right) \tag{6.12}$$

where Φ_k is the spatial Fourier transform of ϕ_n, and $C_k = C_0 d/2$, $L_k = 2dL_0/(k^2\pi^2)$ and d is the length of the transmission line. After quantization using the same procedure as in the quantization of the L-C circuit, the quantized Hamiltonian for a coplanar transmission line is

$$\hat{H} = \hbar \sum_k \omega_k (\hat{a}_k^\dagger \hat{a}_k + 1/2) \tag{6.13}$$

with $\omega_k = k\pi/(\sqrt{C_0 L_0}d)$, $\hat{a}_k^\dagger = (L_k/C_k)^{1/4}\Pi_k/\sqrt{2} - i(C_k/L_k)^{1/4}\Phi_k/\sqrt{2}$ and $\Pi_k = C_k \dot{\Phi}_k$.

If breaks are formed in the central conductor of the transmission line, corresponding to a point-like variation in the capacitance of the waveguide, a resonant structure is formed whose fundamental frequency is determined by its length [23], shown in Fig. 6.3b. The two capacitors act like reflective mirrors in conventional optical cavities and couple the 'external' field into the resonant modes of the resonator. Typical frequency of coplanar waveguides ranges from 2–10 GHz with quality factor of $Q \sim 10^5 - 10^6$ determined by the low internal and external loss rates. Importantly, given the very small resonator volume, microwave fields in the resonators can be very intense, and together with the large dipole of the artificial atom, has enabled the demonstration of various strong coupling phenomena in recent years.

6.2.3 *Amplifiers and Detection Devices*

Conventional microwave detectors are linear: their output is proportional to the *amplitude* of the incoming field. As a result, they are restricted to homodyne and related measurement techniques. This is in contrast to optical photodetectors which are intrinsically sensitive to the photon flux (i.e. power).

To measure microwave power, either the output of a linear detector is squared, or the field passes through a non-linear device (e.g. a rectifier or transistor) and then filtered. In either case, the input field to the detector must be sufficiently large to register above the detector noise floor. For single-photon microwave detectors, the output of a linear detector will be zero on average (since a field in a Fock state has zero average amplitude), ruling out the first approach, making the search for ultra-high nonlinearities critical. In recent experiments, this nonlinearity arises from the fact that an artificial anharmonic atom (e.g. transmon, quantum dot, etc.) will saturate with an intensity of a single photon per lifetime.

Even presuming the existence of such a nonlinearity, there is still a need to amplify the incredibly small powers corresponding to a flux of a few microwave photons. As such, we briefly review some of the main constraints on quantum limited amplifiers. A key paper on quantum limits to amplification by Caves in 1982 showed that a linear, phase-independent amplifier with gain G (i.e. that amplifies both quadratures by G) must necessarily add noise, which when referred to the input signal is $|1 - 1/G|/2$ [24], corresponding to an additional "half-quantum" worth of noise to the input mode. This can be understood heuristically by noting that two nearby coherent states, $|\alpha\rangle$ and $|\beta\rangle$, with substantial overlap in phase space, cannot become more distinguishable by application of any CP map, including amplification. As a consequence, the amplified versions of each state cannot be coherent states of larger amplitude (which would be increasingly distinguishable), but must have additional noise.

In practise, high quality conventional amplifiers based on high-electron mobility transistors (HEMTs) [25] add considerably more noise than the quantum limit. Recent results report that HEMT amplifiers add around 20 quanta worth of noise to the input [26], a factor of 40 times worse than the quantum limit.

It is possible, in principle, to avoid adding additional noise if the amplifier is phase-sensitive, such that one quadrature is amplified by G_1 and the other is attenuated by a corresponding amount $G_2 = G_1^{-1}$ [24]. This situation can in fact be realised using a parametric amplifier [8], in which an idler mode acts as a phase reference to determine which quadrature of a signal mode is amplified (and which is correspondingly attenuated). As a result parametric amplifiers have been increasingly used in low-temperature measurement systems [27]. Indeed, they have been shown to operate very close to their quantum limit, with recent experiments demonstrating the addition of just a few noise quanta to the input signal [26].

6.3 Interaction Between 'Atoms' and Microwaves

6.3.1 Circuit QED

The development of circuit QED (cQED) addresses one of the biggest technical hurdles in conventional cavity QED systems—the difficulty of trapping natural atoms or ions. This field has made impressive advances in the last decade. From the fundamental physics side, quantum circuits have been used to demonstrate the dynamical Casimir effect [28], to mimic the behavior of natural atoms [29], to test the violation of a Bell inequality [30] and so on. In terms of applications in quantum information, quantum circuit technology is a very promising candidate for on-chip quantum computation [31, 32], thanks to system stability, mature techniques of fabrication and integration, long coherence of superconducting qubits [21], the ability to interface with variety of systems [33], and strong Coupling (in cQED) between qubits and field [34, 35]. Hybrids of electronic quantum circuits with atomic systems [36, 37], spins [38, 39], quantum dots [40] and other systems have shown their great promise in connecting optical and microwave frequency, information transfer between solid-state devices to flying qubits and building more controllable quantum systems.

The physics of circuit QED closely parallels that of cQED in optical cavities with natural atoms in Rydberg states with microwave transitions. We consider the interaction between a superconducting qubit and a coplanar transmission line resonator. In light of the fact that the transmon has a well-resolved spectrum, it can be modelled as a two-level system, with energy levels $\{|g\rangle, |e\rangle\}$ coupled to a near-resonant quantum harmonic oscillator described by mode creation and annihilation operators $\hat{a}^\dagger, \hat{a}$, via the dipole moment. In the dipole approximation, this system is described by the well-studied Rabi Hamiltonian ($\hbar = 1$)

$$\hat{H} = \omega_r(\hat{a}^\dagger\hat{a} + 1/2) + \omega_{eg}/2(\hat{\sigma}_{ee} - \hat{\sigma}_{gg}) + g(\hat{a} + \hat{a}^\dagger)(\hat{\sigma}_{ge} + \hat{\sigma}_{eg}), \quad (6.14)$$

where ω_r is the resonator frequency, ω_{eg} is the transmon transition frequency, g is the coupling strength and $\hat{\sigma}_{ij} = |i\rangle\langle j|$. Assuming the transition frequency is much larger than the coupling, $\omega_{eg} \gg g$, we make the rotating wave approximation (RWA) and discard 'counter rotating terms', namely $\hat{a}\hat{\sigma}_{ge}$ and $\hat{a}^\dagger\hat{\sigma}_{eg}$, ending up with the Jaynes-Cummings Hamiltonian

$$\hat{H} = \omega_r\hat{a}^\dagger\hat{a} + \omega_{eg}/2(\hat{\sigma}_{ee} - \hat{\sigma}_{gg}) + g(\hat{a}^\dagger\hat{\sigma}_{ge} + \hat{a}\hat{\sigma}_{eg}). \quad (6.15)$$

The Jaynes-Cummings Hamiltonian commutes with the total excitation number, $\hat{N} = \hat{a}^\dagger\hat{a} + \hat{\sigma}_{ee}$, so the Hamiltonian is block diagonal in eigenstates of $\hat{N}$, with eigenvalues n labelling each block.

We transform to an interaction picture defined by $\omega_r\hat{N}$, the interaction Hamiltonian becomes

$$\hat{H}_I = \Delta|e\rangle\langle e| + g(\hat{a}^\dagger\hat{\sigma}_{ge} + \hat{a}\hat{\sigma}_{eg}), \quad (6.16)$$

where $\Delta = \omega_{eg} - \omega_r$ is the detuning.

On resonance, $\omega_r = \omega_{eg}$ (i.e. $\Delta = 0$), and the system eigenstates are

$$|\psi_{\pm,n}\rangle = (|g, n\rangle \pm |e, n-1\rangle)/\sqrt{2}. \quad (6.17)$$

with eigenenergies $\pm\sqrt{n}g$. The mode splitting in the eigenenergies induced by the interaction with a qubit is a signature of strong atom-field coupling, i.e. $g \gg \gamma, \kappa$, where γ is the decay rate of the atom, and κ is the loss rate of the cavity. When $n = 1$, the mode splitting is referred as the vacuum Rabi splitting. Figure 6.4 shows the transmission through a transmon-cavity system, demonstrating the vacuum Rabi splitting, which is a telltale for strong coupling behavior. In superconducting circuits, vacuum Rabi splitting effects have been observed in a single CPB [34] and in a transmon [41], along with the Mollow triplet in the fluorescence emission spectrum of a driven two-level system [42, 43].

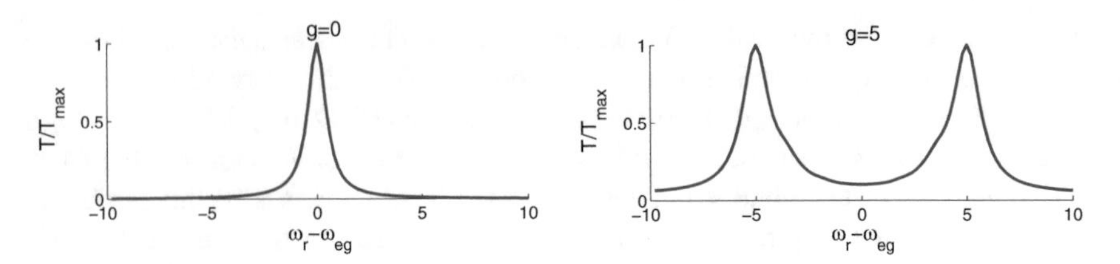

Fig. 6.4 (*left*) Transmission spectrum of the uncoupled system ($g = 0$). (*right*) The vacuum Rabi splitting in the transmission spectrum of a strongly coupled transmon-resonator system. The other system parameters are: $\gamma = \kappa = 1$

In the dispersive limit, $\Delta \gg g$, there is no direct energy exchange between the qubit and the field. In this limit, applying the unitary transformation to the Hamiltonian $e^{-gD/\Delta}He^{gD/\Delta}$, where $D = \hat{a}^\dagger\hat{\sigma}_{ge} - \hat{a}\hat{\sigma}_{eg}$, and expanding to second order in g/Δ we arrive at the effective Hamiltonian

$$H_{eff} \approx (\omega_r + \frac{g^2}{\Delta})(\hat{a}^\dagger\hat{a} + \frac{1}{2})\hat{\sigma}_z + \frac{\omega_{eg}}{2}\hat{\sigma}_z, \tag{6.18}$$

where $\hat{\sigma}_z = \hat{\sigma}_{ee} - \hat{\sigma}_{gg}$ and we have used the relation

$$e^{-xD}He^{xD} = H + x[H, D] + \frac{x^2}{2}[[H, D], D] + \cdots \tag{6.19}$$

From Eq. (6.18) we see that the transition energy depends on the photon number; likewise, the resonator frequency depends on the state of the qubit. In this limit, the dispersive qubit-field interaction can be used for Quantum non-demolition (QND) readout of qubit states or resonator photon number [44]. This important result follows from a very simple model, which nevertheless has led to demonstrations of microwave cavity photon detection [45, 46].

Generalising this to detecting itinerant microwave photons in waveguide is not trivial. Later, we will discuss theoretical approaches which extend these ideas to multi-level transmons coupled to two fields. At large detunings, the model can be reduced to an effective cross-Kerr interaction model, which we will discuss in next section.

6.3.2 *Applications: Measurement of Microwave Photons*

Measuring single quanta of the electromagnetic field is now a standard part of the operating toolbox of optical physics. Numerous commercial photon counters of the 'bucket' variety (i.e. distinguish between 0 and ≥ 1 photons) are available. Several such detectors can be combined to discriminate between different photon numbers [47, 48]. In addition, recent advances in superconducting bolometers [49] and multi-element nanowires [50] has enabled reliable multi-photon counting from near-infra red to the optical domain.

In contrast, due to their extremely low energy, the detection of single microwave photons is very challenging. An early proposal by Brune et al. relied on the interaction of Rydberg atoms with photons in a microwave cavity [44], which was realised some years later (see below) [46]. In circuit QED, there are a number of excellent pioneering theoretical proposals and experimental demonstrations [45, 51–57].

There are several proposed approaches to detecting itinerant microwave photons. One approach is absorptive detection using a Λ-type three-level system [51, 52, 54, 55]. This is conceptually related to conventional optical avalanche photon counters, which produce a macroscopically detectable voltage spike in response to the presence of a photon. In the microwave domain, a single photon excites the system from a ground (or otherwise long-lived) state to an excited state, which rapidly decays to a metastable level or continuum, producing strong measurement signature in the process [45, 53, 55, 57].

Another approach that has received recent attention is to exploit the nonlinear, cross-Kerr interaction between two microwave fields, induced by a strongly anharmonic medium [19, 56, 58]. The secondary field serves as a probe, whose properties are measurably changed in the presence or absence of a photon in the primary field of interest. Notably, experiments in the microwave domain have demonstrated very high single-pass cross-Kerr nonlinearities, approaching tens of degrees cross-phase modulation per photon [19].

After discussing various experimental results, we will give a more detailed theoretical discussion of these two approaches to counting itinerant microwave photons.

For completeness, we note that linear detectors can be used to sample phase space distributions (e.g. the Husimi Q distribution) of a source of light, from which multi-time correlation functions of arbitrary order can be reconstructed [59, 60]. Whilst this does not enable single-shot detection of microwave photons, it does enable the characterisation of photon sources, which we discuss in more detail in Sect. 6.3.3.

The pioneering experiments by Haroche et al. in 2007 demonstrated measurements of microwave fields at the single photon level [46]. In their experiment Rubidium atoms in the circular Rydberg states with radial quantum numbers $n \sim 50$ were used as probe media to measure microwave photons. Velocity selected atoms emanating from an oven initially pass through the first Ramsey cavity, R_1, which coherently rotates the atomic state from the 'ground' Rydberg state ($|g\rangle = |n = 50\rangle$) to an equal superposition of $|g\rangle$ and the 'excited' Rydberg state $|e\rangle = |51\rangle$ (of course, both states are in fact highly excited, and are separated by ~51 GHz), to sit on the equator of the Bloch sphere, i.e. the state $|+\rangle = |g\rangle + |e\rangle$. Following this preparation stage, each atom coherently interacts with the target cavity C, whose occupation number is unknown. The interaction with the cavity adds a photon-number dependent phase $e^{-i\phi(n,\Delta)}$ to the state $|e\rangle$, driving the system to the state $|g\rangle + e^{-i\phi(n,\Delta)}\,|e\rangle$. In the case where $\phi(1, \Delta) - \phi(0, \Delta) = \pi$, the atomic state is unchanged if the target cavity mode is in the vacuum, or evolves coherently to the state $|-\rangle = |g\rangle - |e\rangle$ if the cavity mode is occupied by a single photon. The cavity mode was sufficiently cold that higher photon contributions were negligible (<0.3 %). Importantly, the atom-cavity interaction does not change the photon number within the target cavity. The atoms then pass through the second Ramsey cavity, R_2, which coherently reverses the evolution of R_1, effecting the evolution $|+\rangle \rightarrow |g\rangle$ if the cavity were in the vacuum state, or $|-\rangle = |e\rangle$ if the cavity were in the 1 photon state. Finally the state of the atom is read out in the detector D, and an inference is made about the occupation of the target cavity. Notably, since the cavity mode occupation at the end of the atomic interaction is notionally the same as at the start, this protocol is a QND measurement, ideally leaving the cavity in its original state.

The capacity to correctly measure the cavity photon number is limited by the depth of the Ramsey fringes, which were experimentally reported to have a visibility of 78 % [46]. This leads to probabilities of incorrectly inferring the cavity photon number: $p_{g:1} = 13\,\%$, and $p_{e:0} = 9\,\%$ following each atomic detection. By grouping atomic detection events into blocks of 8 repetitions, a majority vote can be used (at the cost of longer measurement time/smaller bandwidth) to reduce the probability of erroneously estimating the cavity state to $\lesssim 0.1\,\%$, i.e. the measurement fidelity is $>99.9\,\%$.

Rapid progress has been made in chip-based superconducting systems. In 2010, QND detection of the photon number in an on-chip microwave cavity with fidelity $\sim 90\,\%$ was demonstrated [53]. Whilst differing in detail from the results of experiments with bulk resonators and Rydberg atoms, there are some conceptual similarities. In this experiment a single transmon in its ground state is non-resonantly coupled to a high-Q storage cavity, whose photon number is to be measured. The transition frequency, $\omega_{ge}^{(n)}$ between the ground and first-excited state of the transmon is Stark-shifted, depending on the number of photons in the storage cavity, $n \in \{0, 1\}$. Spectroscopy of the transmon transition frequency thus yields information about the photon number in the storage cavity. Furthermore, the state of the transmon undergoes Rabi oscillations when it is driven at the n-dependent frequency corresponding to the state of the storage cavity, i.e. when $\omega_{\text{drive}} = \omega_{ge}^{(n)}$. Thus, if the Rabi drive corresponds to a π-pulse, then the transmon is left in an excited state. Conversely, if the driving frequency applied to the transmon does not match the corresponding Stark-shifted transition frequency, $\omega_{\text{drive}} \neq \omega_{ge}^{(n)}$, then the transmon remains in its ground state. Finally, the state of the transmon can be measured by a probe pulse applied to the far-detuned measurement cavity; the reflected signal acquires a phase shift depending on the state of the transmon. Consequently, observing a phase shift in the signal reflected from the measurement cavity yields information about the occupation of the storage cavity, mediated by the transmon.

QND measurement of itinerant microwave photons has demonstrated using a current-biased Josephson junction [54], and related proposals for cavity photon counting [61]. The junction has a characteristic 'tilted washboard' potential, $U(\phi) = -a\cos(\phi) - I_b\phi$, where I_b is the bias current, a is a constant, and ϕ is the superconducting phase difference across the junction defined earlier. This potential has a series of local minima near $\phi = 2\pi m$, for integer m, for which there are a finite number of bound, metastable states, $|g\rangle, |e\rangle, \ldots$, with well-defined inter-level transition frequencies. The lifetime of the first excited metastable state is substantially shorter than that of the lowest metastable state. As a result, an incoming photon at frequency ω_{ge} will drive the transition from $|g\rangle$ to $|e\rangle$, which will then rapidly tunnel into a continuum of states, corresponding to a voltage pulse across the junction. Thus, the observation of such a pulse indicates the absorption of a microwave photon. This approach was demonstrated to have low dark counts (i.e. false reports of the presence of a photon), but it remains somewhat unclear what the overall quantum efficiency of the scheme is. We note that this proposal followed earlier work which used the same basic mechanism simply as a readout for the state of a qubit encoded in the metastable states of the junction [62].

6.3.2.1 Theory: Λ-type Artificial Atoms

Phase qubits coupled to a microwave field [63] have been used as detectors, with a natural readout mechanism afforded by the application of a bias current [61, 62]. The bias-current induces a 'tilted-washboard' potential, described earlier, which transforms the qubit states into metastable states coupled to a continuum of modes. The large difference in tunnelling rates between the qubit states and the continuum gives a powerful way to measure which state the qubit was in. The same arrangement has been proposed as a microwave photon detector [52], which motivated experimental efforts discussed earlier [54]. The basic idea is for an incoming photon to drive a transition between the qubit states, which are then distinguished by whether there is subsequent voltage spike across the junction. Conceptually, this has some similarities to conventional optical photon counters based on Avalanche photodiodes (semiconductor) in which an incoming photon drives a long-lived metastable state of the detector into a short-lived state, whose decay produces a macroscopic voltage spike.

Recently, Romero et al. proposed a microwave photon detection scheme along these lines [51, 52]. Briefly, one or more phase qubits are embedded in a coplanar transmission line. The qubits are characterised by two metastable states, which decay into a macroscopically detectable continuum of states. They reported 50 % detection efficiency with one phase qubit in the first work and later in an improved work the perfect detection with 100 % efficiency was achieved theoretically by placing the phase qubit in front of a mirror (one end of a coplanar transmission line) [55]. These two proposals are quite promising with the merits of high efficiency, cavity-free and broad bandwidth but they absorb the photon under detection.

Now let us review the basic theory in [52]. The system Hamiltonian for the case of one absorber at position x_0

$$\hat{H} = \omega_1 |1\rangle \langle 1| + i v_g \int [\psi_L^\dagger \partial_x \psi_L - \psi_R^\dagger \partial_x \psi_R] dx \\ + g \int \delta(x - x_0)[(\psi_R + \psi_L) |1\rangle \langle 0| + h.c.] dx, \quad (6.20)$$

where $\psi_L^\dagger (\psi_R^\dagger)$ represents the creation of a photon moving to left (right), v_g is the group velocity in the transmission line, and we take $x_0 = 0$ hereafter. The wave function for the system is

$$|\psi\rangle = \int dx[\phi_L(x)\,|1_L, 0_R, 0\rangle + \phi_R(x)\,|0_L, 1_R, 0\rangle] + c_1(t)\,|0_L, 0_R, 1\rangle\,, \quad (6.21)$$

where $|1_L, 0_R, 0\rangle$ denotes one photon in the left part of the waveguide, while no photon in the right part of the waveguide, and the qubit in the ground state. Substituting the system Hamiltonian and the wave function into the time-independent Schrödinger equation $\hat{H}\psi = E\psi$, we arrive at the following coupled equations for the coefficients:

$$iv_g\frac{\partial}{\partial x}\phi_L + gc_1\delta(x) = E\phi_L, \quad (6.22)$$

$$-iv_g\frac{\partial}{\partial x}\phi_R + gc_1\delta(x) = E\phi_R, \quad (6.23)$$

$$(-i\gamma/2 + \omega_1)c_1 + g(\phi_L(0) + \phi_R(0)) = Ec_1, \quad (6.24)$$

where we phenomenologically introduced the atomic decay rate γ. Assuming that the incident light comes from left with energy $E = v_g k$, we can express the coefficients of the wave function for the microwave field as

$$\phi_L = re^{-ikx}\mathsf{H}(-x), \quad (6.25)$$

$$\phi_R = e^{ikx}\mathsf{H}(-x) + te^{ikx}\mathsf{H}(x), \quad (6.26)$$

where H is the Heaviside-step function, and t and r are the transmission coefficient and the reflection coefficient respectively and they can be solved from coupled equations as:

$$r = \frac{-ig^2/v_g}{(i\gamma/2 + \Delta) + ig^2/v_g} \quad (6.27)$$

$$t = \frac{i\gamma/2 + \Delta}{(i\gamma/2 + \Delta) + ig^2/v_g} \quad (6.28)$$

with the detuning $\Delta = \omega - \omega_1$. The transfer matrix T is given by

$$T = \begin{pmatrix} 1/t^* & -r^*/t^* \\ -r/t & 1/t \end{pmatrix} = \begin{pmatrix} 1 + \frac{g^2}{v_g(\gamma/2+i\Delta)} & \frac{g^2}{v_g(\gamma/2+i\Delta)} \\ \frac{g^2}{v_g(\gamma/2-i\Delta)} & 1 + \frac{g^2}{v_g(\gamma/2-i\Delta)} \end{pmatrix} \quad (6.29)$$

The efficiency of photon detection or the probability of photon absorption is the probability that the photon is neither transmitted nor reflected and it is defined as

$$\eta = 1 - (1 + |T_{01}|^2)|T_{11}|^{-2} = (\gamma_0 + \gamma_0^*)|1 + \gamma_0|^{-2} \quad (6.30)$$

where $\gamma_0 = (\gamma/2 - i\delta)v_g/g^2$. It is easy to check that the up-bound of the efficiency is 50 %. As discussed in [52, 55], the detection efficiency can be significantly improved by adding more absorbers or by placing the absorber in front of a mirror.

6.3.2.2 Theory: Ξ-type Artificial Atoms

When two electromagnetic fields are simultaneously incident on a non-linear medium, an effective interaction between the fields results. In the case of a *cross-Kerr* medium, a probe field experiences a phase shift proportional to the intensity of a control field. On-chip microwave circuits have recently demonstrated giant cross-Kerr phase shift (∼20° cross phase shift per control photon) via the cross-Kerr nonlinearity between microwaves induced by a transmon [19], which is much larger than comparable single-pass optical experiments [64–66]. This experiment, together with theory from the quantum optics literature [67, 68] suggests that QND measurements of single microwave photons using a strong cross-Kerr nonlinearity are possible.

Inspired by these experimental observations, we now discuss the feasibility of using this giant non-linearity to count microwave photons. The basic idea is that an itinerant microwave photon in the control field induces a transient polarisation in the transmon, which in-turn displaces the probe field. The displacement of the probe field is measured by a conventional linear detector, yielding information about the state of the original control field. This approach has been analysed in the literature, and we summarise the results here [56, 58, 69].

Before introducing the full analysis of an anharmonic oscillator (e.g. a transmon) coupled to two itinerant microwave fields, we discuss a limiting case in which the fields are discrete resonator modes, and where the transmon may be adiabatically eliminated to produce an effective cross-Kerr nonlinearity.

We consider a three-level transmon with the energy levels $|0\rangle$, $|1\rangle$ and $|2\rangle$, embedded in a coplanar transmission line, and coupled to two single-mode fields. The Hamiltonian is given by

$$\hat{H} = \omega_2\hat{\sigma}_{22} + \omega_1\hat{\sigma}_{11} + \omega_a\hat{a}^\dagger\hat{a} + \omega_b\hat{b}^\dagger\hat{b} + g_1(\hat{a}\hat{\sigma}_{10} + \hat{a}^\dagger\hat{\sigma}_{01}) + g_2(\hat{b}\hat{\sigma}_{21} + \hat{b}^\dagger\hat{\sigma}_{12}) \tag{6.31}$$

where $\hat{a}$ and $\hat{b}$ are the annihilation operators for the fields, ω_j is a transition frequency, and g_j are coupling strengths. Transforming the Hamiltonian to a rotating frame, we have

$$\hat{H} = (\delta_2 + \delta_1)\hat{\sigma}_{22} + \delta_1\hat{\sigma}_{11} + g_1(\hat{a}\hat{\sigma}_{10} + \hat{a}^\dagger\hat{\sigma}_{01}) + g_2(\hat{b}\hat{\sigma}_{21} + \hat{b}^\dagger\hat{\sigma}_{12}) \tag{6.32}$$

where $\delta_2 = \omega_2 - \omega_1 - \omega_b$ and $\delta_1 = \omega_1 - \omega_a$. The equations of motion for the mode and transmon operators are

$$\dot{\hat{a}} = -ig_1\hat{\sigma}_{01} - \kappa_1\hat{a}/2, \tag{6.33}$$
$$\dot{\hat{b}} = -ig_2\hat{\sigma}_{12} - \kappa_2\hat{b}/2, \tag{6.34}$$
$$\dot{\hat{\sigma}}_{01} = -i\delta_1\hat{\sigma}_{01} - ig_1\hat{a}(\hat{\sigma}_{00} - \hat{\sigma}_{11}) - ig_2\hat{b}^\dagger\hat{\sigma}_{02}, \tag{6.35}$$
$$\dot{\hat{\sigma}}_{12} = -i\delta_2\hat{\sigma}_{12} - ig_2\hat{b}(\hat{\sigma}_{11} - \hat{\sigma}_{22}) + ig_1\hat{a}^\dagger\hat{\sigma}_{02}, \tag{6.36}$$
$$\dot{\hat{\sigma}}_{02} = -i(\delta_1 + \delta_2)\hat{\sigma}_{02} + ig_1\hat{a}\hat{\sigma}_{12} - ig_2\hat{b}\hat{\sigma}_{01} \tag{6.37}$$

where κ_1 and κ_2 are decay rates of the corresponding cavity modes.

In the dispersive regime, $\delta_2, \delta_1 \gg g_1, g_2$, the atomic operators can be adiabatically eliminated. We solve the algebraic equations for the transmon operators, and substitute into the dynamics for the mode operators to obtain effective equations of motion, which are generated by the effective Hamiltonian describing the cross-Kerr interaction between the two input fields

$$\hat{H}_{\text{eff}} \approx -\frac{g_1^2}{\delta_1}\hat{a}^\dagger\hat{a} + \frac{4g_1^4}{\delta_1^3}\hat{a}^{2\dagger}\hat{a}^2 - (\frac{g_1^2g_2^2}{\delta_1^2\delta_2} + \frac{g_1^2g_2^2}{\delta_1\delta_2(\delta_1+\delta_2)})\hat{a}^\dagger\hat{a}\hat{b}^\dagger\hat{b} \tag{6.38}$$

This effective model is derived in the dispersive regime and under the condition of small or negligible dissipation rates of the fields and the atom. The second term in $\hat{H}_{\text{eff}}$ is the desired cross-Kerr interaction. However, the model manifestly neglects any independent dynamics of the transmon system, and breaks down in regimes where this may be significant.

In order to describe the full dynamics close to resonance (where the effective field coupling is largest), it is important to treat the field and transmon dynamics on equal footings. This was done in [56, 58, 69] where it was found that the strong cross-Kerr nonlinearity induced by the transmon is constrained by strong saturation effects.

We now describe a somewhat simpler situation, in which the probe and control fields are simply weak coherent states, to demonstrate both the capacity to function as a power-detector at low control powers, and the way in which saturation becomes significant in the single control-photon regime. This corresponds to the situation in recent experimental results demonstrating giant microwave cross-Kerr nonlinearities [19]. The Hamiltonian for the system of a transmon coupled to two frequency resolved traveling fields in a waveguide (shown in Fig. 6.6a) is given by

$$\hat{H}_s = \omega_1\hat{\sigma}_{11} + \omega_2\hat{\sigma}_{22} + g_2(\alpha_p^* e^{i\omega_p t}\hat{\sigma}_{12} + \alpha_p e^{-i\omega_p t}\hat{\sigma}_{21}) + g_1(\alpha_c^* e^{i\omega_c t}\hat{\sigma}_{01} + \alpha_c e^{-i\omega_c t}\hat{\sigma}_{10}), \tag{6.39}$$

where the coupling coefficients $g_1 = \sqrt{\gamma_{01}/2\pi}$ and $g_2 = \sqrt{\gamma_{12}/2\pi}$, with γ_{01} and γ_{12} being the relaxation rates of the 0–1 and 1–2 transitions of the transmon, respectively. This relation between the coupling coefficient and the "atomic" relaxation rate comes from the Markovian approximation.

Moving to an interaction picture at the frequencies of the driving fields and assuming that the input fields are real, the Hamiltonian becomes

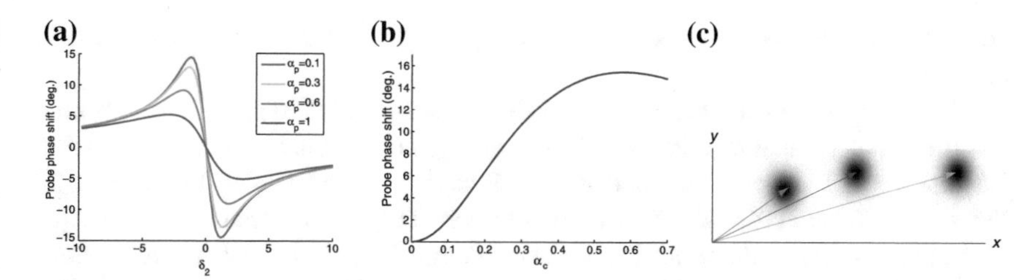

Fig. 6.5 **a** The cross-Kerr phase shift of the probe field induced by the signal field as the detuning δ_2 is varied; **b** The cross-Kerr phase shift of the probe field induced by the signal field as the signal amplitude α_c is varied. The system parameters are: $\gamma_{01} = 1$, $\gamma_{12} = 2$, $\delta_1 = -0.01$ and $\alpha_c = 1/\sqrt{2\pi}$ for the *left* figure and $\alpha_p = 0.1$ and $\delta_2 = -0.8$ for the *right* figure. **c** A phase-space cartoon illustrating the effect of saturation of the probe cross-Kerr phase shift induced by a fixed-amplitude signal field. A weak probe field incident on an unsaturable cross-Kerr medium will be rotated off the x-axis by a fixed angle regardless of its amplitude (*red arrows*). A saturable medium will rotate the signal field by an angle that decreases with amplitude (*blue and green arrows*)

$$\hat{H}_s = \delta_1\hat{\sigma}_{11} + (\delta_1 + \delta_2)\hat{\sigma}_{22} + g_2\alpha_p(\hat{\sigma}_{12} + \hat{\sigma}_{21}) + g_1\alpha_c(\hat{\sigma}_{01} + \hat{\sigma}_{10}) \quad (6.40)$$

where $\delta_1 = \omega_{10} - \omega_c$, $\delta_2 = \omega_{21} - \omega_p$. The unconditional quantum master equation is given by

$$\dot{\rho} = -i[\hat{H}_s, \rho] + \gamma_{01}\mathcal{D}[\hat{\sigma}_{01}]\rho + \gamma_{12}\mathcal{D}[\hat{\sigma}_{12}]\rho \quad (6.41)$$

We solve the dynamics from the transmon given by Eq. (6.41), and then use input-output relations to infer how the probe-field responds to the transmon polarisation, $\hat{a}_{p,\text{out}} = \hat{a}_{p,\text{in}} + \sqrt{\gamma_{12}}\hat{\sigma}_{12}$ [22]. The displacement of the output probe field relative to the input is detectable using a homodyne detector (pictured in Fig. 6.6a). Figure 6.5 presents the phase shift of the probe field induced by the control field via the transmon-induced nonlinearity. Figure 6.5a shows the probe phase shift versus the probe frequency at different probe amplitudes. Clearly, the probe phase shift saturates as the probe amplitude increases. This corresponds to the behavior illustrated in the cartoon Fig. 6.5c, in which the displacement of the probe field does not increase in proportion to its amplitude (blue and green arrows). For comparison, a

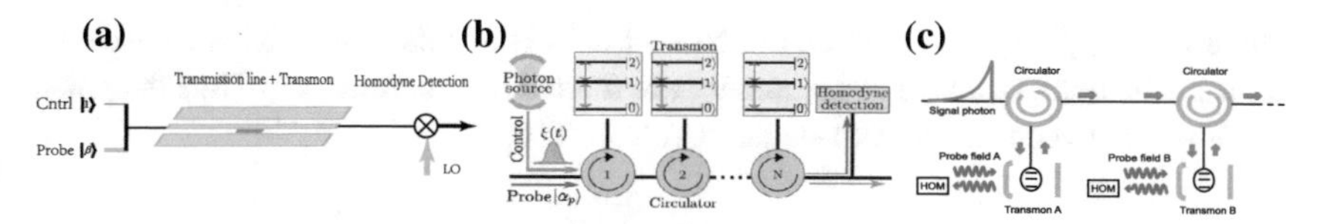

Fig. 6.6 Three detection scheme based on cross-Kerr-like schemes. **a** Photon detection based on nonlinearity induced by a transmon in an open transmission line; **b** Photon detection based on nonlinearity induced by a few cascaded transmons in a transmission line; **c** Photon detection based on nonlinearity induced by two transmons with a probe cavity

non-saturable cross-Kerr medium will yield a probe-independent phase shift (dashed red arrow).

Figure 6.5b shows the probe phase shift versus control field amplitude. We note that the x-axis is denominated in units of "average control photon flux per transmon lifetime"; that is, $\alpha_c = 1$ corresponds to a flux of 1 control photon in a time γ_{01}^{-1}. For small control amplitude there is an approximate quadratic relation between the probe phase shift and the control amplitude, i.e. the phase shift is linear in the control *power*, consistent with the effective cross-Kerr model. This supports the notion that the transmon induced nonlinearity can be used as a power detector of weak control fields in the microwave domain. It also demonstrates saturation of the probe response as the control power increases.

At resonance, the analytical steady-state solution of the model above can be obtained. In this case, the equations of motion for the expectation of the atomic operators are given by

$$\frac{d}{dt}\langle\hat{\sigma}_{00}\rangle = i\alpha_s\sqrt{\gamma_{01}}(\langle\hat{\sigma}_{01}\rangle - \langle\hat{\sigma}_{10}\rangle) + \langle\hat{\sigma}_{11}\rangle\gamma_{01}, \tag{6.42}$$

$$\frac{d}{dt}\langle\hat{\sigma}_{01}\rangle = \frac{-\gamma_{01}\langle\hat{\sigma}_{01}\rangle}{2} + i\sqrt{\gamma_{01}}\alpha_s(\langle\hat{\sigma}_{00}\rangle - \langle\hat{\sigma}_{11}\rangle) + i\sqrt{\gamma_{12}}\alpha_p\langle\hat{\sigma}_{02}\rangle, \tag{6.43}$$

$$\frac{d}{dt}\langle\hat{\sigma}_{02}\rangle = \frac{-\gamma_{12}}{2}\langle\hat{\sigma}_{02}\rangle - i\alpha_s\sqrt{\gamma_{01}}\langle\hat{\sigma}_{12}\rangle + i\sqrt{\gamma_{12}}\alpha_p\langle\hat{\sigma}_{01}\rangle, \tag{6.44}$$

$$\frac{d}{dt}\langle\hat{\sigma}_{12}\rangle = \frac{-(\gamma_{01}+\gamma_{12})}{2}\langle\hat{\sigma}_{12}\rangle - i\sqrt{\gamma_{01}}\alpha_s\langle\hat{\sigma}_{02}\rangle + i\sqrt{\gamma_{12}}\alpha_p(\langle\hat{\sigma}_{11}\rangle - \langle\hat{\sigma}_{22}\rangle), \tag{6.45}$$

$$\frac{d}{dt}\langle\hat{\sigma}_{22}\rangle = -i\alpha_p\sqrt{\gamma_{12}}(\langle\hat{\sigma}_{12}\rangle - \langle\hat{\sigma}_{21}\rangle) - \langle\hat{\sigma}_{22}\rangle\gamma_{12}. \tag{6.46}$$

The transmon induced displacement is given by $\Delta Q = -i\sqrt{\gamma_{12}}(\langle\hat{\sigma}_{12}\rangle - \langle\hat{\sigma}_{21}\rangle)$. We can estimate this displacement in a steady state situation, in which the probe and control fields are held constant. The steady state solution for the transmon polarisation $\langle\hat{\sigma}_{12}\rangle$ in this case is:

$$\langle\hat{\sigma}_{12}\rangle^{on} = \frac{i\alpha_s^2\sqrt{\gamma_{12}}\alpha_p\,(1+\gamma_{01}/\gamma_{12})}{\left[\left(\frac{\gamma_{01}}{2}+4\alpha_s^2\right)+2\alpha_p^2\right]\left[\left(\Gamma+2\alpha_p^2\right)/2+\alpha_s^2\gamma_{01}/\gamma_{12}\right]}. \tag{6.47}$$

where $\Gamma = (\gamma_{01}+\gamma_{12})/2$. When the control field is off, the transmon is transparent to the probe field, which is consequently unaffected by the transmon. Thus the phase difference between the control-on and control-off cases is

$$\Delta Q \sim \left|\langle\hat{\sigma}_{12}\rangle^{on} - 0\right| \simeq \frac{\alpha_c^2\sqrt{\gamma_{12}}\alpha_p\,(1+\gamma_{01}/\gamma_{12})}{\left[\left(\frac{\gamma_{01}}{2}+4\alpha_c^2\right)+2\alpha_p^2\right]\left[\left(\Gamma+2\alpha_p^2\right)/2+\alpha_c^2\gamma_{01}/\gamma_{12}\right]}. \tag{6.48}$$

At high probe intensity $\alpha_p \gg \alpha_s, \gamma_{12}, \gamma_{01}$, this reduces to

$$\Delta Q \sim \frac{\alpha_s^2 \sqrt{\gamma_{12}}\,(1 + \gamma_{01}/\gamma_{12})}{2\alpha_p^3} \tag{6.49}$$

The phase displacement is inversely proportional to cube of the probe amplitude.

This illustrates that for weak probe and control fields, the displacement of the probe is both linear in α_p, and quadratic in α_c (i.e. linear in the power in the control field). It further illustrates that saturation effects become important when the control and probe field amplitudes are comparable to $\sqrt{\gamma_{01}}$, i.e. the photon flux per transmon lifetime is $\sim$1.

Unsurprisingly, the saturation of the transmon is also important when the control field is a Fock state, with a single photon in a wave packet of temporal width $\sim 1/\gamma_{01}$ [56, 58, 69]. As a consequence, it was shown that a single transmon cannot be used to displace a coherent probe-field by more than the quantum noise of the field, when the control field has a single photon [69]. That is, a probe field displaced by a single control photon is not well-resolved from the undisplaced probe. It was concluded that a single anharmonic resonator (e.g. a transmon) in a microwave waveguide cannot be used as the basis for a reliable counter of itinerant microwave photons.

To overcome the limitations imposed by the saturation of a single transmon, it is desirable to cascade several in a chain, so that either a single probe field may be displaced by each successive interaction, or so that a collection of independent probes at each transmon may be measured and synthesised into a single signal better able to resolve the control photon number. Naively cascading transmons in a waveguide was shown not to work, since there is a necessary connection between the strength of the cross-Kerr interaction (the dispersive part), and the fact that the control photon is reflected from the transmon (the dissipative part). These are connected by a Kramers-Kronig type relation, so that the more powerful is the induced cross-Kerr interaction, the greater the chance of the control photon being reflected back along the waveguide [69].

Instead, the transmons must be cascaded in such a way that back-reflection is strongly suppressed. One scheme for implementing this is shown in Fig. 6.6b. Here, transmons are interleaved with circulators—uni-directional couplers that allow a field to propagate in only a single direction. In the configuration show a control photon entering the circulator is directed to the transmon, which being located at the end of a waveguide stub, deterministically sends the photon back to the circulator, which directs the control photon into the next circulator. In this way, probe fields interacting with each transmon collectively integrate information about the control field, without the damaging effects of back scattering. This approach was investigated in detail in [56], where it was shown that it is indeed a viable approach to microwave photon counting. With several transmons it is possible to achieve photon counting fidelities and quantum efficiencies >90 %.

A further extension of the protocol includes a probe field cavity (albeit with the control field still itinerant), as shown in Fig. 6.6c. This enhances the probe-transmon coupling, and was shown to enhance the fidelity and efficiency for a given number of transmons/circulators, such that just two transmons, and a single circulator are sufficient to achieve photon counting fidelities and quantum efficiencies >90 % [58]. In any case, circulators are currently one of the technical bottlenecks in superconducting circuit QED systems. We note passing that the development of compact, on-chip circulators will be crucial for future large scale circuit QED systems, not least for microwave photon counting.

There are two common merits of these three schemes discussed above:

1. Single microwave photons are not stored in a cavity so that this type of detection devices are suitable for a relatively wide frequency range of photons, determined by the coupling bandwidth of the transmon.
2. The detection process is non-absorbing so that photons can be repeatedly measured or used for further applications.

As such, this approach lends itself to QND photon counting. Importantly however, there is a necessary trade-off with the pulse-envelope distortion of a detected photon [56, 58, 69].

6.3.3 Applications: Generation of Microwave Photons

Arguably, the generation of Fock states of microwave photons has proven rather more straightforward than their detection. Again, there are two distinct regimes that have been probed experimentally, firstly in cavity QED systems, and secondly in freely propagating modes of a waveguide. Here we discuss very briefly some recent advances in microwave photon generation.

The reliable production of microwave-cavity photons in bulk cavities was pioneered by Haroche's experimental group. Their approach to generating Fock states of microwave cavities was to initially prepare the cavity in a weak coherent field, consisting of a superposition of Fock states, then use measurements effected by the flux of Rydberg atoms [46] to project onto the number eigenbasis. By this means, microwave cavity Fock states with up to 7 photons have been prepared [45], along with other non-classical states of light [70]. Furthermore, the quasi-continuous measurement record of the photon number (provided by the Rydberg detection apparatus) can be used in a control system to drive the cavity occupation to a desired photon number [71].

Alternatively, one could engineer sequential, controlled interactions between a two-level "atom", repeatedly initialised into its excited state, and a cavity to deposit quanta one-by-one, building up a Fock state to a desired photon number. This approach was adopted in the context of on-chip microwave devices by Hofheinz et al. [72]. A superconducting phase qubit [73] was used as the two-level system, coupled to a microwave stripline resonator with resonant frequency ~6.6 GHz. This kind of system is well described by the Jaynes-Cummings Hamiltonian, which predicts that the Rabi frequency at which energy is exchanged between a two-level system and a cavity depends on the cavity occupation via $\Omega_n = \Omega_0\sqrt{n}$ [74], where $\Omega_0 \approx 30$ to 40 MHz is the vacuum Rabi splitting. Notably, this is 3 orders of magnitude larger than the vacuum Rabi splitting achieved between Rydberg atoms and bulk resonator modes [46]. Thus to effect perfect energy transfer from the two-level system to the cavity (i.e. a π-pulse), requires that the interaction time between the two-level system and the cavity needs to be reduced with increasing n. Through this sequential approach, Fock states of up to $n = 6$ photons have been reliably produced [72, 75].

Linear detectors, which record the phase and amplitude of a signal, can be used to sample phase space distributions (e.g. the Huisimi Q distribution) of a source of light. In the context of microwave electronics these are known as IQ-mixers, named because they record both In-phase and Quadrature components of some input mode, $\hat{a}$. As mentioned earlier, it is also possible to reconstruct multi-time correlation functions of arbitrary order from the output record of an IQ detector [59, 60]. Essentially, the output of the IQ mixer, $\hat{S}_a$ is related to the input mode via $\hat{S}_a = \hat{a} + \hat{h}^\dagger$, where $\hat{h}^\dagger$ is an ancillary input mode that accounts for noise in the mixer. At best, this ancillary mode may be quantum limited, but in typical experiments it is characterised by some effective noise temperature, or noise-equivalent photon flux. The output of the IQ mixer is a (classical) complex number $S_a = \langle \hat{S}_a \rangle$, so moments such as $\langle (\hat{a}^\dagger)^j \hat{a}^k \rangle$ can be expressed in terms of moments of $\hat{S}_a$. These can be reconstructed from the measurement record S_r, together with moments of $\hat{h}$, which can be measured independently. Importantly, this includes higher order correlation functions including $g^{(2)}(0) = \langle (\hat{a}^\dagger)^2 \hat{a}^2 \rangle / \langle \hat{a}^\dagger \hat{a} \rangle^2$, which vanishes for e.g. single photon sources.

While this technique is not suitable for single-shot photon counting, it is extremely useful for characterising the output field of a microwave photon source, and to observe tell-tale anti-bunching that distinguishes a single-photon source from a coherent or thermal source. It has been used to characterise the microwave output of a pair of transmon superconducting qubits into a common cavity mode enabling the reconstruction of the density matrix of the two-qubit system [76]. More recently, the same approach was used to characterise the emission profile and statistics from a coherently controlled transmon in a microwave cavity, and to confirm that the output was indeed consistent with the emission of single microwave photons [77]. Similarly, an artificial atom driven by a weak coherent source acts as a non-linear mirror [78], with the reflected field showing substantial anti-bunching. Likewise, a microwave resonator strongly coupled to an artificial atom acts as a photon filter, transmitting single photons one-at-a-time [42, 43]. In both case, the resulting field exhibits the signature of non-classical, antibunched light consistent with strong single-photon character.

6.4 Conclusions

The last decade has seen remarkable advances in microwave photonics with few quanta. The production, detection and manipulation of Fock states (and other non-classical states) of the field in a microwave cavity has been demonstrated in a variety of systems, including the Nobel-prize winning work of Haroche using bulk cavities and Rydberg atoms, as well as in micro-fabricated devices using a range of superconducting artificial atoms. Indeed, a number of canonical quantum- and atom-optics phenomena have been demonstrated in these systems with much less experimental difficulty than in systems using real atoms. Furthermore, these techniques have enabled true QND detection of photons in microwave cavities, a feat which remains to be achieved in the optical domain.

In contrast, there remains an outstanding problem of detecting guided, itinerant microwave photons. This is in contrast to the optical domain, where high-quality photo-detectors have existed for many decades. We have discussed various protocols for filling in this last piece of the microwave-photon toolkit. However the proposals we have discussed suffer limitations related to their efficiency, fidelity or technical complexity. The latter is largely limited by the bulky, off-chip microwave circulators, and we view the development of on-chip circulators as a key technology that will enable development of reliable, high-efficiency microwave photon detectors.

References

1. A. Blais, R.-S. Huang, A. Wallraff, S. Girvin, R.J. Schoelkopf, Phys. Rev. A **69**, 062320 (2004)
2. R. Schoelkopf, S. Girvin, Nature **451**, 664 (2008)
3. Y. Makhlin, G. Schön, A. Shnirman, Rev. Mod. Phys. **73**, 357 (2001)
4. G. Burkard, R. Koch, D. DiVincenzo, Phys. Rev. B **69**, 064503 (2004)
5. M.H. Devoret, A. Wallraff, J. Martinis, arXiv preprint cond-mat/0411174 (2004)
6. A. Houck, J. Koch, M. Devoret, S. Girvin, R. Schoelkopf, Quantum Inf. Process. **8**, 105 (2009)
7. J. Clarke, F.K. Wilhelm, Nature **453**, 1031 (2008)
8. B. Yurke, JOSA B **4**, 1551 (1987)
9. J. Clarke, A.I. Braginski, *The SQUID Handbook* (Wiley, New York, 2004)
10. V. Bouchiat, D. Vion, P. Joyez, D. Esteve, M. Devoret, Phys. Scr. **1998**, 165 (1998)
11. Y. Nakamura, Y.A. Pashkin, J. Tsai, Nature **398**, 786 (1999)
12. J.R. Friedman, V. Patel, W. Chen, S. Tolpygo, J.E. Lukens, Nature **406**, 43 (2000)
13. C.H. Van der Wal, A. Ter Haar, F. Wilhelm, R. Schouten, C. Harmans, T. Orlando, S. Lloyd, J. Mooij, Science **290**, 773 (2000)
14. J.M. Martinis, S. Nam, J. Aumentado, C. Urbina, Phys. Rev. Lett. **89**, 117901 (2002)
15. J. Koch, T.M. Yu, J. Gambetta, A.A. Houck, D.I. Schuster, J. Majer, A. Blais, M.H. Devoret, S.M. Girvin, R.J. Schoelkopf, Phys. Rev. A **76**, 042319 (2007)
16. V.E. Manucharyan, J. Koch, L.I. Glazman, M.H. Devoret, Science **326**, 113 (2009)
17. Y. Makhlin, G. Scohn, A. Shnirman, Nature **398**, 305 (1999)
18. L.S. Bishop, J. Chow, J. Koch, A. Houck, M. Devoret, E. Thuneberg, S. Girvin, R. Schoelkopf, Nat. Phys. **5**, 105 (2009)
19. I.-C. Hoi, A.F. Kockum, T. Palomaki, T.M. Stace, B. Fan, L. Tornberg, S.R. Sathyamoorthy, G. Johansson, P. Delsing, C. Wilson, Phys. Rev. Lett. **111**, 053601 (2013)
20. A. Fedorov, L. Steffen, M. Baur, M. Da Silva, A. Wallraff, Nature **481**, 170 (2011)
21. C. Rigetti, J.M. Gambetta, S. Poletto, B. Plourde, J.M. Chow, A. Córcoles, J.A. Smolin, S.T. Merkel, J. Rozen, G.A. Keefe et al., Phys. Rev. B **86**, 100506 (2012)
22. C. Gardiner, P. Zoller, *Quantum Noise: A Handbook of Markovian and Non-Markovian Quantum Stochastic Methods with Applications to Quantum Optics*, vol. 56 (Springer, Berlin, 2004)
23. M. Göppl, A. Fragner, M. Baur, R. Bianchetti, S. Filipp, J. Fink, P. Leek, G. Puebla, L. Steffen, A. Wallraff, J. Appl. Phys. **104**, 113904 (2008)
24. C.M. Caves, Phys. Rev. D **26**, 1817 (1982)
25. T. Mimura, IEEE Trans. Microw. Theory Tech. **50**, 780 (2002)

26. B. Ho Eom, P.K. Day, H.G. LeDuc, J. Zmuidzinas, Nat. Phys. **8**, 623 (2012)
27. J. Teufel, T. Donner, M. Castellanos-Beltran, J. Harlow, K. Lehnert, Nat. Nanotech. **4**, 820 (2009)
28. C. Wilson, G. Johansson, A. Pourkabirian, M. Simoen, J. Johansson, T. Duty, F. Nori, P. Delsing, Nature **479**, 376 (2011)
29. J. You, F. Nori, Nature **474**, 589 (2011)
30. M. Ansmann, H. Wang, R.C. Bialczak, M. Hofheinz, E. Lucero, M. Neeley, A. O'Connell, D. Sank, M. Weides, J. Wenner et al., Nature **461**, 504 (2009)
31. T. Yamamoto, Y.A. Pashkin, O. Astafiev, Y. Nakamura, J.-S. Tsai, Nature **425**, 941 (2003)
32. J. Plantenberg, P. De Groot, C. Harmans, J. Mooij, Nature **447**, 836 (2007)
33. Z.-L. Xiang, S. Ashhab, J. You, F. Nori, Rev. Mod. Phys. **85**, 623 (2013)
34. A. Wallraff, D.I. Schuster, A. Blais, L. Frunzio, R.-S. Huang, J. Majer, S. Kumar, S.M. Girvin, R.J. Schoelkopf, Nature **431**, 162 (2004)
35. T. Niemczyk, F. Deppe, H. Huebl, E. Menzel, F. Hocke, M. Schwarz, J. Garcia-Ripoll, D. Zueco, T. Hümmer, E. Solano et al., Nat. Phys. **6**, 772 (2010)
36. J. Verdú, H. Zoubi, C. Koller, J. Majer, H. Ritsch, J. Schmiedmayer, Phys. Rev. Lett. **103**, 043603 (2009)
37. Z. Deng, Q. Xie, C. Wu, W. Yang, Phys. Rev. A **82**, 034306 (2010)
38. X. Zhu, S. Saito, A. Kemp, K. Kakuyanagi, S.-I. Karimoto, H. Nakano, W.J. Munro, Y. Tokura, M.S. Everitt, K. Nemoto et al., Nature **478**, 221 (2011)
39. Y. Kubo, C. Grezes, A. Dewes, T. Umeda, J. Isoya, H. Sumiya, N. Morishita, H. Abe, S. Onoda, T. Ohshima et al., Phys. Rev. Lett. **107**, 220501 (2011)
40. L. Childress, A. Sørensen, M. Lukin, Phys. Rev. A **69**, 042302 (2004)
41. J. Fink, M. Göppl, M. Baur, R. Bianchetti, P. Leek, A. Blais, A. Wallraff, Nature **454**, 315 (2008)
42. O. Astafiev, A.M. Zagoskin, A.A. Abdumalikov, Y.A. Pashkin, T. Yamamoto, K. Inomata, Y. Nakamura, J.S. Tsai, Science **327**, 840 (2010)
43. C. Lang, D. Bozyigit, C. Eichler, L. Steffen, J. Fink, A. Abdumalikov, M. Baur, S. Filipp, M. da Silva, A. Blais et al., Phys. Rev. Lett. **106**, 243601 (2011)
44. M. Brune, S. Haroche, V. Lefevre, J.M. Raimond, N. Zagury, Phys. Rev. Lett. **65**, 976 (1990)
45. C. Guerlin, J. Bernu, S. Deleglise, C. Sayrin, S. Gleyzes, S. Kuhr, M. Brune, J.-M. Raimond, S. Haroche, Nature **448**, 889 (2007)
46. S. Gleyzes, S. Kuhr, C. Guerlin, J. Bernu, S. Deleglise, U.B. Hoff, M. Brune, J.-M. Raimond, S. Haroche, Nature **446**, 297 (2007)
47. K.E., Y.L., and S.J., Nat. Photon. **2**, 425 (2008)
48. O. Thomas, Z.L. Yuan, J.F. Dynes, A.W. Sharpe, A.J. Shields, Appl. Phys. Lett. **97**, 031102 (2010)
49. A.E. Lita, A.J. Miller, S.W. Nam, Opt. Express **16**, 3032 (2008)
50. E.A. Dauler, A.J. Kerman, B.S. Robinson, J.K. Yang, B. Voronov, G. Goltsman, S.A. Hamilton, K.K. Berggren, J. Mod. Opt. 56, 364 (2009)
51. G. Romero, J.J. García-Ripoll, E. Solano, Phys. Rev. Lett. **102**, 173602 (2009a)
52. G. Romero, J.J. García-Ripoll, E. Solano, Phys. Scr. **2009**, 014004 (2009)
53. B. Johnson, M. Reed, A. Houck, D. Schuster, L.S. Bishop, E. Ginossar, J. Gambetta, L. DiCarlo, L. Frunzio, S. Girvin et al., Nat. Phys. **6**, 663 (2010)
54. Y.-F. Chen, D. Hover, S. Sendelbach, L. Maurer, S.T. Merkel, E.J. Pritchett, F.K. Wilhelm, R. McDermott, Phys. Rev. Lett. **107**, 217401 (2011)
55. B. Peropadre, G. Romero, G. Johansson, C.M. Wilson, E. Solano, J.J. García-Ripoll, Phys. Rev. A **84**, 063834 (2011)
56. S.R. Sathyamoorthy, L. Tornberg, A.F. Kockum, B.Q. Baragiola, J. Combes, C.M. Wilson, T.M. Stace, G. Johansson, Phys. Rev. Lett. **112**, 093601 (2014)
57. B. Peaudecerf, T. Rybarczyk, S. Gerlich, S. Gleyzes, J.M. Raimond, S. Haroche, I. Dotsenko, M. Brune, Phys. Rev. Lett. **112**, 080401 (2014)
58. B. Fan, G. Johansson, J. Combes, G.J. Milburn, T.M. Stace, Phys. Rev. B **90**, 035132 (2014)
59. N. Grosse, T. Symul, M. Stobińska, T. Ralph, P. Lam, Phys. Rev. Lett. **98**, 153603 (2007)

60. M. da Silva, D. Bozyigit, A. Wallraff, A. Blais, Phys. Rev. A **82**, 043804 (2010)
61. L.C. Govia, E.J. Pritchett, C. Xu, B. Plourde, M.G. Vavilov, F.K. Wilhelm, R. McDermott, Phys. Rev. A **90**, 062307 (2014)
62. D. Vion, A. Aassime, A. Cottet, P. Joyez, H. Pothier, C. Urbina, D. Esteve, M.H. Devoret, Science **296**, 886 (2002)
63. M.A. Sillanpää, J. Li, K. Cicak, F. Altomare, J.I. Park, R.W. Simmonds, G.S. Paraoanu, P.J. Hakonen, Phys. Rev. Lett. **103**, 193601 (2009)
64. C. Perrella, P.S. Light, J.D. Anstie, F. Benabid, T.M. Stace, A.G. White, A.N. Luiten, Phys. Rev. A **88**, 013819 (2013)
65. V. Venkataraman, K. Saha, A.L. Gaeta, Nat. Photonics **7**, 138 (2013)
66. J. Volz, M. Scheucher, C. Junge, A. Rauschenbeutel, Nat. Photon. **8**, 965 (2014)
67. V. Braginsky, F. Khalili, *Quantum Measurement* (Cambridge University Press, Cambridge, 1995)
68. W.J. Munro, K. Nemoto, T.P. Spiller, S.D. Barrett, P. Kok, R.G. Beausoleil, J. Opt. B: Quantum Semiclass. Opt. **7**, S135 (2005)
69. B. Fan, G. Johansson, I.-C. Hoi, T.M. Stace, Phys. Rev. Lett. **110**, 053601 (2013)
70. S. Deleglise, I. Dotsenko, C. Sayrin, J. Bernu, M. Brune, J.-M. Raimond, S. Haroche, Nature **455**, 510 (2008)
71. C. Sayrin, I. Dotsenko, X. Zhou, B. Peaudecerf, T. Rybarczyk, S. Gleyzes, P. Rouchon, M. Mirrahimi, H. Amini, M. Brune et al., Nature **477**, 73 (2011)
72. M. Hofheinz, E.M. Weig, M. Ansmann, R.C. Bialczak, E. Lucero, M. Neeley, A.D. O'Connell, H. Wang, J.M. Martinis, A.N. Cleland, Nature **454**, 310 (2008)
73. M.H. Devoret, J.M. Martinis, Quantum Inf. Process. **3**, 163 (2004)
74. D.F. Walls, G.J. Milburn, *Quantum Optics* (Springer, New York, 2007)
75. A. Houck, D. Schuster, J. Gambetta, J. Schreier, B. Johnson, J. Chow, L. Frunzio, J. Majer, M. Devoret, S. Girvin et al., Nature **449**, 328 (2007)
76. S. Filipp, P. Maurer, P. Leek, M. Baur, R. Bianchetti, J. Fink, M. Göppl, L. Steffen, J. Gambetta, A. Blais et al., Phys. Rev. Lett. **102**, 200402 (2009)
77. M. Pechal, L. Huthmacher, C. Eichler, S. Zeytinoğlu, A.A. Abdumalikov, S. Berger, A. Wallraff, S. Filipp, Phys. Rev. X **4**, 041010 (2014)
78. I.-C. Hoi, T. Palomaki, J. Lindkvist, G. Johansson, P. Delsing, C. Wilson, Phys. Rev. Lett. **108**, 263601 (2012)

Chapter 7
Weak Measurement and Feedback in Superconducting Quantum Circuits

Kater W. Murch, Rajamani Vijay and Irfan Siddiqi

Abstract We describe the implementation of weak quantum measurements in superconducting qubits, focusing specifically on transmon type devices in the circuit quantum electrodynamics architecture. To access this regime, the readout cavity is probed with on average a single microwave photon. Such low-level signals are detected using near quantum-noise-limited superconducting parametric amplifiers. Weak measurements yield partial information about the quantum state, and correspondingly do not completely project the qubit onto an eigenstate. As such, we use the measurement record to either sequentially reconstruct the quantum state at a given time, yielding a quantum trajectory, or to close a direct quantum feedback loop, stabilizing Rabi oscillations indefinitely.

7.1 Introduction

Measurement-based feedback routines are commonplace in modern electronics, including the thermostats regulating the temperature in our homes and sophisticated motion stabilization hardware needed for the autonomous operation of aircraft. The basic elements present in such classical feedback loops include a sensor element which provides information to a controller, which in turn steers the system of interest toward a desired target. In this paradigm, the act of sensing itself does not a priori perturb the system in a significant way. Moreover, extracting more or less

K.W. Murch (✉)
Department of Physics, Washington University, St. Louis, MO, USA
e-mail: murch@physics.wustl.edu

R. Vijay
Tata Institute of Fundamental Research, Department of Condensed Matter Physics & Materials Science, Mumbai, India
e-mail: rvijay26@gmail.com

I. Siddiqi
Quantum Nanoelectronics Laboratory, Department of Physics, University of California, Berkeley, CA, USA
e-mail: irfan@berkeley.edu

© Springer International Publishing Switzerland 2016

R.H. Hadfield and G. Johansson (eds.), *Superconducting Devices in Quantum Optics*, Quantum Science and Technology,
DOI 10.1007/978-3-319-24091-6_7

information during the measurement process does not factor into the control algorithm. For quantum coherent circuits, these basic assumptions do not hold. The act of measurement is invasive, and the so-called backaction drives a system toward an eigenstate of the measurement operator. Furthermore, the information content of the measurement determines the degree of backaction imparted. For example, strong measurements completely project a superposition state into a definite eigenstate while extracting enough information to unambiguously allow an observer to determine which eigenstate has been populated. Weak measurements on the other hand, accrue less information, indicating which eigenstate is more likely to be populated, and have proportionally weaker backaction that does not completely collapse a quantum superposition. This type of measurement provides a natural route to implement active feedback in quantum systems with the weak measurement outcome providing the input for a quantum controller that takes into account the measurement induced backaction.

In this chapter, we first discuss a general formalism based on positive operator-valued measures (POVMs) which can describe both weak and strong measurements. We then apply this formalism to a superconducting two-level system coupled to a microwave frequency cavity, which is well approximated by the Jaynes-Cummings model commonly used in cavity quantum electrodynamics. In such a system, the measurement strength for a given integration time can be adjusted by varying the number of photons in the cavity. To access the weak measurement regime, the cavity is typically populated, on average, with less than one photon, requiring an ultra-low-noise amplifier for efficient detection. In contemporary experiments, this function is realized using superconducting parametric amplifiers. After briefly describing these devices, we discuss two basic types of experiments. In the first set, the result of a sequence of weak measurements is used to reconstruct individual quantum trajectories using a Bayesian update procedure. The statistical distribution of an ensemble of many such trajectories is then analyzed to experimentally and theoretically obtain the most likely path. In these experiments, the backaction results either solely from the measurement process or from the combination of measurement and unitary evolution under a coherent drive. In principle, the reconstructed state and the detector values corresponding to the most likely path can be used for arbitrary feedback protocols and optimal control. In the second type of experiment, we use the weak measurement outcome to directly complete an analog feedback circuit. In particular, we demonstrate the real-time stabilization of Rabi oscillations resulting in the suppression of their ensemble decay. Finally, we close with a discussion of future directions for hardware improvement and additional experimentation, particularly in multi-qubit systems.

7.2 Generalized Measurements

In quantum mechanics, predictions about the outcome of experiments are given by Born's rule which for a state vector $|\psi_i\rangle$ provides the probability $P(a) = |\langle a|\psi_i\rangle|^2$ that a measurement of an observable described by an operator $\hat{A}$ with eigenstates $|a\rangle$

yields one of the eigenvalues a. As a consequence of the measurement, the quantum state is projected into the state $|a\rangle$. Here we consider a qubit, with states $|0\rangle$ and $|1\rangle$, which is conveniently described by the Pauli matrix algebra with pseudo-spin operators σ_x, σ_y, and σ_z. For example, if we prepare the qubit in an initial state,

$$|\psi_i\rangle = |+x\rangle \equiv \frac{1}{\sqrt{2}}(|0\rangle + |1\rangle) \tag{7.1}$$

Born's rule tells us the probability of finding the qubit in the state $|0\rangle$, $P(+z) = 1/2$. Here we have introduced the notation $\pm x$, $\pm z$ to indicate the eigenstate of the σ_x and σ_z pseudo-spin operators. After the measurement, the qubit will remain in the eigenstate corresponding to the eigenvalue and subsequent measurements will yield the same result. In this way the measurement changes $\langle\sigma_z\rangle$ from 0 to 1, while changing $\langle\sigma_x\rangle$ from 1 to 0. This change, associated with a projective measurement of σ_z is the backaction of measurement and also referred to as the collapse of the wavefunction.

The fact that the measurement takes $\langle\sigma_x\rangle$ from 1 to 0 is a consequence of the Heisenberg uncertainty principle. Because σ_x and σ_z do not commute, if one component of the pseudo-spin is known then the others must be maximally uncertain. Thus some amount of backaction occurs in any measurement, and measurements (such as the projective measurement discussed above) that cause no more backaction than the amount mandated by the Heisenberg uncertainty principle are said to be quantum non-demolition (QND) measurements.

In this chapter, we are concerned with a more general type of measurement that can be performed on the system [1–6]. Real measurements take place over a finite amount of time and we are interested in describing the evolution of the system along the way. As such, a projective measurement is composed of a sequence of partial measurements. Formally, we can describe these measurements by the theory of positive operator-valued measures [1–4], (POVMs) which yields the probability $P(m) = \mathrm{Tr}(\Omega_m \rho \Omega_m^\dagger)$ for outcome m, and the associated backaction on the quantum state, $\rho \to \Omega_m \rho \Omega_m^\dagger / P(m)$ for a system described by a density matrix ρ. POVMs are "positive" simply because they describe outcomes with positive probabilities and "operator-valued" because they are expressed as operators. These operators Ω_m obey $\sum_m \Omega_m^\dagger \Omega_m = \hat{I}$ as is necessary for POVMs. When $\Omega_a = |a\rangle\langle a|$ is a projection operator and $\rho = |\psi\rangle\langle\psi|$, the theory of POVMs reproduces Born's rule. For the case described above, the projective measurement of σ_z is described by the POVM operators $\Omega_{+z} = (\hat{I} + \sigma_z)/2$ and $\Omega_{-z} = (\hat{I} - \sigma_z)/2$ and straightforward application of the theory reproduces the effects of projective measurement discussed above.

7.2.1 Indirect Measurements

While it is tempting to connect a quantum system directly to a classical measuring apparatus, such an arrangement often results in far more backaction on the quantum system than is dictated by the Heisenberg uncertainty principle. For example, an avalanche photodiode can be used to detect the presence of a single photon, but it achieves this detection by absorbing the photon, thereby completely destroying the quantum state. In practice, classical devices are composed of too many noisy degrees of freedom to perform QND measurements.

To circumvent this problem, we break the measurement apparatus into three parts to conduct an indirect measurement. The three parts of the apparatus are the quantum system of interest, a quantum pointer system that can be coupled to the quantum system, and then a classical measurement apparatus that can be used to record the pointer system. To execute a measurement, the pointer system is coupled to the quantum system and after sufficient interaction the two systems become entangled. Then, a classical measurement apparatus, such as a photodetector or an ammeter, is used to make measurements on the pointer system. These measurements on the pointer system are projective and depending on their outcome, the quantum system's state is accordingly changed.

To illustrate this process, consider a system consisting of two qubits [3]. We initialize one qubit (the system qubit) in the state in (7.1) and allow a second (environment) qubit to interact with the primary qubit for a brief period of time such that the two qubits are in the entangled state,

$$|\Psi\rangle \propto \big((1+\epsilon)|0\rangle_{\text{sys}} + (1-\epsilon)|1\rangle_{\text{sys}}\big) \otimes |0\rangle_{\text{env}} + \big((1-\epsilon)|0\rangle_{\text{sys}} + (1+\epsilon)|1\rangle_{\text{sys}}\big) \otimes |1\rangle_{\text{env}}. \tag{7.2}$$

We then make a projective measurement of the second qubit (using the POVM $\Omega_{\pm z}$), and if the result is the $|0\rangle$ state, then the system qubit is left in the state,

$$|\Psi\rangle \propto (1+\epsilon)|0\rangle_{\text{sys}} + (1-\epsilon)|1\rangle_{\text{sys}}. \tag{7.3}$$

For $\epsilon \ll 1$ this partial measurement slightly drives the system qubit to the ground state. However, what happens if we instead measure the environment qubit in the σ_y basis, $|y_\pm\rangle = \frac{1}{\sqrt{2}}(|0\rangle \pm i|1\rangle)$? We can express the joint state in this basis,

$$|\Psi\rangle \propto \big((1+\epsilon-i(1-\epsilon))|0\rangle_{\text{sys}} + (1-\epsilon-i(1+\epsilon))|1\rangle_{\text{sys}}\big) \otimes |y_+\rangle_{\text{env}} + \big((1-\epsilon-i(1+\epsilon))|0\rangle_{\text{sys}} + (1+\epsilon-i(1-\epsilon))|1\rangle_{\text{sys}}\big) \otimes |y_-\rangle_{\text{env}}. \tag{7.4}$$

By factoring out global phase factors $\phi_+ = \pi/4+\epsilon$ and $\phi_- = \pi/4-\epsilon$ and assuming that $\epsilon \ll 1$, the joint state is,

$$|\Psi\rangle \propto e^{i\phi_+}\big(|0\rangle_{\text{sys}} + e^{-2i\epsilon}|1\rangle_{\text{sys}}\big) \otimes |y_+\rangle_{\text{env}} + e^{i\phi_-}\big(|0\rangle_{\text{sys}} + e^{2i\epsilon}|1\rangle_{\text{sys}}\big) \otimes |y_-\rangle_{\text{env}} \tag{7.5}$$

which shows that the system qubit state obtains a slight rotation depending on the detected state of the environment qubit. This simple model indicates two salient features of partial measurements. First, if the environment and the system are weakly entangled, then projective measurements on the environment cause only small changes in the qubit state, and second, the choice of measurement basis for the environment gives different conditional evolution for the system. After the projective measurement on the environment qubit the system is in a product state with the environment qubit and no entanglement remains.

We now imagine many such successive interactions between the system qubit and a set of environment qubits. In this case each entangling interaction and projective measurement of the environment causes a small amount of random backaction on the qubit, which results in a diffusive trajectory for the state of the qubit. Such weak measurements have been recently implemented using coupled superconducting qubits [7].

7.2.2 Continuous Measurement

The measurements that are realized in the circuit Quantum Electrodynamics (cQED) architecture described later in Sect. 7.3 can be represented by a set of POVMs $\{\Omega_V\}$, where V is the dimensionless measurement result which is scaled so that it takes on average values ± 1 for the qubit in states $\pm z$ respectively,

$$\Omega_V = \left(2\pi a^2\right)^{-1/4} e^{(-(V-\sigma_z)^2/4a^2)}. \tag{7.6}$$

Here, $1/4a^2 = k\eta\Delta t$, and k parametrizes the strength of the measurement, η is the quantum efficiency, and Δt is the duration of the measurement. The operators Ω_V satisfy $\int \Omega_V^\dagger \Omega_V dV = \hat{I}$ as expected for POVMs. The probability of each measurement yielding a value V is $P(V) = \mathrm{Tr}(\Omega_V \rho_t \Omega_V^\dagger)$, which is the sum of two Gaussian distributions with variance a^2 centered at $+1$ and -1 and weighted by the populations ρ_{00} and ρ_{11} of the two qubit states. The σ_z term in Ω_V causes the back action on the qubit degree of freedom, $\rho \rightarrow \Omega_V \rho \Omega_V^\dagger / P(V)$, due to the readout of the measurement result V. Equation (7.6) can also describe a stronger measurement, ultimately yielding the limit where the two Gaussian distributions are disjoint, and the readout projects the qubit onto one of its σ_z eigenstates, with probabilities ρ_{00} or ρ_{11}. The strength of the measurement is controlled by the parameter k and we will see later in Sect. 7.3 how this quantity is related to experimental parameters.

For weak measurements, $\Delta t \ll \tau_c$, where the characteristic measurement time $\tau_c = 1/4k\eta$ describes the time it takes to separate the Gaussian measurement histograms by two standard deviations. The distribution of measurement results is then approximately given by a single Gaussian that is centered on the expectation value of σ_z,

$$P(\Omega_V) \simeq e^{-4k\eta\Delta t (V - \langle\sigma_z\rangle)^2}, \tag{7.7}$$

which highlights the fact that V is simply a noisy estimate of $\langle\sigma_z\rangle$. This allows the measurement signal to be written as the sum of $\langle\sigma_z\rangle$ and a zero mean Gaussian random variable [2].

The time evolution of the quantum state following a sequence of measurements described by Eq. (7.6) in the limit $\Delta t \to 0$ is given by the stochastic master equation [1, 2]:

$$\frac{d\rho}{dt} = k(\sigma_z\rho\sigma_z - \rho) + 2\eta k(\sigma_z\rho + \rho\sigma_z - 2\langle\sigma_z\rangle\rho)(V(t) - \langle\sigma_z\rangle). \quad (7.8)$$

Here, the first term is the standard master equation in Lindblad form and the second term is the stochastic term that updates the state based on the measurement result. In the case of unit detector efficiency ($\eta = 1$) the stochastic master equation (perhaps surprisingly) yields pure state dynamics. This happens because the decoherence term with rate k in the equation is exactly compensated by the stochastic term to yield a random pure state evolution; the first term shrinks the Bloch vector towards the z-axis, while the stochastic term adds a random transverse component, putting it back to the surface of the Bloch sphere again. In contrast, if we set $\eta = 0$, all random backaction is suppressed, and we instead obtain a conventional, deterministic master equation, where probing of the system causes extra Lindblad decoherence on the system, but yields no information.

7.3 Quantum Measurements in the cQED Architecture

The cQED architecture [8] provides both excellent coherence properties and high fidelity measurement. In particular, it is an ideal test bed for studying quantum measurements because it enables one to implement text book quantum measurements relatively easily. The two key aspects which make this possible is the applicability of an ideal measurement Hamiltonian [8] and the availability of quantum limited parametric amplifiers [9, 10].

Circuit QED is essentially the implementation of the cavity QED architecture [11] using superconducting circuits. Instead of an atom interacting with the electromagnetic field inside a Fabry-Perot cavity, the cQED architecture uses a superconducting qubit as an 'artificial' atom which interacts coherently with the electromagnetic field in an on-chip waveguide resonator [12] or a 3D waveguide cavity [13]. Figure 7.1 shows the basic cQED setup. Dispersive readout in cQED was introduced in Chap. 6 and we briefly reintroduce it here to make this chapter more self contained. The Hamiltonian describing this interaction is the Jaynes-Cummings Hamiltonian,

$$H = \frac{\hbar\omega_{01}}{2}\sigma_z + \hbar\omega_c(a^\dagger a + \frac{1}{2}) + \hbar g(a\sigma^+ + a^\dagger\sigma_-) \quad (7.9)$$

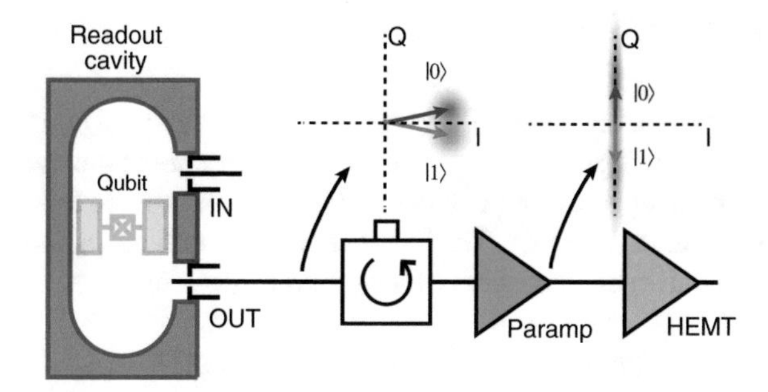

Fig. 7.1 Circuit QED setup consisting of a qubit dispersively coupled to a readout cavity. A signal transmitting the cavity at the cavity frequency acquires a qubit-state-dependent phase shift, shown as two phasors in the IQ plane. Phase sensitive amplification amplifies one quadrature exclusively

where the first term represents the qubit as a pseudo-spin, the second term is the cavity mode, and the third term represents the interaction between the qubit and the electromagnetic field in the rotating wave approximation. Here ω_{01} is the transition frequency between the qubit levels, ω_c is the cavity mode frequency, g is the coupling strength between the qubit and cavity mode and $\sigma_\pm = (\sigma_x \pm i\sigma_y)/2$ are the qubit raising and lowering operators. Typically, the qubit frequency is far detuned from the cavity frequency to protect the qubit from decaying into the mode to which the cavity is strongly coupled. In this dispersive regime ($|\Delta| = |\omega_{01} - \omega_c| \gg g$), the effective Hamiltonian reduces to,

$$H = \frac{\hbar\omega_{01}}{2}\sigma_z + \hbar(\omega_c + \chi\sigma_z)\left(a^\dagger a + \frac{1}{2}\right) \tag{7.10}$$

where χ is called the dispersive shift. It is now possible to see how the measurement is implemented. The cavity mode frequency is a function of the qubit state with χ, which depends on g and Δ, setting the magnitude of this shift. The measurement proceeds by probing the cavity with a microwave signal and detecting the qubit state dependent phase shift of the scattered microwave signal. In this architecture, one can clearly see the process of indirect measurement introduced earlier. The pointer system is the microwave field which interacts with the qubit and gets Entanglement, entangled states with it. One then measures the output microwave field which in turn determines the qubit state. As mentioned earlier, the exact measurement on the pointer determines the backaction on the qubit and that can be controlled by the choice of amplification method used to detect the output microwave field. We will however come to that a little later.

7.3.1 Dispersive Measurements

We will now formalize the interaction between the qubit and the pointer state using the language of coherent states to describe the quantum state of the microwave field. Let $|\alpha\rangle$ represent the coherent state of the microwave field sent to the cavity with an average photon number $\bar{n} = |\alpha|^2$. The initial state of the qubit is $a_0|0\rangle + a_1|1\rangle$. After interaction, the final entangled state of the system is given by

$$|\psi_f\rangle = a_0|e^{-i\theta_d}\alpha\rangle|0\rangle + a_1|e^{+i\theta_d}\alpha\rangle|1\rangle \tag{7.11}$$

where θ_d is the dispersive phase shift of the scattered microwave signal. The two coherent states can be represented in the quadrature plane as shown in Fig. 7.2. For a fixed dispersive phase shift θ_d, the larger the average number of photons in the coherent states, the easier it is to distinguish them. Here, the distinguishability of the output coherent states is directly related to the distinguishability of the qubit state i.e. the strength of the measurement can be controlled by the microwave power.

One can now choose to measure the output microwave field by performing a homodyne measurement and choosing one of the two possible quadratures. From Fig. 7.2, it is clear that the qubit state information is encoded in the 'Q' quadrature while the 'I' quadrature has the same value for both qubit states. We will first describe the single quadrature measurement of the 'Q' quadrature using a phase-sensitive amplifier and the associated backaction on the qubit. Since the coherent state is not a two-level state, a measurement on it yields a continuous range of values and corresponds to the case discussed in Sect. 7.2.2.

Typically, experiments with microwave signals employ mixers operating at room temperature to implement homodyne detection. However, the microwave signals used to probe the cavity are extremely weak and need to be amplified before they can be processed by room temperature electronics. The challenge is that commercial amplifiers add significant noise which results in imperfect correlation between the measured output and the measurement backaction on the qubit. We use superconducting phase-sensitive parametric amplifiers to implement near noiseless amplification of a single quadrature of the microwave signal.

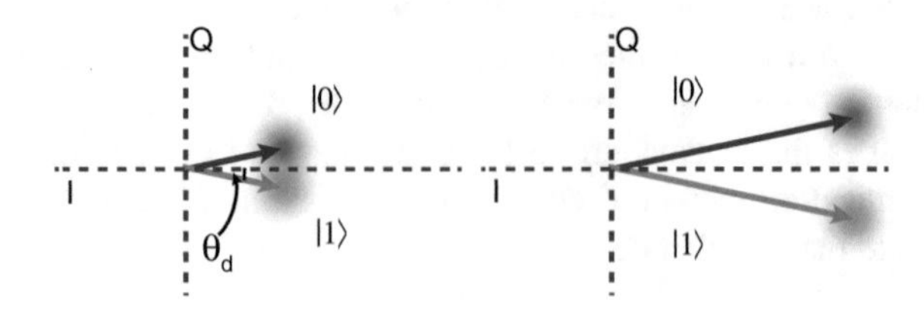

Fig. 7.2 Output coherent states represented in the IQ plane. The length of the vector increases with increasing photon number in the coherent state and leads to better distinguishability between the qubit states

7.3.2 Parametric Amplification

There are several parametric amplifier designs [9, 10, 14] but we will restrict our discussion to one based on a lumped element non-linear oscillator [10]. The basic physics of parametric amplification in such a system can be understood by considering a driven, damped Duffing oscillator model [15] whose classical equation of motion is given by

$$\frac{d^2\delta(t)}{dt^2} + 2\Gamma\frac{d\delta(t)}{dt} + \omega_0^2\left(\delta(t) - \frac{\delta(t)^3}{6}\right) = F\cos(\omega_d t). \tag{7.12}$$

Here ω_0 is the linear resonant frequency for small oscillations, Γ is the amplitude damping coefficient, δ is the gauge-invariant phase difference across the junction which is the dynamical variable of oscillator, and ω_d is the frequency of the harmonic driving term often referred to as the pump.

Figure 7.3a shows the basic circuit diagram of a Josephson junction based non-linear oscillator. The phase diagram of such a non-linear oscillator as a function of drive frequency and drive power is shown in Fig. 7.3b. A characteristic feature of a non-linear oscillator is that its effective resonant frequency is no longer independent of the amplitude of its oscillations. This is shown schematically by the black solid line in Fig. 7.3b where the effective resonant frequency decreases with increasing driving power for Josephson junction based nonlinear oscillators. Beyond a critical drive frequency and power (ω_c and P_c), the driven response becomes bistable and we are not interested in that regime. The relevant regime for parametric amplification is marked in Fig. 7.3b and is just before the onset of bistability. Because of the power dependence of the effective resonant frequency, it is possible to cross the resonance (black line) both in frequency and amplitude. Figure 7.3c shows the phase of the

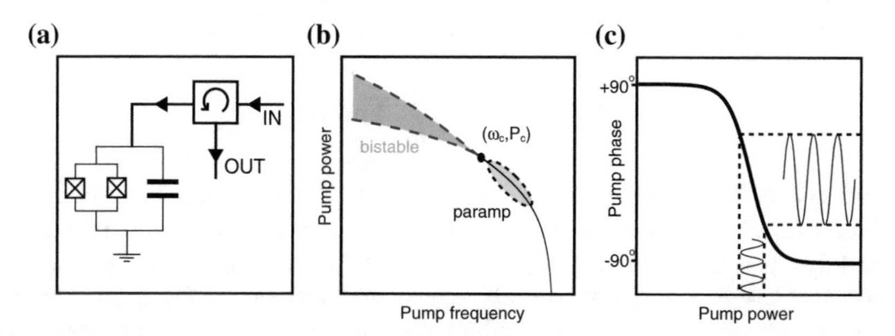

Fig. 7.3 **a** The basic circuit diagram of a Josephson junction based non-linear oscillator. A circulator is used to separate the amplified reflected signal from the incident signal. **b** Phase diagram of a driven non-linear oscillator. The *solid line* indicates the effective resonant frequency as a function of drive power. The *dashed lines* enclose the bistable regime of the non-linear oscillator. The oval marks the paramp biasing regime. **c** The reflected pump signal phase is plotted as a function of pump power and defines the transfer function of the paramp

reflected pump signal from the oscillator as a function of pump amplitude. This is essentially the transfer function of the amplifier and the pump power is chosen to bias the system in the steep part of this curve.

The signal to be amplified has the same frequency ω_d with an amplitude which is typically less than 1 % of the pump amplitude, is combined with the pump signal and sent to the nonlinear resonator. Since the signal and pump are at the same frequency, one can define a phase difference between them. If the signal is in phase with the pump, then the combined amplitude changes significantly with the input signal which moves the bias point and consequently the reflected pump phase. This is essentially the mechanism of amplification, i.e. a small signal brings about a large phase shift in the large pump signal. However, if the signal is 90° out of phase with the pump, then to first order, it has no effect on the bias point and consequently no change in the reflected pump phase. So the signals which are in phase with the pump get amplified while the quadrature phase signals get de-amplified. This allows one to selectively amplify the 'Q' quadrature of the microwaves scattered from the cavity in a cQED measurement. In principle, such a phase-sensitive amplifier can amplify one quadrature without adding any additional noise to the signal [10], resulting in a signal-to-noise-ratio that is maintained after amplification and an output noise level that is entirely set by amplified zero point fluctuations associated with the coherent state. In practice, due to signal losses between the cavity and the amplifier and due to inefficiencies in the amplifier itself, one typically obtains an efficiency $\eta \sim 0.5$. In other words, the output noise is only about double the unavoidable quantum noise [16, 17].

7.3.3 Weak Measurement and Backaction

A single weak measurement V_m is obtained by integrating the homodyne signal $V_Q(t)$ corresponding to the 'Q' quadrature for a measurement time τ,

$$V_m(\tau) = \frac{2}{\tau \Delta V} \int_0^\tau V_Q(t)dt \tag{7.13}$$

where ΔV is the difference in the mean value of $V_Q(t)$ corresponding to the two qubit states. This implies that the mean value of V_m is ± 1 for the two qubit states. Measuring the 'Q' quadrature of the scattered microwave signal in cQED using a phase-sensitive amplifier corresponds to a σ_z measurement of the qubit. The corresponding backaction pushes the qubit state towards one of the eigenstates of σ_z. This is shown as the solid black line in Fig. 7.4a where $Z^Z = \langle \sigma_z \rangle |_{V_m}$ is plotted as a function of weak measurement result V_m for an initial qubit state $(|0\rangle + |1\rangle)/\sqrt{2}$. Similarly, the expectation values of σ_x and σ_y are also shown as solid blue and red lines respectively. The dashed line corresponds to the theoretical prediction given by [17, 18],

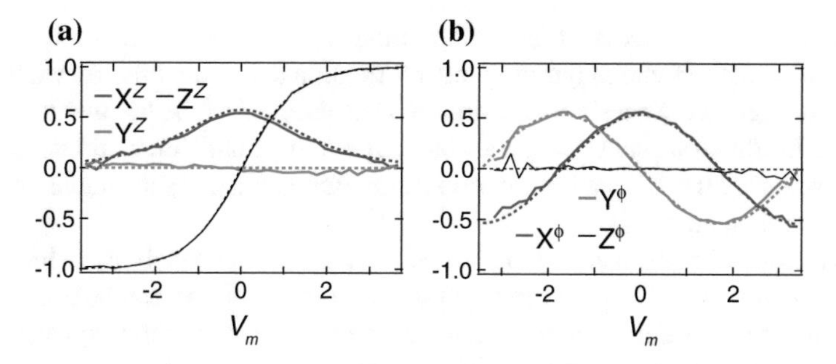

Fig. 7.4 Backaction of a weak measurement. **a** When the 'Q' quadrature of the scattered microwave signal is amplified a σ_z measurement of the qubit occurs. The *dashed lines* indicate the theoretical prediction from (7.14) and the *solid lines* are the tomographic results conditioned on the measurement result V_m. **b** A 'ϕ' measurement of the qubit is obtained by amplifying the 'I' quadrature. The *dashed lines* indicate the predictions from (7.15) and the *solid lines* are the conditioned tomography. The data correspond to $\eta = 0.49$ and $S = 3.15$ and correspond to the experimental setup in [17]

$$Z^Z = \tanh\left(\frac{V_m S}{4}\right), X^Z = \sqrt{1-(Z^Z)^2}e^{-\gamma\tau}, Y^Z = 0 \tag{7.14}$$

where τ is the measurement duration, $S = 64\tau\chi^2\bar{n}\eta/\kappa$ is the dimensionless measurement strength and γ is the environmental dephasing rate of the qubit. To connect with the nomenclature used in Sect. 7.2.2, we note that V_m is equal to the dimensional measurement result V defined earlier whereas $S = 4/a^2$ where a^2 is the variance of Gaussian measurement distributions. If $\eta = 1$, the state remains pure during the entire measurement process. Here, the data corresponds to $\eta = 0.49$ and $S = 3.15$.

It is important to note that even though the other quadrature 'I' does not contain any information about σ_z, it does contain information about the photon number fluctuations in the coherent state [18]. This implies that a measurement of the 'I' quadrature will result in the qubit state rotating about the Z axis in the Bloch sphere [17] and will not push the state towards one of the eigenstates of σ_z. This can be seen in Fig. 7.4b where the expectation values of $\sigma_{x,y,z}$ are plotted (solid lines) after a weak measurement with the paramp amplifying the 'I' quadrature. Here V_m is obtained by integrating the homodyne signal $V_I(t)$ corresponding to the 'I' quadrature for a measurement time τ. Note that this is not the same as measuring the σ_x or the σ_y operator and we label it as a 'ϕ' measurement since it results in rotation about the Z axis in Bloch sphere. Consequently, there is no asymptotic value for X^ϕ or Y^ϕ and they evolve sinusoidally as,

$$X^\phi = \cos\left(\frac{V_m S}{4}\right)e^{-\gamma\tau}, \quad Y^\phi = -\sin\left(\frac{V_m S}{4}\right)e^{-\gamma\tau}, \quad Z^\phi = 0. \tag{7.15}$$

and Z^ϕ doesn't evolve and remains zero.

A measurement using a phase-sensitive amplifier truly enables one to measure one quadrature only while erasing the information in the other quadrature so that no observer can get access to that information. In fact, that information no longer exists and is not the same as ignoring the information in that quadrature which would lead to decoherence. So a single quadrature measurement of the 'Q' quadrature only provides information about σ_z and there are no photon number fluctuations. Similarly, when measuring the 'I' quadrature, no information about σ_z is available and hence $\langle\sigma_z\rangle$ does not change [17].

It is also possible to measure both quadratures simultaneously by using a phase preserving amplifier. The parametric amplifier described above can be used in phase preserving mode by detuning the signal frequency from the pump frequency [10] or one could use a two mode amplifier like the Josephson Parametric Converter (JPC) [14]. In this case, each measurement gives two outputs corresponding to the 'I' and 'Q' quadrature [19]. This implies that one learns about the qubit state as well as the photon number fluctuations and consequently the measurement backaction results in the Bloch vector rotating around the Z axis while it approaches one of the eigenstates of σ_z. Somewhat surprisingly, even in this case the qubit state can remain pure [19] throughout the measurement process provided the measurement efficiency $\eta = 1$.

7.4 Quantum Trajectories

Quantum trajectories were first introduced as a theoretical tool to study open quantum systems [20–24]. Rather than describe an open quantum system by a density matrix, which for a N-dimensional Hilbert space requires N^2 real numbers and requires solving the master equation, the evolution of a pure state (which requires only N complex numbers), can be repeatedly calculated to determine the evolution of $\rho(t)$. The quantum trajectory formalism thus assumes that the evolution of an open (and therefore mixed) quantum system can be expressed as the evolution of several, individual, pure quantum trajectories.

7.4.1 Continuous Quantum Measurement

The process of continuous quantum measurement can be built out of a sequence of discrete weak measurements as sketched in Fig. 7.5. As we have already established, each of these measurements induces a specific conditional backaction on the quantum state. If several of these weak measurements are conducted in series then the discrete time evolution of the quantum state can be determined. The cQED measurement apparatus that we consider forms a continuous probe of the quantum system, however since the cavity has a finite bandwidth κ, the measurement signal is correlated at times less than $\sim 1/\kappa$ and it makes sense to bin the measurement signal in time steps $\Delta t \sim 1/\kappa$. For a cavity with $\kappa/2\pi = 10$ MHz, this correlation time is roughly 16 ns, so the continuous measurement record is discretized in similar time steps. These discrete, weak measurements can be considered to be continuous in the limit where

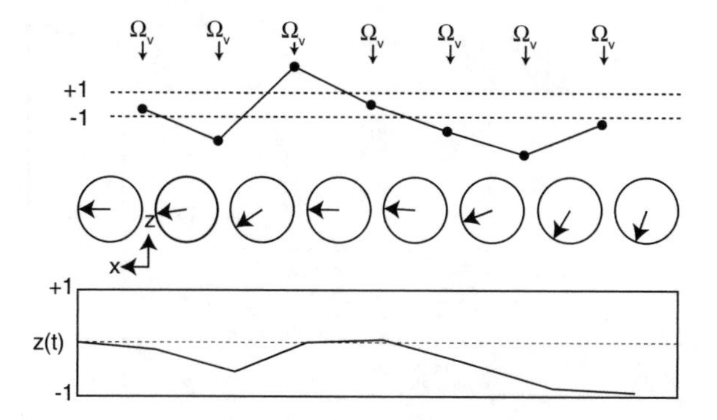

Fig. 7.5 A sequence of weak measurements leading to a quantum trajectory of the qubit state on the Bloch sphere. The qubit is initially along $x = 1$ and each measurement imparts conditional dynamics on the state

the discretization steps are much smaller than the characteristic measurement time $\tau_c = \kappa/16\pi\chi^2\bar{n}\eta$ which was defined in Sect. 7.2.2. Typically, experiments operate with $\tau_c \simeq 1\,\mu s$, such that $\Delta t \ll \tau_c$.

A single experimental iteration results in a continuous measurement signal $V(t)$ which is subsequently binned into a string of measurement results (V_i, V_{i+1}, V_{i+2}). Given the initial state of the qubit, the state is updated at each time step t_i, either based on the stochastic master equation [2, 25], or a Bayesian argument [17]. This leads to a discrete time trajectory $x(t)$, $y(t)$, $z(t)$. Figure 7.6 displays several of these trajectories for the qubit initialized in the state $x = +1$. Because each experiment results in a different measurement signal $V(t)$, each trajectory is different.

To verify that these trajectories, which are conditional on a single measurement signal, are correct, we have to prove that at every point along the trajectory the conditional state makes correct predictions for the outcome of any measurements that can be performed on the system. To accomplish this, we perform conditional quantum state tomography at discrete times along the trajectory (Fig. 7.7). We denote a single trajectory $\tilde{x}(t)$, $\tilde{z}(t)$ as a "target" trajectory. This trajectory makes predictions for the

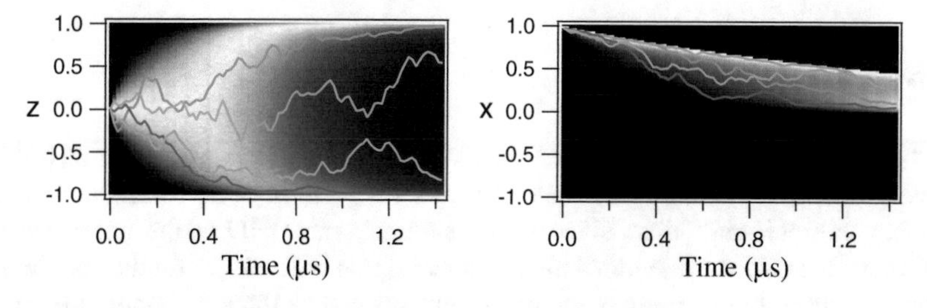

Fig. 7.6 Individual quantum trajectories of the qubit state as given by $z = \langle\sigma_z\rangle$ and $x = \langle\sigma_x\rangle$ from an initial state $+x$. Four different trajectories are shown in *color* on top of a *greyscale* histogram indicates the relative occurrence of different states at different times

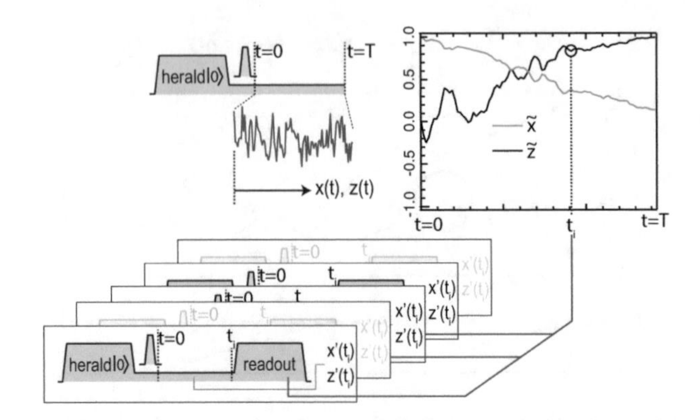

Fig. 7.7 Tomographic reconstruction of the trajectory. To reconstruct the target trajectory at time $\tilde{z}(t_i)$ several experiments are performed with a projective measurement of σ_z at time t_i. For each of these, the state $(x'(t_i), z'(t))$ is calculated and if matches the target trajectory the outcome of the projective measurement is included in the average

mean values of measurements of σ_x and σ_z versus time. To create a tomographic validation of these predictions at a specific time t_i we perform several experiments with a weak measurement duration of t_i that conclude with one of three tomographic measurements. For each of these experiments, we calculate the trajectory $x'(t_i)$, $z'(t_i)$ and if $x'(t_i) = \tilde{x}(t_i) \pm \epsilon$ and $z'(t_i) = \tilde{z}(t_i) \pm \epsilon$ the outcomes of the projective tomographic measurements are included in the average trajectory $x(t)$, $z(t)$. We find that the target trajectory and tomographic trajectory are in close agreement.

If we did not condition our trajectories on the measurement signal, for example if $\eta = 0$, then the trajectories would simply follow the ensemble evolution as described by a standard Lindblad master equation. In this case, each trajectory would be the same, yet would still make correct predictions for the outcome of projective measurements performed on the system. While this unconditioned evolution makes correct predictions, it also quickly takes an initial pure state to a mixed state. However, if $\eta \sim 1$, the conditional quantum state retains substantial purity for all time and makes correct predictions for measurements on the system.

7.4.2 Unitary Evolution

So far in this section we have considered the case of quantum non-demolition (QND) measurement, in that the weak measurements we perform $\propto \sigma_z$ commute with the qubit evolution Hamiltonian. Since the measurements are QND, all the measurements commute and only the integrated measurement signal is necessary for the conditional state evolution. In this regime, the measurements can be treated simply in terms of classical probabilities, since the evolution only depends on the state populations, and the measurement signal $\propto \sigma_z$ also depends only on the populations.

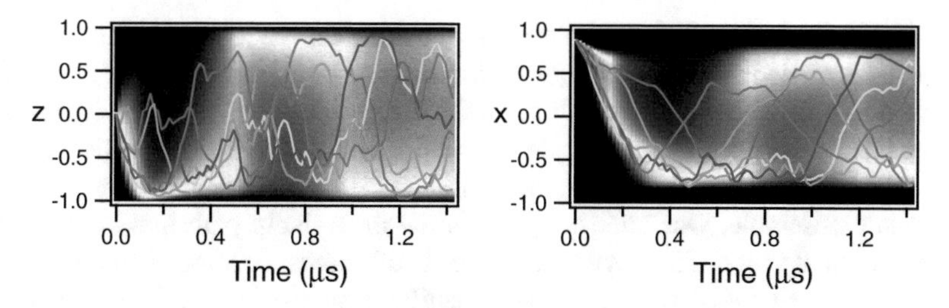

Fig. 7.8 Quantum trajectories in the presence of unitary driven evolution. Individual quantum trajectories of the qubit state as given by $z = \langle\sigma_z\rangle$ and $x = \langle\sigma_x\rangle$ from an initial state $+x$. Six different trajectories are shown in *color* on top of a *greyscale* histogram indicates the relative occurrence of different states at different times

The situation becomes much more rich if we allow for unitary evolution of the qubit state that does not commute with the measurement operators. To accomplish this, we drive the qubit resonantly to induce Rabi oscillations. This drive is described by the Hamiltonian, $H_R = \hbar\Omega\sigma_y/2$. In this case, the unitary evolution is included in the stochastic master equation,

$$\frac{d\rho}{dt} = -\frac{\mathbf{i}}{\hbar}[H_{\mathrm{R}}, \rho] + k(\sigma_z\rho\sigma_z - \rho) + 2\eta k(\sigma_z\rho + \rho\sigma_z - 2\mathrm{Tr}(\sigma_z\rho)\rho)V_t. \quad (7.16)$$

The Rabi drive turns coherences into populations and vice versa, causing the measurement signal to depend on the coherences of the qubit. Such evolution is fully quantum and reveals interesting features regarding the competition between unitary dynamics and measurement dynamics. Figure 7.8 displays several trajectories which exhibit this competition between measurement and driven evolution. The trajectories are oscillatory, but distorted by the stochastic backaction of the measurement, approaching jump-like behavior expected for the regime of quantum jumps. This figure also highlights how state tracking maintains the purity of the state in comparison to the full ensemble evolution.

7.4.3 *The Statistics of Quantum Trajectories*

We have so far demonstrated the ability to track individual quantum trajectories which evolve in response to measurement dynamics (wave function collapse) and also in competition with unitary driven evolution. These trajectories are stochastic in the sense that the evolution of each trajectory is different. But, what statements can we make about these trajectories in general? Clearly, the evolution associated with weak measurement would be in some way different than the dynamics of quantum jumps [16], but how do we quantify and characterize trajectories?

In order to characterize the general properties of trajectories we need a method to look at ensembles of trajectories but that does so in a way that depends on the individual trajectories. To accomplish this, we consider the sub-ensemble of trajectories that start and end at certain points in quantum phase space. This sub-ensemble is pre- and post-selected in that all the trajectories start from the same initial state (pre-selection) and we then post-select a sub-ensemble that ends in a specific final state. Given these pre- and post-selected trajectories we can now examine aspects of the sub-ensemble. One such property is the most-likely path that connects the initial and final states. This particular choice is of interest because such most-likely paths can be calculated with a stochastic action principle for continuous quantum measurement which maximizes the total path probability connecting quantum states [26]. Experiments [27] show good agreement with the theoretical most-likely path and the predicted path from theory, thus validating the theory which may be applicable in other quantum control problems. This analysis gives insight into the dynamics associated with open quantum systems, with applications in quantum control and state and parameter estimation.

7.4.4 Time-Symmetric State Estimation

The examination of pre- and post-selected quantum trajectories raises the notion of time symmetry in quantum evolution and quantum measurement [28–31]. We have so far used the quantum state as predictive tool, that is, at a time t the quantum state described by $\rho(t)$ makes correct predictions for the probabilities of the outcomes of measurements performed at time t and the associated mean values of these observables. However, after a measurement is performed, the quantum state may continue to evolve due to further probing and unitary driving, and we may ask at some later time $T > t$ what is the probability for the outcome of that measurement *in the past* given the results of later probing.

Consider a measurement scenario where two experimenters monitor the evolution of a qubit and track its quantum state $\rho(t)$. At time t one experimenter makes a measurement of the qubit but locks the result "in a safe". The second experimenter then continues to monitor the qubit and at a later time T the second experimenter wants to guess the outcome of the measurement whose result is locked in the safe. Clearly more information is available if the second experimenter accounts for information about the qubit obtained after the first measurement, and if this experimenter can correctly account for those results, he will be able to make more confident predictions for the result in the safe. Stated simply, the second experimenter must determine what result is most likely to be locked in the safe given the subsequent measurement signal.

One can show [32] that at time T, the second experimenter's hindsight prediction for the measurement performed in the past is given by,

$$P_p(m) = \frac{\mathrm{Tr}(\Omega_m \rho(t) \Omega_m^\dagger E(t))}{\sum_m \mathrm{Tr}(\Omega_m \rho(t) \Omega_m^\dagger E(t))}, \tag{7.17}$$

which describes the probability of obtaining outcome m from the POVM measurement Ω_m performed by the first experimenter. Here $\rho(t)$ is the usual quantum state propagated forward in time until time t and $E(t)$ is a similar matrix which is propagated backwards from time T to time t using a similar method for the calculation of $\rho(t)$. The matrix $E(t)$ has recently been calculated for experimental data and demonstrated that Eq. (7.17) makes correct and indeed more confident predictions for the outcome of measurements performed in the past [25, 33].

7.5 Analog Feedback Stabilization: Rabi Oscillations

In this section, we will explore how one can use the continuous measurement record obtained using weak measurements to modify the behavior of the quantum system being monitored with the help of feedback. Unlike feedback in classical systems, one has to worry about the random backaction associated with the measurement of a quantum system. However, as explained in previous sections, the state of a quantum system can be monitored perfectly using weak measurements if the initial state is known and the measurement efficiency is unity. Even though the evolution of the quantum system is random and unpredictable, it is still knowable by monitoring the weak measurement output. This can also be done in the presence of any additional unitary evolution as explained in Sect. 7.4.2. However, decoherence processes which are always present will tend to take the system away from a desired state one might want. We will now describe how one can use quantum feedback to prevent the quantum system from deviating from a desired state. We will consider the particular case of feedback control in a resonantly driven qubit undergoing Rabi oscillations [34].

7.5.1 *Weak Monitoring of Rabi Oscillations*

Let us first look at Rabi oscillations more carefully. The state of a resonantly driven qubit evolves sinusoidally between its two states with a rate Ω_R which depends on the strength of the resonant drive. To be specific, for a qubit state $\alpha(t)|0\rangle + \beta(t)|1\rangle$, $|\alpha(t)|^2 = \sin^2(\Omega_R t/2)$ and $|\beta(t)|^2 = \cos^2(\Omega_R t)$. These oscillations in the qubit state probability are called Rabi oscillations. In the absence of any decoherence, given the initial state and the Rabi frequency Ω_R, we can predict the qubit state at any future time. One can equivalently say that the phase of the Rabi oscillations is known and remains unchanged with time. However, decoherence processes will introduce errors in this deterministic evolution and over some characteristic time scale, the phase of the Rabi oscillations will diffuse.

In a typical Rabi oscillation experiment, the qubit is initialized in the ground state and resonantly driven for a fixed duration of time (τ_R) followed by a projective measurement. This process is repeated many times to obtain the ensemble averaged qubit state $\langle\sigma_z\rangle$. A plot of $\langle\sigma_z\rangle$ as a function of τ_R yields decaying sinusoidal oscillations where the decay constant depends on the decoherence in the system. The decaying oscillations are an indication of the diffusion of the Rabi oscillation phase with time. Here, qubit driving and projective measurement are never done simultaneously. However, one can drive the qubit while measuring it weakly and still obtain ensemble averaged Rabi oscillations. As discussed in Sect. 7.2.2, since the weak measurement output can be thought of as a noisy estimate of $\langle\sigma_z\rangle$, an ensemble average of the weak measurement signal in the presence of Rabi drive also yields decaying oscillations. Figure 7.9 shows Rabi oscillations obtained using weak measurements. However, one important difference is that the decay constant now depends on both environmental decoherence and measurement strength. This additional measurement induced decoherence is a consequence of the ensemble averaging where we ignore the individual results of the weak measurements. This is in contrast to the oscillatory quantum trajectories shown in Fig. 7.8 with simultaneous Rabi driving and weak measurement.

We will now discuss a feedback protocol [34] which corrects for the phase diffusion of Rabi oscillations and prevents the decay of Rabi oscillations. In principle, one can do a full reconstruction of the quantum state [35] to estimate the feedback signals required. However, to do that in real-time is experimentally challenging. Instead, we use classical intuition in this feedback protocol motivated by phase-locked loops (PLL) used to stabilize classical oscillators. In a PLL, one compares the phase of an oscillator with that of a reference signal. Any phase error is then corrected by creating a feedback signal proportional to the error which controls the oscillator frequency. Essentially, if the oscillator is lagging in phase, then the feedback signal increases the frequency and vice-versa.

We can now apply the same idea to our weakly monitored qubit in the presence of Rabi driving [16]. The basic feedback setup is shown in Fig. 7.10. A reference signal at the Rabi frequency is multiplied with the weak measurement signal and low pass filtered to create the error signal. Since the weak measurement signal is a noisy oscillatory signal corresponding to the oscillating qubit state, the error signal is proportional to the deviation in phase of the Rabi oscillations with respect to the reference signal. The error signal is used to control the amplitude of the Rabi drive

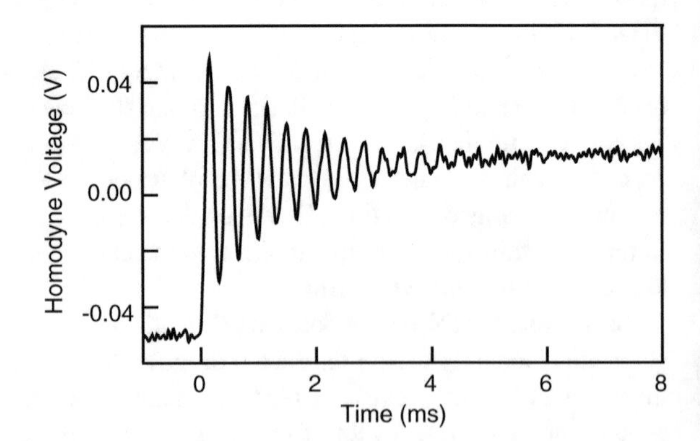

Fig. 7.9 Rabi oscillations obtained using ensemble averaged weak measurements. The decay constant is set by both environmental decoherence and measurement induced decoherence

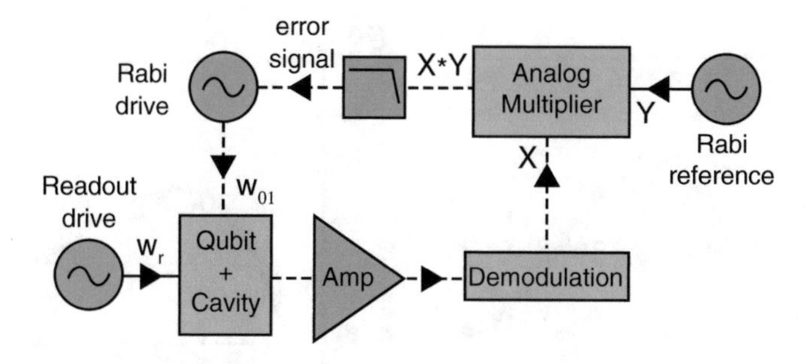

Fig. 7.10 Simplified feedback setup for stabilization of Rabi oscillations

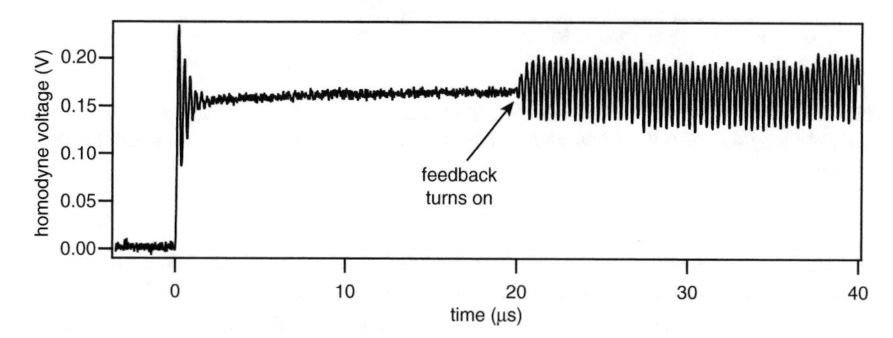

Fig. 7.11 Feedback stabilized Rabi oscillations. Initially the ensemble averaged Rabi oscillations decay, primarily due to measurement induced decoherence. Feedback is turned on later and results in the oscillations recovering as the phase of Rabi oscillations synchronize with the reference signal

which in turn controls the Rabi frequency just as in a PLL. Figure 7.11 shows the effect of such a feedback signal which is turned on after a time much greater than the decay constant of the Rabi oscillation in the absence of feedback. One can clearly see that the ensemble averaged oscillations recover when the feedback is turned on and stabilize to a fixed amplitude. As long as the feedback is on, the ensemble averaged oscillations will never decay. This implies that the Rabi oscillations have synchronized with the reference signal and the phase diffusion has been reduced due to feedback though not completely eliminated. Note that the slow drift in the mean level of the signal is due to finite probability of getting excited into the second excited state of the transmon qubit [16].

To ensure that the stabilized oscillations are not an artifact of the measurement setup, quantum state tomography was used to verify the quantum nature of the stabilized oscillations. Figure 7.12a shows a plot of $\langle\sigma_X\rangle$, $\langle\sigma_Y\rangle$, and $\langle\sigma_Z\rangle$ for one full stabilized Rabi oscillation. The magnitude of these oscillations do not reach ± 1 indicating that the synchronization is not perfect and the phase diffusion of the Rabi oscillations has not been completely eliminated. The data shown is the best synchronization we obtained in this experiment corresponding to a feedback efficiency $D = 0.45$ which is approximately given by the amplitude of the oscillations in $\langle\sigma_Z\rangle$

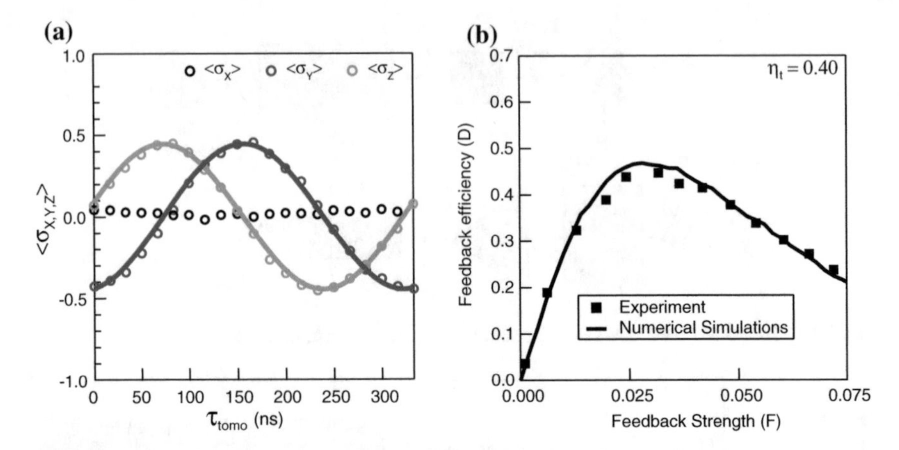

Fig. 7.12 **a** Quantum state tomography of feedback stabilized Rabi oscillations showing $\langle\sigma_X\rangle$, $\langle\sigma_Y\rangle$, and $\langle\sigma_Z\rangle$ for one full Rabi oscillation. **b** Feedback efficiency is plotted a function of feedback strength

or $\langle\sigma_X\rangle$. The best synchronization is obtained for an optimal feedback strength F as is evident from Fig. 7.12b which shows a plot of D versus F (solid squares). The dimensionless feedback strength F is essentially the feedback loop gain and is controlled by the amplitude of the reference signal.

There are two main factors in this experiment which result in $D < 1$. The first one is the total measurement efficiency η_t which had a value of 0.4 in this experiment. This efficiency has two contributions given by $\eta_t = \eta\,\eta_{\mathrm{env}}$. Here η is the measurement efficiency due to the detector as introduced in Sect. 7.2.2 and is set by the excess noise introduced by the amplification chain. The other term $\eta_{\mathrm{env}} = (1 + \Gamma_{\mathrm{env}}/\Gamma_m)^{-1}$ takes into account the environmental dephasing Γ_{env}, where Γ_m is the measurement induced dephasing. In order for feedback stabilization to work well, one needs to ensure that the environmental dephasing is small compared to the measurement induced dephasing i.e. $\eta_{\mathrm{env}} \to 1$. In other words, you want the measurement induced disturbance to dominate over the environmental disturbance since the measurement output can then be used to correct for those. The second factor affecting the feedback efficiency is the loop delay. Since the qubits are operating inside a dilution refrigerator while the feedback electronics are operating at room temperature, there is a delay in the feedback signal which results in inefficient synchronization. The solid line in Fig. 7.12b is obtained using numerical simulations including the effect of feedback delay and shows good agreement with the experimental data. While one might be tempted to increase measurement strength arbitrarily to approach $\eta_t = 1$, feedback delay and finite feedback bandwidth leads to an optimal value of measurement strength for maximizing feedback efficiency. This is because stronger measurement requires faster feedback which is limited by the bandwidth of the feedback loop.

This experiment demonstrates the use of continuous measurement and feedback to stabilize Rabi oscillations in a qubit. The simple feedback protocol which is based on classical intuition works successfully because the feedback signal achieves near perfect cancellation of the random measurement backaction for an optimal value of F. This technology can provide another route for quantum error correction based on weak continuous measurements and feedback [36, 37] with a potential advantage in situations where strong measurements can cause qubit state mixing [38]. Such techniques also offer the possibility of measurement based quantum control for solid-state quantum information processing [39–45].

7.6 Conclusion

Weak measurements realize a flexible method of implementing active feedback in quantum systems. With superconducting circuits that operate at microwave frequencies, feedback fidelity is currently limited by the overall measurement efficiency of the amplification chain. In typical setups, the amplifier is housed in a separate cryo-package and inefficiencies result from losses in cable connectors and passive components such as circulators which add directionality to the signal path. One promising avenue to overcome this limitation is to use parametric devices that can be directly integrated on-chip with the qubit. Such types of circuits use a combination of complex pumping techniques at different frequencies [46] or multiple cavity modes to isolate the amplifier bias form the qubit circuit [47]. Additionally, as more sophisticated feedback routines are developed, particularly sequences which involve digital processing of the measurement data, then loop delays resulting from long data paths and latencies in classical electronics must also be taken into account. Wiring complexity and dead time may be minimized by integrating quantum circuits with cryogenic classical logic, either superconducting or semiconducting.

Another frontier to be explored in superconducting circuits is the use of feedback and control in multi-qubit arrays. For example, in such an architecture, one can imagine simultaneously performing weak measurements on each qubit. The resulting joint-state information can potentially be used to reconstruct the initial state of the array from families of quantum trajectories, realizing another form of state tomography that capitalizes on the dense information embedded in the correlations of an analog weak measurement. Such techniques can also be extended for Hamiltonian parameter estimation, parity measurements, and measurement based error correction. In essence, simultaneous probing of a quantum many body system parallelizes both 'read' and 'write' operations.

References

1. H.M. Wiseman, G.J. Milburn, *Quantum Measurement and Control* (Cambridge University Press, Cambridge, 2010)
2. K. Jacobs, D.A. Steck, Contemp. Phys. **47**, 279 (2006)
3. T. Brun, Am. J. Phys. **70**, 719 (2002)

4. M. Nielsen, I. Chuang, *Quantum Computation and Quantum Information* (Cambridge University Press, Cambridge, 2000)
5. J.P. Groen, D. Ristè, L. Tornberg, J. Cramer, P.C. de Groot, T. Picot, G. Johansson, L. DiCarlo, Phys. Rev. Lett. **111**, 090506 (2013)
6. G.A. Smith, A. Silberfarb, I.H. Deutsch, P.S. Jessen, Phys. Rev. Lett. **97**, 180403 (2006)
7. J.P. Groen, D. Ristè, L. Tornberg, J. Cramer, P.C. de Groot, T. Picot, G. Johansson, L. DiCarlo, Phys. Rev. Lett. **111**, 090506 (2013)
8. A. Blais, R.S. Huang, A. Wallraff, S.M. Girvin, R.J. Schoelkopf, Phys. Rev. A **69**(6), 062320 (2004)
9. M.A. Castellanos-Beltran, K.D. Irwin, G.C. Hilton, L.R. Vale, K.W. Lehnert, Nat. Phys. **4**, 929 (2008)
10. M. Hatridge, R. Vijay, D.H. Slichter, J. Clarke, I. Siddiqi, Phys. Rev. B **83**, 134501 (2011)
11. P. Berman (ed.), *Cavity Quantum Electrodynamics* (Academic Press, Boston, 1994)
12. A. Wallraff, D.I. Schuster, A. Blais1, L. Frunzio, R.S. Huang, J. Majer, S. Kumar, S.M. Girvin, R.J. Schoelkopf, Nature **431**, 162 (2004)
13. H. Paik, D.I. Schuster, L.S. Bishop, G. Kirchmair, G. Catelani, A.P. Sears, B.R. Johnson, M.J. Reagor, L. Frunzio, L.I. Glazman, S.M. Girvin, M.H. Devoret, R.J. Schoelkopf, Phys. Rev. Lett. **107**, 240501 (2011)
14. N. Bergeal, F. Schackert, M. Metcalfe, R. Vijay, V.E. Manucharyan, L. Frunzio, D.E. Prober, R.J. Schoelkopf, S.M. Girvin, M.H. Devoret, Nature **465**(7294), 64 (2010)
15. R. Vijay, M.H. Devoret, I. Siddiqi, Rev. Sci. Instrum. **80**(11), 111101 (2009)
16. R. Vijay, C. Macklin, D.H. Slichter, S.J. Weber, K.W. Murch, R. Naik, A.N. Korotkov, I. Siddiqi, Nature **490**, 77 (2012)
17. K.W. Murch, S.J. Weber, C. Macklin, I. Siddiqi, Nature **502**, 211 (2013)
18. A.N. Korotkov (2011). arXiv:1111.4016
19. M. Hatridge, S. Shankar, M. Mirrahimi, F. Schackert, K. Geerlings, T. Brecht, K.M. Sliwa, B. Abdo, L. Frunzio, S.M. Girvin, R.J. Schoelkopf, M.H. Devoret, Science **339**(6116), 178 (2013)
20. H. Carmichael, *An Open Systems Approach to Quantum Optics* (Springer, Berlin, 1993)
21. C. Gardiner, P. Zoller, *Quantum Noise* (Springer, Berlin, 2004)
22. J. Dalibard, Y. Castin, K. Mølmer, Wave-function approach to dissipative processes in quantum optics. Phys. Rev. Lett. **68**(5), 580–583 (1992)
23. C. Gardiner, A. Parkins, P. Zoller, Wave-function quantum stochastic differential equations and quantum-jump simulation methods. Phys. Rev. A **46**(7), 4363–4381 (1992)
24. R. Schack, T.A. Brun, I.C. Percival, J. Phys. A **28**, 5401 (1995)
25. D. Tan, J. Weber, S.I. Siddiqi, K. Mølmer, W. Murch, K. Phys, Rev. Lett. **114**, 090403 (2015)
26. A. Chantasri, J. Dressel, A.N. Jordan, Phys. Rev. A **88**, 042110 (2013)
27. S.J. Weber, A. Chantasri, J. Dressel, A.N. Jordan, K.W. Murch, I. Siddiqi, Nature **511**, 570–573 (2014)
28. S. Watanabe, Rev. Mod. Phys. **27**, 179 (1955)
29. Y. Aharonov, P.G. Bergmann, J.L. Lebowitz, Phys. Rev. **134**, B1410 (1964)
30. Y. Aharonov, S. Popescu, J. Tollaksen, Phys. Today **63**, 27 (2010)
31. Y. Aharonov, S. Popescu, J. Tollaksen, Phys. Today **64**, 62 (2011)
32. S. Gammelmark, B. Julsgaard, K. Mølmer, Phys. Rev. Lett. **111**, 160401 (2013)
33. T. Rybarczyk, S. Gerlich, B. Peaudecerf, M. Penasa, B. Julsgaard, K.M. lmer, S. Gleyzes, M. Brune, J.M. Raimond, S. Haroche, I. Dotsenko (2014) arXiv:1409.0958
34. R. Ruskov, A.N. Korotkov, Phys. Rev. B **66**, 041401 (2002)
35. C. Sayrin, I. Dotsenko, X. Zhou, B. Peaudecerf, T. Rybarczyk, G. Sebastien, P. Rouchon, M. Mirrahimi, H. Amini, M. Brune, J. Raimond, S. Haroche, Nature **477**, 73 (2011)
36. C. Ahn, A.C. Doherty, A.J. Landahl, Phys. Rev. A **65**(4, Part A), 042301 (2002)
37. L. Tornberg, G. Johansson, Phys. Rev. A **82**, 012329 (2010)
38. D.H. Slichter, R. Vijay, S.J. Weber, S. Boutin, M. Boissonneault, J.M. Gambetta, A. Blais, I. Siddiqi, Phys. Rev. Lett. **109**, 153601 (2012)
39. H.F. Hofmann, G. Mahler, O. Hess, Phys. Rev. A **57**, 4877 (1998)
40. J. Wang, H.M. Wiseman, Phys. Rev. A **64**, 063810 (2001)

41. G.G. Gillett, R.B. Dalton, B.P. Lanyon, M.P. Almeida, M. Barbieri, G.J. Pryde, J.L. O'Brien, K.J. Resch, S.D. Bartlett, A.G. White, Phys. Rev. Lett. **104**, 080503 (2010)
42. R. Ruskov, A.N. Korotkov, Phys. Rev. B **67**, 241305 (2003)
43. J. Combes, K. Jacobs, Phys. Rev. Lett. **96**, 010504 (2006)
44. K. Jacobs, Quantum Inf. Comput. **7**(1), 127 (2007)
45. R.L. Cook, P.J. Martin, J.M. Geremia, Nature **446**, 774 (2007)
46. L. Ranzani, J. Aumentado, New J. Phys. **17**(2), 023024 (2015)
47. S. Khan, R. Vijay, I. Siddiqi, A.A. Clerk, New J. Phys. **16**(11), 113032 (2014)

Chapter 8
Digital Feedback Control

Diego Ristè and Leonardo DiCarlo

This chapter covers the development of feedback control of superconducting qubits using projective measurement and a discrete set of conditional actions, here referred to as digital feedback. We begin with an overview of the applications of digital feedback in quantum computing. We then introduce an implementation of high-fidelity projective measurement of superconducting qubits. This development lays the ground for closed-loop control based on the binary measurement result. A first application of digital feedback control is fast and deterministic qubit reset, allowing the repeated initialization of a qubit more than an order of magnitude faster than its relaxation rate. A second application employs feedback in a multi-qubit setting to convert the generation of entanglement by parity measurement from probabilistic to deterministic, targeting an entangled state with the desired parity every time.

Part of this chapter appeared in Refs. [1–3].

D. Ristè
Raytheon BBN Technologies, 10 Moulton Street, Cambridge, MA 02138, USA
e-mail: driste@bbn.com

L. DiCarlo (✉)
QuTech Advanced Research Center and Kavli Institute of Nanoscience, Delft University of Technology, Lorentzweg 1, 2628 CJ Delft, The Netherlands
e-mail: l.dicarlo@tudelft.nl

© Springer International Publishing Switzerland 2016

R.H. Hadfield and G. Johansson (eds.), *Superconducting Devices in Quantum Optics*, Quantum Science and Technology,
DOI 10.1007/978-3-319-24091-6_8

8.1 Digital Feedback Control in Quantum Computing

Moving from proof-of-principle demonstrations of quantum gates and algorithms to fully fledged quantum hardware requires closing the loop between qubit measurement and control. There are different categories of quantum feedback control, depending on the type of measurement and feedback law used. For clarity, we first offer a classification of quantum feedback, similarly to that used in classical feedback. Then, we focus on the particular class of discrete-time, digital feedback.

8.1.1 Classification of Quantum Feedback

A first distinction is between continuous-time and discrete-time feedback. In the first case, measurement and control are continuous in time and concurrent. An example is the stabilization of a qubit state using continuous partial measurement, as discussed in Chap. 8 of this volume as well as in Refs. [4–8]. In discrete-time feedback, instead, the conditional control is applied only after a measurement has been performed and processed. Here, we focus exclusively on discrete-time implementations. This class can be further divided into two categories, analog and digital. We speak of analog feedback when the measurement result assumes a continuum of values and the feedback law is a continuous function of the result. An example is the experiment in Ref. [9], where the feedback controller first integrates the signal produced by a weak measurement and then applies the resulting coherent operation on the qubit. If the measurement has a finite set of possible results, instead, the possible feedback actions are also finite. We refer to this as digital feedback. The simplest example is qubit reset (Sect. 8.4.2), in which a strong projective measurement collapses the qubit into either the ground or excited state. Here, a π rotation brings the qubit to ground. Another interesting example is digital feedback using ancilla-based partial measurement [10, 11]. In this case, the measurement output is discrete, showing that partial measurement is not necessarily associated with analog feedback. In many applications, digital feedback forces determinism into one of the most controversial aspects of quantum mechanics, namely the measurement, whose result is intrinsically probabilistic. Looking at the action of digital feedback as a black box, we expect to see a definite output qubit state for a given input. In an ideal feedback scheme, measurement results and the conditioned operations vary at every run of the protocol, but the overall process is deterministic and the output state is always the same. For example, one can project a two-qubit superposition to a specific Bell state by combining a parity measurement with digital feedback (Sect. 8.5).

8.1.2 Protocols Using Digital Feedback

Several quantum information processing (QIP) protocols call for digital feedback. One of the requirements for a quantum computer is efficient qubit initialization [12]. Often, the steady state of a qubit does not correspond to a pure computational state $|0\rangle$ or $|1\rangle$, bur rather to a mixture of the two. Therefore, active initialization methods have been used in many QIP architectures. Examples are laser or microwave initialization [13–16] and initialization by control of the qubit relaxation rate [17, 18]. An alternative method, recently used with NV centers in diamond [19] and superconducting qubits (Sect. 8.2), relies on projective measurement to initialize the qubits into a pure state. However, measurement alone cannot produce the desired state with certainty, since the measurement result is probabilistic. Closing a feedback loop based on this measurement turns the unwanted outcomes into the desired state. A qubit register must not only be initialized in a pure state at the beginning of computation, but often also during the computation. For example, performing multiple rounds of error correction is facilitated by resetting ancilla qubits to their ground state after each parity check [20]. When using a qubit as a detector (e.g. of charge [21] or photon parity [22]), a fast reset can be used to increase the sampling rate without keeping track of past measurement outcomes.

Similarly, in the multi-qubit setting, digital feedback is key to turning measurement-based protocols from probabilistic to deterministic. An example is the generation of entanglement by parity measurement [23]. A parity measurement projects an initial maximal superposition state into an entangled state with a well-defined parity, i.e., with either even or odd total number of qubit excitations (Sect. 8.5). However, once again, the outcome of the parity measurement is random. When running the protocol open-loop multiple times, the average final state has no specific parity and is unentangled. Only by forcing a definite parity using feedback can one generate a target entangled state deterministically.

A variation of closed-loop control, named feedforward, applies control on qubits different from those measured. Feedforward schemes have already found application in quantum communication, where the main objective is the secure transmission of quantum information over a distance. In quantum teleportation, a measurement on the Bell basis of two qubits projects a third qubit, at any distance, into the state of the first, to within a single-qubit rotation [24]. The measurement result determines which qubit rotation, if any, must be applied to teleport the original state. An extension of teleportation is entanglement swapping [24]. This protocol transfers entanglement to two qubits which never interact, and forms the basis for quantum repeaters [25], aiming to distribute entanglement across larger distances than allowed by a lossy communication channel. Here, measurement and feedback are used in every step to first purify [26] and then deterministically transfer entangled pairs to progressively farther nodes.

In quantum computing, feedforward operations are at the basis of the first schemes devised to protect a qubit state from errors. The simplest protocol is the bit-flip code [27], which encodes the quantum state of one qubit into a an entangled state of three, and uses measurement of two-qubit operators (syndromes) in combination with feedback to correct for σ_x (bit-flip) errors. Of similar structure is the phase-flip code, which protects against σ_z (phase-flip) errors. To protect against errors on any axis, the minimum size of the encoding is five qubits. In quantum error correction, projective measurement is more than a tool to detect discrete errors that have already occurred. In fact, the measurement serves to discretize the set of possible errors. Measuring the error syndromes forces one and only one of these errors to happen. This greatly simplifies the feedback step, which is now restricted to a finite set of correcting actions.

While few-qubit error correction schemes are capable of correcting any single error, they require currently inaccessible measurement and gate fidelities. A more realistic approach is offered by topologically protected circuits such as surface codes [28], where errors as high as 1 % are tolerated at the expense of a larger number of physical qubits required [29]. One cycle in a surface code, aimed at maintaining a logical state encoded in a square lattice of qubits, includes the projective measurements of 4-qubit operators as error syndromes. When an error is detected on a data qubit, the corrective, coherent feedback operation is replaced by a change of sign in the operators for the following syndrome measurements involving that qubit. In other words, errors are kept track of by the classical controller rather than fixed [30, 31]. Beyond protecting a state from external perturbations, performing fault-tolerant quantum computing will require robustness to gate errors. In surface codes, single- and two-qubit gates on logical qubits are also based on projective measurements and in some cases require digital feedback to apply conditional rotations [32].

In addition to the gate model [12], digital feedback is central to the paradigm of measurement-based quantum computing [33]. In this approach, also called one-way computation, the initial state is an entangled state of a large number of qubits. All logical operations are performed by projective measurements. To make computation deterministic, feedback selects the measurement bases at each computational step, conditional on the measurement results.

8.1.3 Experimental Realizations of Digital Feedback

Experimentally, digital feedback has been employed for entanglement swapping with trapped ions [34] and for the unconditional teleportation of photonic [35], ionic [36, 37], and atomic [38, 39] qubits. In linear optics, feedforward has been used to implement segments of one-way quantum computing [40–45] and for photon multiplexing [46]. In the solid state, the first approach to feedback, of the analog type, was used to stabilize Rabi oscillations of a superconducting qubit [47]. Soon after, digital feedback with high-fidelity projective measurement was introduced in the solid state, also using superconducting circuits [2, 48]. Recently, digital feedback has been extended to multi-qubit protocols with superconducting qubits (Sect. 8.5 and Ref. [49]) and NV centers in diamond [50].

8.1.4 Concepts in Digital Feedback

The basic ingredients for a digital feedback loop are: (1) projective qubit readout and (2) control conditional on the measurement result (see Fig. 8.1a for the simplest single-qubit loop). The main challenge for (1) is to obtain a high-fidelity readout which is also nondemolition, thus leaving the qubits in a state consistent with the measurement result. A mismatch between measurement result and post-measurement qubit state will trigger the wrong feedback action (Fig. 8.1b). The requirement for (2) is to minimize the time, or latency, between measurement and conditional action. Various sources contribute to latency: the time for the signal to travel from the sample to the feedback controller, the time for the feedback controller to process the signal and discretize it, and the delay to the execution of the conditional qubit gates. If a transition between levels occurs in one of the measured qubits during this interval, for instance because of spontaneous relaxation, its state becomes inconsistent with the chosen feedback action, resulting in the wrong final state (Fig. 8.1c). In feedforward protocols, such as error correction or teleportation, the feedback action is applied to data qubits, which are different from the measured ancilla qubits. In this case, the loop must also be fast compared to the coherence times of data qubits.

The simplest example of digital feedback is single-qubit reset. Here, the qubit is projected by measurement onto $|0\rangle$ or $|1\rangle$ and, depending on the targeted state, a π pulse is applied conditional on the measurement result. In this example, we consider the effect of the errors in Fig. 8.1b, c, modeling the qubit as a classical three-level system, where the third level includes the possibility of transitions out of the qubit subspace. This is relevant in the case of transmon qubits with a sizeable steady-state excitation [2, 48]. We indicate with p_{ij}^M the probability of obtaining the measurement result M with initial state $|i\rangle$ and post-measurement state $|j\rangle$. With Γ_{ij} we indicate the transition rates from $|i\rangle$ to $|j\rangle$, and with τ_{Fb} the time between the end of measurement and the end of the conditional operation. For perfect pulses, the combined errors P_{err}^{θ} for initial state $\cos(\theta)|0\rangle + \sin(\theta)|1\rangle$ are, to first order:

$$
\begin{aligned}
P_{\text{err}}^{\theta=0} &= p_{00}^{L} + p_{01}^{H} + \Gamma_{01}\tau_{\text{Fb}}, \\
P_{\text{err}}^{\theta=\pi} &= p_{11}^{H} + p_{10}^{L} + p_{12} + (\Gamma_{10} + \Gamma_{12})\tau_{\text{Fb}},
\end{aligned}
\tag{8.1}
$$

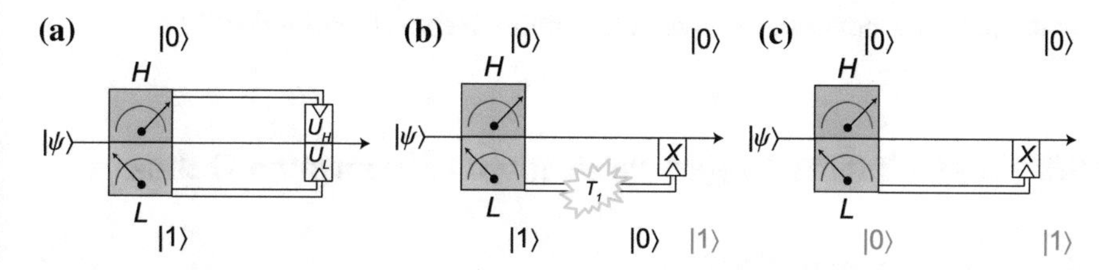

Fig. 8.1 Concept of a single-qubit digital feedback loop and possible errors. a The measurement is digitized into either H or L for qubit declared in $|0\rangle$ or $|1\rangle$. A different unitary rotation is applied for each result. Errors occurring in case of qubit relaxation between measurement and action (**b**) or wrong measurement assignment (**c**). *Top* (*bottom*) *row* indicates the actual qubit state corresponding to result H (L)

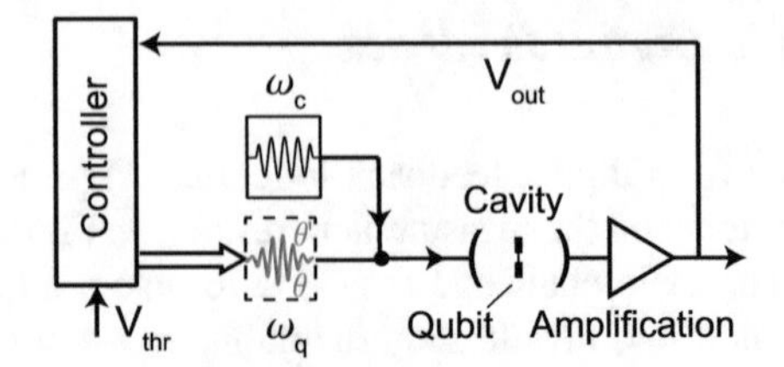

Fig. 8.2 Simplified schematic of a single-qubit feedback loop in cQED. Upon application of a measurement tone at ω_C, the signal V_{out} obtained from processing of the cavity output, carrying information on the qubit state, is input to the feedback controller and compared to a preset threshold V_{th}. If $V_{out} > V_{th}$ (or $V_{th} < V_{th}$), the conditional rotation θ (θ') is applied to the qubit

and weighted combinations thereof for other θ. A simple way to improve feedback fidelity is to perform two cycles back to back. While the dominant error for $\theta = 0$ remains unchanged, for $\theta = \pi$ it decreases to $P_{err}^{\theta=0} + p_{12} + \Gamma_{12}\tau_{Fb}$. The second cycle compensates errors arising from relaxation to $|0\rangle$ between measurement and pulse in the first cycle. However, it does not correct for excitation from $|1\rangle$ to $|2\rangle$. For this reason, adding more cycles does not significantly reduce the error, unless the population in $|2\rangle$ is brought back to the qubit subspace. This can be done [2, 48] by a deterministic π pulse returning the population from $|2\rangle$ to $|1\rangle$, or with a more complex feedback loop capable of resolving and manipulating all three states.

8.1.5 Closing the Loop in cQED

Until recently, the coherence times of superconducting qubits bottlenecked both achievable readout fidelity and required feedback speed. The development of circuit quantum electrodynamics [51, 52] with 3D cavities (3D cQED) [53] constitutes a watershed. The new order of magnitude in qubit coherence times ($>10\,\mu$s), combined with Josephson parametric amplification [54, 55], allows projective readout with fidelities ~99 % and realizing feedback control with off-the-shelf electronics. In the following section, we detail our implementation of high-fidelity projective readout of a transmon qubit in 3D cQED. We then shift focus to the real-time signal processing by the feedback controller, and on the resulting feedback action (Fig. 8.2).

8.2 High-Fidelity Projective Readout of Transmon Qubits

8.2.1 Experimental Setup

Our system consists of an Al 3D cavity enclosing two superconducting transmon qubits, labeled Q_A and Q_B, with transition frequencies $\omega_{A(B)}/2\pi = 5.606(5.327)$ GHz, relaxation times $T_{1A(B)} = 23$ (27) μs. The fundamental mode of the cavity

(TE101) resonates at $\omega_r/2\pi = 6.548$ GHz (for qubits in ground state) with $\kappa/2\pi =$ 430 kHz linewidth, and couples with $g/2\pi \sim 75$ MHz to both qubits. The dispersive shifts [52] $\chi_{A(B)}/\pi = -3.7\ (-2.6)$ MHz, both large compared to $\kappa/2\pi$, place the system in the strong dispersive regime of cQED [56].

Qubit readout in cQED typically exploits dispersive interaction with the cavity. A readout pulse is applied at or near resonance with the cavity, and a coherent state builds up in the cavity with amplitude and phase encoding the multi-qubit state [52, 57]. We optimize readout of Q_A by injecting a microwave pulse through the cavity at $\omega_{RF} = \omega_r - \chi_A$, the average of the resonance frequencies corresponding to qubits in $|00\rangle$ and $|01\rangle$, with left (right) index denoting the state of Q_B (Q_A) (Fig. 8.3a, d). This choice maximizes the phase difference between the pointer coherent states. Homodyne detection of the output signal, itself proportional to the intra-cavity state, is overwhelmed by the noise added by the semiconductor amplifier (HEMT), precluding high-fidelity single-shot readout (Fig. 8.3c). We introduce a Josephson parametric amplifier (JPA) [54] at the front end of the amplification chain to boost the readout signal by exploiting the power-dependent phase of reflection at the JPA (see Fig. 8.3a, b). Depending on the qubit state, the weak signal transmitted through the cavity is either added to or subtracted from a much stronger pump tone incident on the JPA, allowing single-shot discrimination between the two cases (Fig. 8.3c).

8.2.2 *Characterization of JPA-Backed Qubit Readout and Initialization*

The ability to better discern the qubit states with the JPA-backed readout is quantified by collecting statistics of single-shot measurements. The sequence used to benchmark the readout includes two measurement pulses, M_0 and M_1, each 700 ns long, with a central integration window of 300 ns (Fig. 8.4a). Immediately before M_1, a π pulse is applied to Q_A in half of the cases, inverting the population of ground and excited state (Fig. 8.4b). We observe a dominant peak for each prepared state, accompanied by a smaller one overlapping with the main peak of the other case. We hypothesize that the main peak centered at positive voltage corresponds to state $|00\rangle$, and that the smaller peaks are due to residual qubit excitations, mixing the two distributions. To test this hypothesis, we first digitize the result of M_0 with a threshold voltage V_{th}, chosen to maximize the contrast between the cumulative histograms for the two prepared states (Fig. 8.4c), and assign the value $H(L)$ to the shots falling above (below) the threshold. Then we only keep the results of M_1 corresponding to $M_0 = H$. Indeed, we observe that postselecting 91 % of the shots reduces the overlaps from ~6 to 2 % and from ~9 to 1 % in the H and L regions, respectively (Fig. 8.4d). This supports the hypothesis of partial qubit excitation in the steady state, lifted by restricting to a subset of measurements where M_0 declares the register to be in $|00\rangle$. Further evidence is obtained by observing that moving the threshold substantially decreases the fraction of postselected measurements without signif-

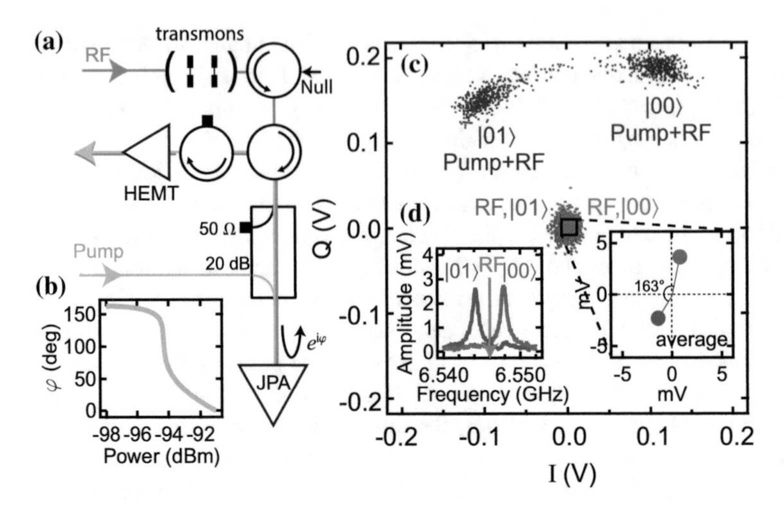

Fig. 8.3 **JPA-backed dispersive transmon readout**. **a** Simplified diagram of the experimental setup, showing the input path for the readout signal carrying the information on the qubit state (RF, *green*) and the stronger, degenerate tone (Pump, *grey*) biasing the JPA. Both microwave tones are combined at the JPA and their sum is reflected with a phase dependent on the total power (**b**), amplifying the small signal. An additional tone (Null) is used to cancel any pump leakage into the cavity. The JPA is operated at the low-signal gain of ∼25 dB and 2 MHz Bandwidth (of coupling to superconducting qubit). **c** Scatter plot in the $I - Q$ plane for sets of 500 single-shot measurements. *Light red* and *blue*: readout signal obtained with an RF tone probing the cavity for qubits in $|00\rangle$ and $|01\rangle$, respectively. *Dark red* and *blue*: the Pump tone is added to the RF. **d** Spectroscopy of the cavity fundamental mode for qubits in $|00\rangle$ and $|01\rangle$. The RF frequency is chosen halfway between the two resonance peaks, giving the maximum phase contrast (163°, see *inset* on the *right*). Reproduced with permission of the American Physical Society from Ref. [1]

icantly improving the contrast [$\sim + 0.1$ (0.2) % keeping 85 (13) % of the shots] (Fig. 8.5b). Postselection is effective at suppressing the residual excitation of any one qubit, since the $|01\rangle$ and $|10\rangle$ distributions are both highly separated from $|00\rangle$, and the probability that both qubits are excited is only ∼0.2 %.

The performance of JPA-backed readout and the effect of initialization by measurement are quantified by the optimum readout contrast, defined as the maximum difference between the cumulative probabilities for the two prepared states (Fig. 8.5a). Without initialization, the use of the JPA gives an optimum contrast of 84.9 %, a significant improvement over the 26 % obtained without the pump tone. Comparing the deviations from unity contrast without and with initialization, we can extract the parameters for the error model shown in Fig. 8.5c. The model [1] takes into account the residual steady-state excitation of both qubits, found to be ∼4.7 % each, and the error probabilities for the qubits prepared in the four basis states. Although the projection into $|00\rangle$ occurs with 99.8 ± 0.1 % fidelity, this probability is reduced to 98.8 % in the time $\tau = 2.4\,\mu$s between M_0 and M_1, chosen to fully deplete the cavity of photons before the π pulse preceding M_1. We note that τ could be reduced by increasing κ by at least a factor of two without compromising T_{1A} by the Purcell

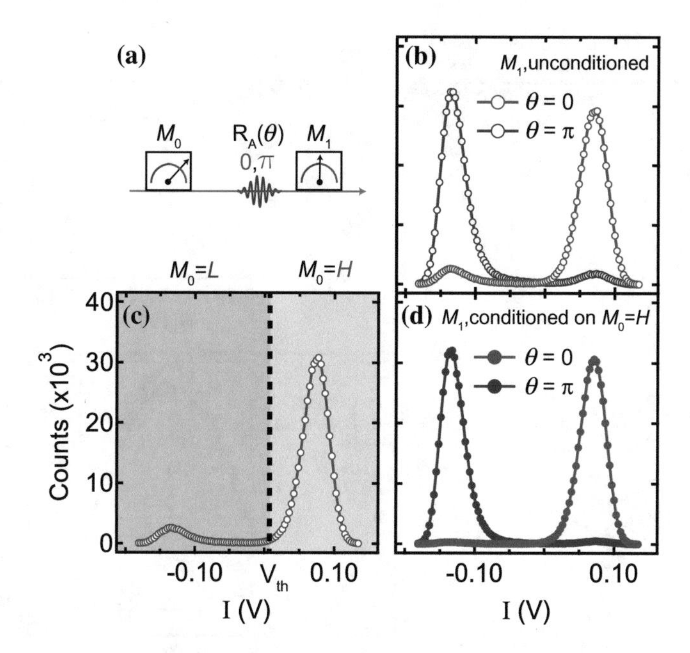

Fig. 8.4 Ground-state initialization by measurement. **a** Pulse sequence used to distinguish between the qubit states (M_1), upon conditioning on the result of an initialization measurement M_0. The sequence is repeated every 250 μs. **b** Histograms of 500 000 shots of M_1, without (*red*) and with (*blue*) inverting the population of Q_A with a π pulse. **c** Histograms of M_0, with V_{th} indicating the threshold voltage used to digitize the result. **d** M_1 conditioned on $M_0 = H$ to initialize the system in the ground state, suppressing the residual steady-state excitation. The conditioning threshold, selecting 91 % of the shots, matches the value for optimum discrimination of the state of Q_A. Reproduced with permission of the American Physical Society from Ref. [1]

effect [58]. By correcting for partial equilibration during τ, we calculate an actual readout fidelity of 98.1 ± 0.3 %. The remaining infidelity is mainly attributed to qubit relaxation during the integration window.

As a test for readout fidelity, we performed single-shot measurements of a Rabi oscillation sequence applied to Q_A, with variable amplitude of a resonant 32 ns Gaussian pulse preceding M_1, and using ground-state initialization as described above (Fig. 8.5d). The density of discrete dots reflects the probability of measuring H or L depending on the prepared state. By averaging over ~10 000 shots, we recover the sinusoidal Rabi oscillations without (white) and with (black) ground-state initialization. As expected, the peak-to-peak amplitudes (85.2 and 96.7 %, respectively) equal the optimum readout contrasts in Fig. 8.5a, within statistical error.

8.2.3 Repeated Quantum Nondemolition Measurements

In an ideal projective measurement, there is a one-to-one relation between the outcome and the post-measurement state. We perform repeated measurements to assess

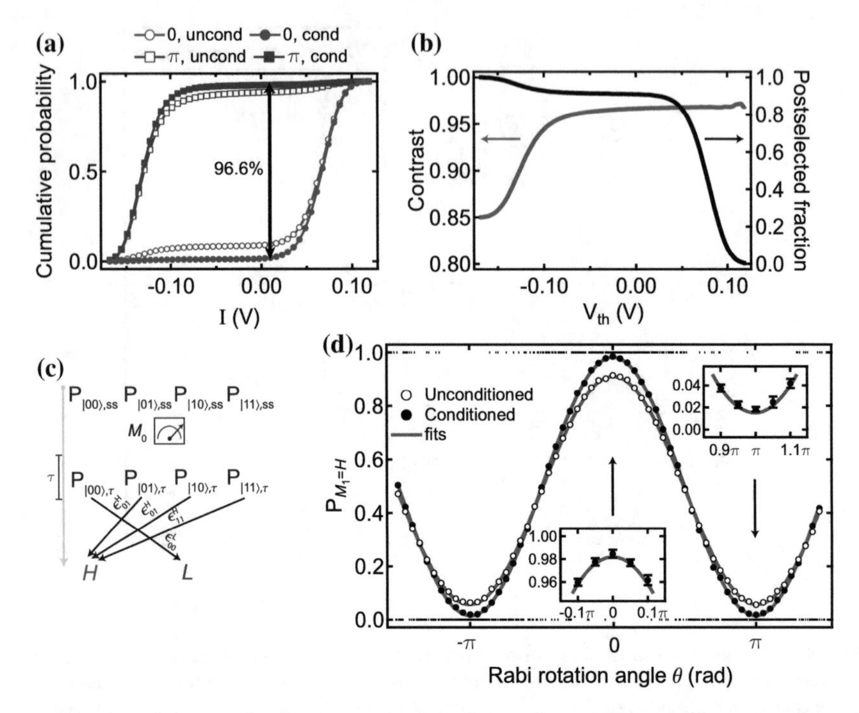

Fig. 8.5 Analysis of readout fidelity. **a** Cumulative histograms for M_1 without and with conditioning on $M_0 = H$, obtained from data in Fig. 8.4c, d. The optimum threshold maximizing the contrast between the two prepared states is the same in both cases. Deviations of the outcome from the intended prepared state are: 8.9 % (1.3 %) for the ground state, 6.2 % (2.1 %) for the excited state without (with) conditioning. Therefore, initialization by measurement and postselection increases the readout contrast from 84.9 to 96.6 %. **b** Readout contrast (*purple*) and postselected fraction (*black*) as a function of V_{th}. **c** Schematics of the readout error model, including the qubit populations in the steady state and at $\tau = 2.4\,\mu\text{s}$ after M_0. Only the arrows corresponding to readout errors are shown. **d** Rabi oscillations of Q_A without (*empty*) and with (*full dots*) initialization by measurement and postselection. In each case, data are taken by first digitizing 10 000 single shots of M_1 into H or L, then averaging the results. Error bars on the average values are estimated from a subset of 175 measurements per point. For each angle, 7 randomly-chosen single-shot outcomes are also plotted (*black dots* at 0 or 1). The visibility of the averaged signal increases upon conditioning M_1 on $M_0 = H$. Figure adapted from Ref. [1]

the nondemolition nature of the readout, following Refs. [59, 60]. The correlation between two consecutive measurements, M_1 and M_2, is found to be independent of the initial state over a large range of Rabi rotation angles θ (see Fig. 8.6a). A decrease in the probabilities occurs when the chance to obtain a certain outcome on M_1 is low (for instance to measure $M_1 = H$ for a state close to $|01\rangle$) and comparable to readout errors or to the partial recovery arising between M_1 and M_2. We extend the readout model of Fig. 8.5c to include the correlations between each outcome on M_1 and the post-measurement state. The deviation of the asymptotic levels from unity, $P_{H|H} = 0.99$ and $P_{L|L} = 0.89$, is largely due to recovery during τ, as demon-

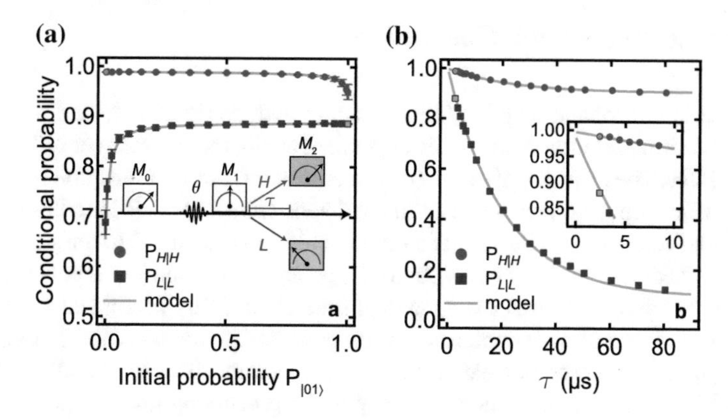

Fig. 8.6 Projectiveness of the measurement. a Conditional probabilities for two consecutive measurements M_1 and M_2, separated by $\tau = 2.4\,\mu\text{s}$. Following an initial measurement pulse M_0 used for initialization into $|00\rangle$ by the method described, a Rabi pulse with variable amplitude rotates Q_A by an angle θ along the x-axis of the Bloch sphere, preparing a state with $P_{|01\rangle} = \sin^2(\theta/2)$. *Red* (*blue*): probability to measure $M_2 = H(L)$ conditioned on having obtained the same result in M_1, as a function of the initial excitation of Q_A. Error bars are the standard error obtained from 40 repetitions of the experiment, each one having a minimum of 250 postselected shots per point. Deviations from an ideal projective measurement are due to the finite readout fidelity, and to partial recovery after M_1. The latter effect is shown in (**b**), where the conditional probabilities converge to the unconditioned values, $P_H = 0.91$ and $P_L = 0.09$ for $\tau \gg T_1$, in agreement with Fig. 8.4, taking into account relaxation between the π pulse and M_2. Error bars are smaller than the dot size. Reproduced with permission of the American Physical Society from Ref. [1]

strated in Fig. 8.6b. From the model, we extrapolate the correlations for two adjacent measurements, $P_{H|H}(\tau = 0) = 0.996 \pm 0.001$ and $P_{L|L}(\tau = 0) = 0.985 \pm 0.002$, corresponding to the probabilities that pre- and post-measurement state coincide. In the latter case, mismatches between the two outcomes are mainly due to qubit relaxation during M_2. Multiple measurement pulses, as well as a long pulse, do not have a significant effect on the qubit state, supporting the nondemolition character of the readout at the chosen power.

Josephson parametric amplification has become a standard technique for the high-fidelity readout of qubits in cQED. Since this experiment and the parallel work in Ref. [61], projective readout of transmon [49, 62, 63] and flux [64] qubits has been performed using different varieties of Josephson junction-based amplifiers. The technology for these amplifiers continuously evolves to meet the needs of quantum circuits of growing complexity. One approach to high-fidelity readout of multiple qubits is to increase the amplifier bandwidth to include several resonators, each coupled to a distinct qubit [11]. Recent implementations in this direction included Josephson junctions in a transmission line [65], in low-Q resonators [66, 67], or in a circuit realizing a superconducting low-inductance undulatory galvanometer (SLUG) [68]. Another approach for multi-qubit readout uses dedicated, on-chip Josephson bifurcation amplifiers [69].

8.3 Digital Feedback Controllers

The input to a feedback loop in cQED is the homodyne signal obtained by amplification and demodulation of the qubit-dependent cavity transmission or reflection, as shown above. The response of the feedback controller is one or more qubit microwave pulses, which are generated and sent to the device (Fig. 8.2). This loop has a significant spatial extension, as the qubits sit in the coldest stage of a dilution refrigerator, while the feedback controller is at room temperature. A round trip involves 5–10 m of cable, which translates to a propagation time of 25–50 ns without accounting for delays due to filters and other microwave components. This physical limitation, which would require fast cryogenic electronics to be overcome, is only a small fraction of the total latency. A major source of delay is the processing time in the controller, combined with the generation or triggering of the microwave pulses for the conditional qubit rotations. The details of this process depend on the type of controller. We describe the first implementations below.

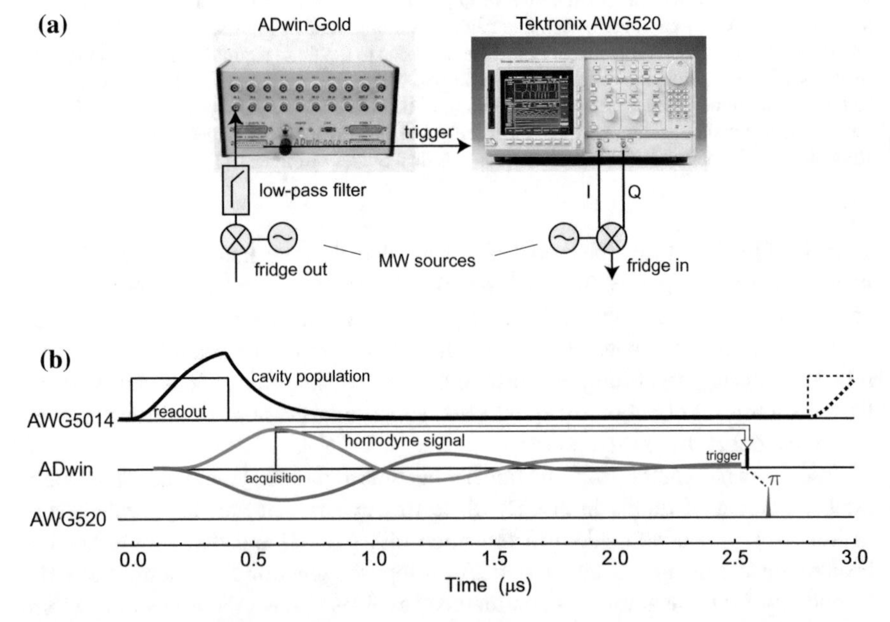

Fig. 8.7 Digital feedback loop with an ADwin controller. **a** Schematic of the feedback loop, consisting of an ADwin, sampling the signal, and a Textronix AWG520, conditionally generating a qubit π pulse. **b** Timings of the feedback loop. The measurement pulse, here 400 ns long, reaches the cavity at $t = 0$. The ADwin, triggered by an AWG5014, measures one channel of the output homodyne signal (*red*: qubit in $|0\rangle$, *blue*: $|1\rangle$), delayed by $\sim$200 ns due to a low-pass filter at its input side. After comparison of the measured voltage at $t = 0.6\,\mu$s to the reference threshold, the AWG520 is conditionally triggered at $t = 2.54\,\mu$s, resulting in a π pulse reaching the cavity at 2.62 μs. Figure adapted from Ref. [2]

The first realization of a digital feedback controller used commercial components for data sampling, processing, and conditional operations [3]. The core of the controller is an ADwin-Gold, a processor with a set of analog inputs and configurable analog and digital outputs. The ADwin samples the readout signal once, at a set delay following a trigger from an arbitrary waveform generator (Tektronix AWG5014). This delay is optimized to maximize readout fidelity. A routine determines the optimum threshold for digitizing the readout signal. This voltage is then used to assign H or L to the measurement. For the reset function in Sect. 8.4.2, the ADwin triggers another arbitrary waveform generator (Tektronix AWG520) to produce a π pulse when the outcome is L. Pulse timings and signal delays in the feedback cycle are illustrated in Fig. 8.7. The total time between start of the measurement and end of the feedback pulse is $\approx$2.6 μs, mainly limited by the processing time of the ADwin.

To shorten the loop time, our second generation of digital feedback used a complex programmable logic device (CPLD, Altera MAX V), acquiring the signal following a 8-bit ADC, in place of the ADwin. This home-assembled feedback controller offers two advantages over the first: a programmable integration window and a response time of 0.11 μs (Fig. 8.8), an order of magnitude faster than the ADwin. As the feedback response time is now comparable or faster than the typical cavity decay

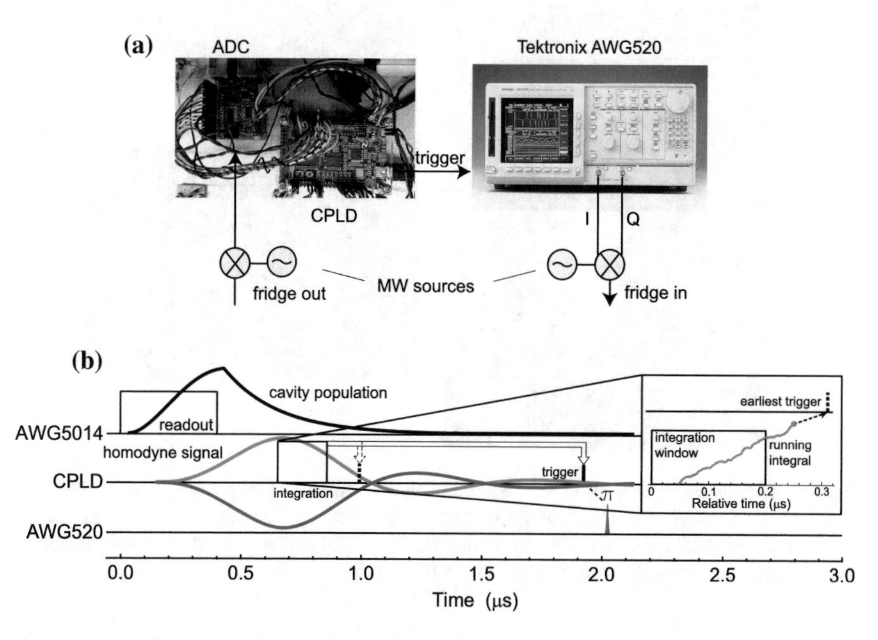

Fig. 8.8 Digital feedback loop with a CPLD-based controller. **a** Schematics of the feedback loop, with an ADC and a CPLD (or FPGA) board replacing the ADwin in Fig. 8.7. **b** Timings of the feedback loop. The CPLD samples the signal at every clock cycle (10 ns) and then integrates it over a window set by a marker of an AWG5014. The internal delay of the CPLD breaks down into the analog-to-digital conversion (60 ns) and the processing to compare the integrated signal to a calibrated threshold, determining the binary output (50 ns). These timings are multiples of the clock (reduced to 4 ns in a recent FPGA-based implementation [70]). The total delay in Ref. [21] is increased to 2 μs to let the cavity return to the ground state before the conditional π pulse

time, active depletion of the cavity [71] will be required to take full advantage of the CPLD speed and further shorten the feedback loop.

Further developments in the feedback controller replaced the CPLD with a field-programmable-gate-array (FPGA) to increase the on-board memory and enable more complex signal processing. For example, the FPGA allows different weights for the measurement record and maximal correlation with the qubit evolution. A FPGA-based controller has also been employed for digital feedback at ETH Zürich [49]. Recent developments at TU Delft and at Yale [72] include the pulse generation on a FPGA board, eliminating the need of an additional AWG. For comparison, Fig. 8.9 shows the setup that would be required for the 3-qubit repetition code [27] using our first generation of feedback (**a**) and the most recent one based on FPGAs (**b**, **c**).

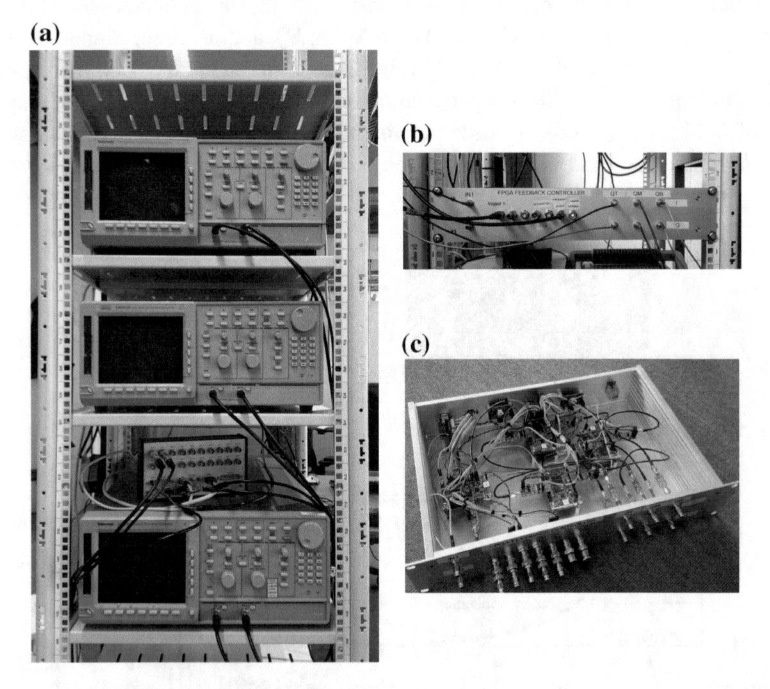

Fig. 8.9 Hardware comparison for feedback control in the bit-flip code. The bit-flip code requires a two-bit digital feedback, acting on three qubits. Scaling the system in Fig. 8.7 would take an AWG520 for each qubit (**a**). A recent implementation [70] performs readout signal processing and pulse generation on FPGA boards, resulting in the compact controller shown in (**b**), (**c**)

8.4 Fast Qubit Reset Based on Digital Feedback

8.4.1 Passive Qubit Initialization to Steady State

Our first application of feedback is qubit initialization, also known as reset [12]. The ideal reset for QIP is deterministic (as opposed to heralded or postselected, see previous section) and fast compared to qubit coherence times. Obviously, the passive method of waiting several times T_1 does not meet the speed requirement. Moreover, it can suffer from residual steady-state qubit excitations [1, 47, 61, 73], whose cause in cQED remains an active research area. The drawbacks of passive initialization are evident for our qubit, whose ground-state population $P_{|0\rangle}$ evolves from states ρ_0 and ρ_1 as shown in Fig. 8.10. With ρ_0 and ρ_1 we indicate our closest realization (~99 % fidelity) of the ideal pure states $|0\rangle$ and $|1\rangle$. $P_{|0\rangle}$ at variable time after preparation is obtained by comparing the average readout homodyne voltage to calibrated levels, as in standard three-level tomography [74, 75]. These populations dynamics are captured by a master equation model for a three-level system:

$$\begin{pmatrix} \dot{P}_{|0\rangle} \\ \dot{P}_{|1\rangle} \\ \dot{P}_{|2\rangle} \end{pmatrix} = \begin{pmatrix} -\Gamma_{01} & \Gamma_{10} & 0 \\ \Gamma_{01} & -\Gamma_{10}-\Gamma_{12} & \Gamma_{21} \\ 0 & \Gamma_{12} & -\Gamma_{21} \end{pmatrix} \begin{pmatrix} P_{|0\rangle} \\ P_{|1\rangle} \\ P_{|2\rangle} \end{pmatrix}. \tag{8.2}$$

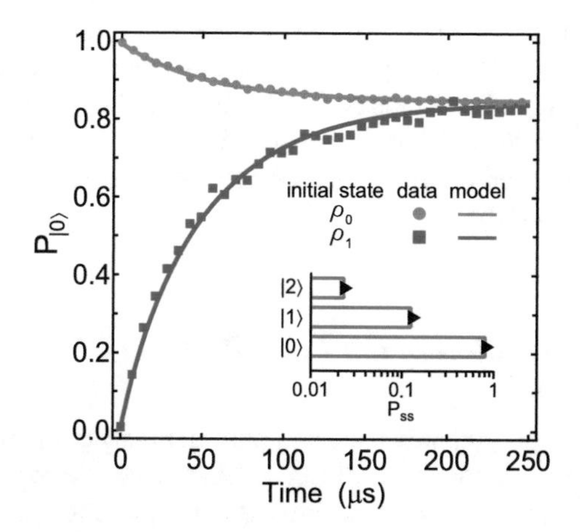

Fig. 8.10 Transmon equilibration to steady state. Time evolution of the ground-state population $P_{|0\rangle}$ starting from states ρ_0 and ρ_1 (notation defined in the text). *Solid curves* are the best fit to Eq. (8.2), giving the inverse transition rates $\Gamma_{10}^{-1} = 50 \pm 2\,\mu\text{s}$, $\Gamma_{21}^{-1} = 20 \pm 2\,\mu\text{s}$, $\Gamma_{01}^{-1} = 324 \pm 32\,\mu\text{s}$, $\Gamma_{12}^{-1} = 111 \pm 25\,\mu\text{s}$. From the steady-state solution, we extract residual excitations $P_{|1\rangle,\text{ss}} = 13.1 \pm 0.8\,\%$, $P_{|2\rangle,\text{ss}} = 2.4 \pm 0.4\,\%$. *Inset*: steady-state population distribution (bars). Markers correspond to a Boltzmann distribution with best-fit temperature 127 mK, significantly higher than the dilution refrigerator base temperature (15 mK). Reproduced with permission of the American Physical Society from Ref. [2]

The best fit to the data gives the qubit relaxation time $T_1 = 1/\Gamma_{10} = 50 \pm 2\,\mu\text{s}$ and the asymptotic 15.5 % residual total excitation.

8.4.2 Qubit Reset Based on Digital Feedback

Previous approaches to accelerate qubit equilibration include coupling to dissipative resonators [17] or two-level systems [18]. However, these are also susceptible to spurious excitation, potentially inhibiting complete qubit relaxation. Feedback-based reset circumvents the equilibration problem by not relying on coupling to a dissipative medium. Rather, it works by projecting the qubit with a measurement (M_1, performed by the controller) and conditionally applying a π pulse to drive the qubit to a targeted basis state (Fig. 8.11). A final measurement (M_2) determines the qubit state immediately afterwards. In both measurements, the result is digitized into levels H or L, associated with $|0\rangle$ and $|1\rangle$, respectively. The digitization threshold voltage V_{th} maximizes the readout fidelity at 99 %. The π pulse is conditioned on $M_1 = L$ to target $|0\rangle$ (scheme Fb_0) or on $M_1 = H$ to target $|1\rangle(\text{Fb}_1)$. In a QIP context, reset is typically used to reinitialize a qubit following measurement, when it is in a computational basis state. Therefore, to benchmark the reset protocol, we first quantify its action on ρ_0 and ρ_1. This step is accomplished with a preliminary measurement M_0 (initializing the qubit in ρ_0 by postselection), followed by a calibrated pulse resonant on the transmon $0 \leftrightarrow 1$ transition to prepare ρ_1. The overlap of the M_2 histograms with the targeted region (H for Fb_0 and L for Fb_1) averages at 96 %, indicating the success of reset. Imperfections are more evident for $\theta = \pi$ and mainly due to equilibration of the transmon during the feedback loop. A detailed error analysis is presented below. We emphasize that qubit initialization by postselection is here only used to prepare nearly pure states useful for characterizing the feedback-based reset, which is deterministic.

8.4.3 Characterization of the Reset Protocol

An ideal reset function prepares the same pure qubit state regardless of its input. To fully quantify the performance of our reset scheme, we measure its effect on our closest approximation to superposition states $|\theta\rangle = \cos(\theta/2)|0\rangle + \sin(\theta/2)|1\rangle$ (Fig. 8.12). Without feedback, $P_{|0\rangle}$ is trivially a sinusoidal function of θ, with near unit contrast. Feedback highly suppresses the Rabi oscillation, with $P_{|0\rangle}$ approaching the ideal value 1 (0) for Fb_0 (Fb_1) for any input state. However, a dependence on θ remains, with $P_{\text{err}} = 1 - P_{|0\rangle}$ for Fb_0 ($1 - P_{|1\rangle}$ for Fb_1) ranging from 1.2 % (1.4 %) for $\theta = 0$ to 7.8 % (8.4 %) for $\theta = \pi$. The remaining errors are discussed

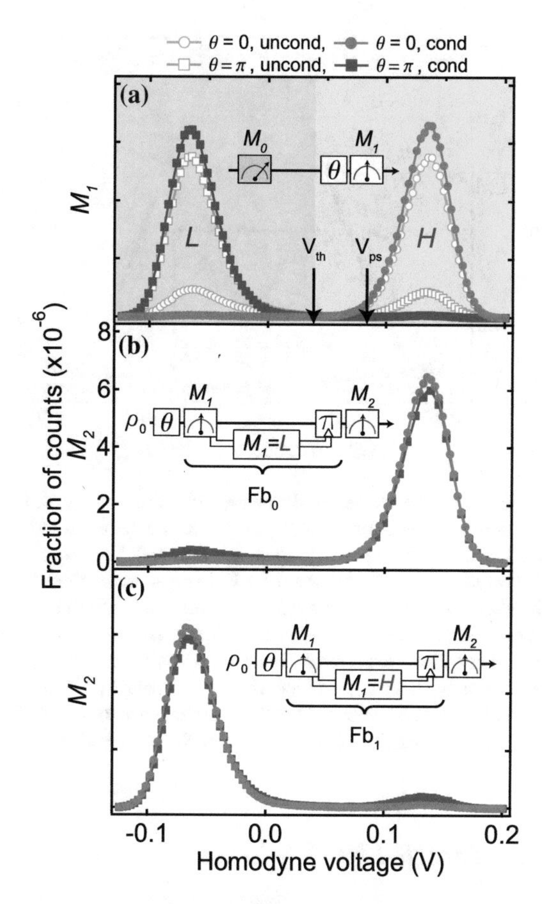

Fig. 8.11 Reset by measurement and feedback. **a** Before feedback: histograms of 300 000 shots of M_1, with (*squares*) and without (*circles*) inverting the qubit population with a π pulse. Each shot is obtained by averaging the homodyne voltage over the second half (200 ns) of a readout pulse. H and L denote the two possible outcomes of M_1, digitized with the threshold V_{th}, maximizing the contrast, analogously to Sect. 8.2. *Full* (*empty*) *dots* indicate (no) postselection on $M_0 > V_{\text{ps}}$. This protocol is used to prepare ρ_0 and ρ_1, which are the input states for the feedback sequences in (**b**) and (**c**). **b** After feedback: histograms of M_2 after applying the feedback protocol Fb_0, which triggers a π pulse when $M_1 = L$. Using this feedback, ~99 % (92 %) of measurements digitize to H for $\theta = 0$ (π), respectively. **c** Feedback with opposite logic Fb_1 preparing the excited state. In this case, ~98 % (94 %) of measurements digitize to L for $\theta = 0$ (π). Reproduced with permission of the American Physical Society from Ref. [2]

in Sect. 8.1.4. From Eq. (8.1), using the best-fit Γ_{ij} and $\tau_{\text{Fb}} = 2.4\,\mu\text{s}$, errors due to equilibration sum to 0.7 % (6.9 %) for $\theta = 0$ (π), while readout errors account for the remaining 0.4 % (1.4 %). In agreement with these values, concatenating two feedback cycles suppresses the error for $\theta = \pi$ to 3.4 %, while there is no benefit for $\theta = 0$ (1.3 %).

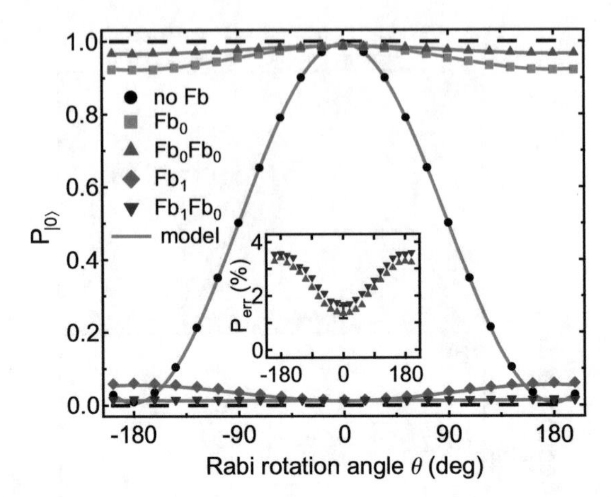

Fig. 8.12 Deterministic reset (for supercondicting qubits) from any qubit state. Ground-state population $P_{|0\rangle}$ as a function of the initial state ρ_θ, prepared by coherent rotation after initialization in ρ_0, as in Fig. 8.11. The cases shown are: no feedback (*circles*), Fb_0 (*squares*), Fb_1 (*diamonds*), twice Fb_0 (*upward triangles*), and Fb_0 followed by Fb_1 (*downward triangles*). The vertical axis is calibrated with the average measurement outcome for the reference states ρ_0, ρ_1, and corrected for imperfect state preparation. The curve with no feedback has a visibility of 99 %, equal to the average preparation fidelity. Each experiment is averaged over 300 000 repetitions. *Inset*: error probabilities for two rounds of feedback, defined as $1 - P_{|t\rangle}$, where $|t\rangle \in \{0, 1\}$ is the target state. The systematic ~0.3 % difference between the two cases is attributed to error in the π pulse preceding the measurement of $P_{|1\rangle}$ following Fb_1. *Curves*: model including readout errors and equilibration (Sect. 8.1.4)

8.4.4 Speed-Up Enabled by Fast Reset

The key advantage of reset by feedback is the ability to ready a qubit for further computation fast compared to coherence times available in 3D cQED [53, 76]. This will be important, for example, when refreshing ancilla qubits in multi-round error correction [20]. We now show that reset suppresses the accumulation of initialization error when a simple experiment is repeated with decreasing in-between time τ_{init}. The simple sequence in Fig. 8.13 emulates an algorithm that leaves the qubit in $|1\rangle$ [case (a)] or $|0\rangle$ [case (b)]. A measurement pulse follows τ_{init} to quantify the initialization error P_{err}. Without feedback, P_{err} in case (a) grows exponentially as $\tau_{init} \to 0$. This accruement of error, due to the rapid succession of π pulses, would occur even at zero temperature, where residual excitation would vanish (i.e., $\Gamma_{i+1,i} = 0$), in which case $P_{err} \to 50\,\%$ as $\tau_{init} \to 0$. In case (b), P_{err} matches the total steady-state excitation for all τ_{init}. Using feedback significantly improves initialization for both long and short τ_{init}. For $\tau_{init} \gg T_1$, feedback suppresses P_{err} from the 16 % residual excitation to 3 % (black symbols and curves),[1] cooling the transmon. Crucially, unlike passive initialization, reset by feedback is also effective at short τ_{init}, where it limits the other-

[1] We note that $P_{|1\rangle} \approx P_{|2\rangle} = 1.6\,\%$ is a non-thermal distribution.

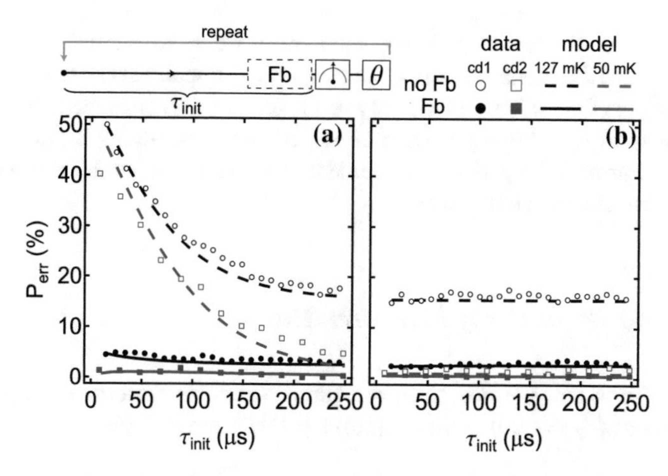

Fig. 8.13 Fast qubit reset. Initialization errors as a function of initialization time τ_{init} under looped execution of a simple experiment leaving the qubit ideally in $|1\rangle$ (**a**, measurement and π pulse) or $|0\rangle$ (**b**, measurement only). *Empty symbols*: initialization by waiting (no feedback). *Solid symbols*: initialization by feedback, with three rounds of Fb_0 and a π pulse on the $1 \leftrightarrow 2$ transition. Two data sets correspond to two different cooldowns: the one corresponding to Figs. 8.10, 8.11 and 8.12 (*black*) and a following one with improved thermalization (*blue*). Curves correspond to a master equation simulation assuming perfect pulses and measured transition rates Γ_{ij} (dashed, no feedback; solid, triple Fb_0 with a π pulse on $1 \leftrightarrow 2$). Feedback reset successfully bounds the otherwise exponential accruement of P_{err} in case **a** as $\tau_{\text{init}} \to 0$. The reduction of P_{err} in **b** reflects the cooling of the transmon by feedback (see text for details). Figure adapted from Ref. [2]

wise exponential accruement of error in (a), bounding P_{err} to an average of 3.5 % over the two cases. Our scheme combines three rounds of Fb_0 with a pulse on the $1 \leftrightarrow 2$ transition before the final Fb_0 to partially counter leakage to the second excited state, which is the dominant error source [see Eq. (8.1)]. The remaining leakage is proportional to the average $P_{|1\rangle}$, which slightly increases in **a** and decreases in **b** as $\tau_{\text{init}} \to 0$. In a following cooldown, with improved thermalization and a faster feedback loop (Fig. 8.8), reset constrained $P_{\text{err}} \lesssim 1\,\%$ (blue), quoted as the fault-tolerance threshold for initialization in modern error correction schemes [29]. In addition to the near simultaneous implementation at ENS [48], similar implementations of qubit reset have followed at Yale [72] and at Raytheon BBN Technologies using a FPGA-based feedback controller.

8.5 Deterministic Entanglement by Parity Measurement and Feedback

In this section, we extend the use of digital feedback to a multi-qubit experiment, targeting the deterministic generation of entanglement by measurement. We first turn the cavity into a parity meter to measure the joint state of two coupled qubits. By carefully engineering the cavity-qubit dispersive shifts, we make the cavity transmission only sensitive to the excitation parity, but unable to distinguish states within each parity. Binning the final states on the parity result generates an entangled state in either case, with up to 88 % fidelity to the closest Bell state. Integrating the demonstrated feedback control in the parity measurement, we turn the entanglement generation from probabilistic to deterministic.

8.5.1 Two-Qubit Parity Measurement

In a two-qubit system, the ideal parity measurement transforms an unentangled superposition state $|\psi^0\rangle = (|00\rangle + |01\rangle + |10\rangle + |11\rangle)/2$ into Bell states

$$|\Phi^+\rangle = \frac{1}{\sqrt{2}}(|01\rangle + |10\rangle) \text{ and } |\Psi^+\rangle = \frac{1}{\sqrt{2}}(|00\rangle + |11\rangle) \tag{8.3}$$

for odd and even outcome, respectively. Beyond generating entanglement between non-interacting qubits [23, 77–80], parity measurements allow deterministic two-qubit gates [81, 82] and play a key role as syndrome detectors in quantum error correction [24, 83]. A heralded parity measurement has been recently realized for nuclear spins in diamond [84]. By minimizing measurement-induced decoherence at the expense of single-shot fidelity, highly entangled states were generated with 3 % success probability. Here, we realize the first solid-state parity meter that produces entanglement with unity probability.

8.5.2 Engineering the Cavity as a Parity Meter

Our parity meter realization exploits the dispersive regime [51] in two-qubit cQED. Qubit-state dependent shifts of a cavity resonance (here, the fundamental of a 3D cavity enclosing transmon qubits Q_A and Q_B) allow joint qubit readout by homodyne detection of an applied microwave pulse transmitted through the cavity (Fig. 8.14a). The temporal average V_{int} of the homodyne response $V_P(t)$ over the time interval $[t_i, t_f]$ constitutes the measurement needle, with expectation value

$$\langle V_{\text{int}}\rangle = \text{Tr}(\mathcal{O}\rho),$$

where ρ is the two-qubit density matrix and the observable $\mathcal{O}$ has the general form

$$\mathcal{O} = \beta_0 + \beta_A\sigma_z^A + \beta_B\sigma_z^B + \beta_{BA}\sigma_z^B\sigma_z^A.$$

The coefficients β_0, β_A, β_B, and β_{BA} depend on the strength ϵ_p, frequency f_p and duration τ_P of the measurement pulse, the cavity linewidth κ, and the frequency shifts

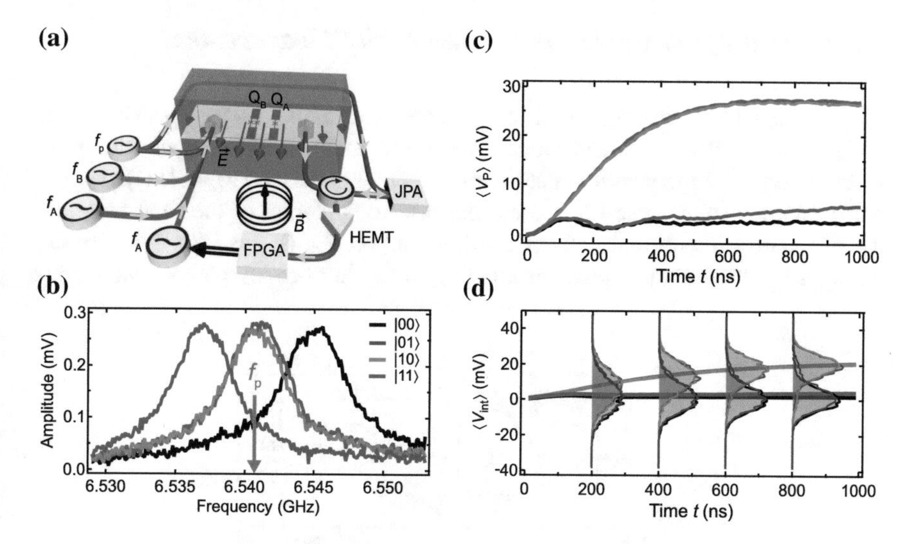

Fig. 8.14 Cavity-based two-qubit parity readout in cQED. **a** Simplified diagram of the experimental setup. Single- and double-junction transmon qubits (Q_A and Q_B, respectively) dispersively couple to the fundamental mode of a 3D copper cavity enclosing them. Parity measurement is performed by homodyne detection of the qubit state-dependent cavity response [51] using a JPA [54]. Following further amplification at 4 K (HEMT) and room temperature, the signal is demodulated and integrated. A FPGA controller closes the feedback loop that achieves deterministic entanglement by parity measurement (Fig. 8.17). **b** Matching of the dispersive cavity shifts realizing a parity measurement. **c** Ensemble-averaged homodyne response $\langle V_P \rangle$ for qubits prepared in the four computational basis states. **d** *Curves*: corresponding ensemble averages of the running integral $\langle V_{\mathrm{int}} \rangle$ of $\langle V_P \rangle$ between $t_i = 0$ and $t_f = t$. Single-shot histograms (5 000 counts each) of V_{int} are shown in 200 ns increments. Figure adapted from Ref. [3]

$2\chi_A$ and $2\chi_B$ of the fundamental mode when Q_A and Q_B are individually excited from $|0\rangle$ to $|1\rangle$. The necessary condition for realizing a parity meter is $\beta_A = \beta_B = 0$ (β_0 constitutes a trivial offset). A simple approach [85, 86], pursued here, is to set f_p to the average of the resonance frequencies for the four computational basis states $|ij\rangle$ $(i, j \in \{0, 1\})$ and to match $\chi_A = \chi_B$. We engineer this matching by targeting specific qubit transition frequencies f_A and f_B below and above the fundamental mode during fabrication and using an external magnetic field to fine-tune f_B in situ. We align χ_A to χ_B to within $\sim 0.06\,\kappa = 2\pi \times 90\,\mathrm{kHz}$ (Fig. 8.14b). The ensemble-average $\langle V_P \rangle$ confirms nearly identical high response for odd-parity computational states $|01\rangle$ and $|10\rangle$, and nearly identical low response for the even-parity $|00\rangle$ and $|11\rangle$ (Fig. 8.14c). The transients observed are consistent with the independently measured κ, χ_A and χ_B values, and the 4 MHz bandwidth of the JPA at the front end of the output amplification chain. Single-shot histograms (Fig. 8.14d) demonstrate the increasing ability of V_{int} to discern states of different parity as t_f grows (keeping $t_i = 0$), and its inability to discriminate between states of the same parity. The histogram separations at $t_f = 400\,\mathrm{ns}$ give $|\beta_A|, |\beta_B| < 0.02\,|\beta_{BA}|$.

8.5.3 Two-Qubit Evolution During Parity Measurement

Moving beyond the description of the measurement needle, we now investigate the collapse of the two-qubit state during parity measurement. We prepare the qubits in the maximal superposition state $|\psi^0\rangle = \frac{1}{2}(|00\rangle + |01\rangle + |10\rangle + |11\rangle)$, apply a parity measurement pulse for τ_P, and perform tomography of the final two-qubit density matrix ρ with and without conditioning on V_{int} (Fig. 8.15a). We choose a weak parity measurement pulse exciting $\bar{n}_{\text{ss}} = 2.5$ intra-cavity photons on average

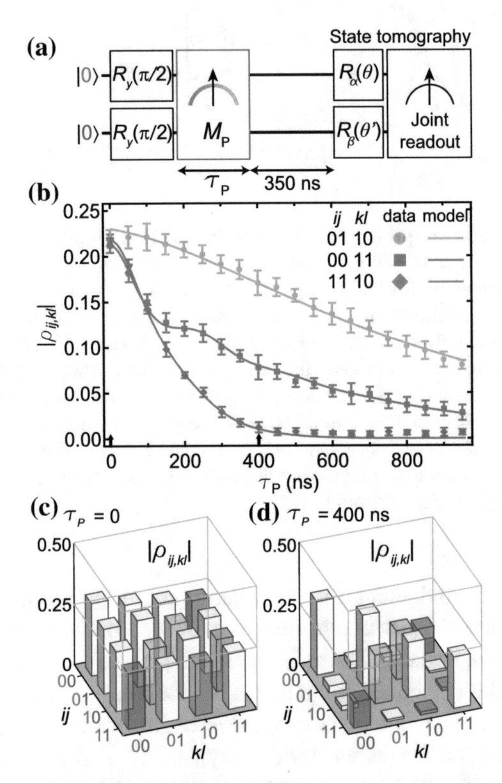

Fig. 8.15 Unconditioned two-qubit evolution under continuous parity measurement. a Pulse sequence including preparation of the qubits in the maximal superposition state $\rho^{(0)} = |\psi^0\rangle\langle\psi^0|$, parity measurement and tomography of the final two-qubit state ρ using joint readout. **b** Absolute coherences $|\rho_{11,10}|$, $|\rho_{01,10}|$, $|\rho_{00,11}|$ following a parity measurement with variable duration τ_P. Free parameters of the model are the steady-state photon number on resonance $\bar{n}_{\text{ss}} = 2.5 \pm 0.1$, the difference $(\chi_A - \chi_B)/\pi = 235 \pm 4$ kHz, and the absolute coherence values at $\tau_P = 0$ to account for few-percent pulse errors in state preparation and tomography pre-rotations. Note that the frequency mismatch differs from that in Fig. 8.14b due to its sensitivity to measurement power. **c**, **d** Extracted density matrices for $\tau_P = 0$ (**c**) and $\tau_P = 400$ ns (**d**), by which time coherence across the parity subspaces (*grey*) is almost fully suppressed, while coherence persists within the odd-parity (*orange*) and even-parity (*green*) subspaces. Error bars correspond to the standard deviation of 15 repetitions. Figure reproduced with permission of Nature Publishing Group from Ref. [3]

in the steady-state, at resonance. A delay of $3.5/\kappa = 350$ ns is inserted to deplete the cavity of photons before performing tomography. The tomographic joint readout is also carried out at f_p, but with 14 dB higher power, at which the cavity response is weakly nonlinear and sensitive to both single-qubit terms and two-qubit correlations ($\beta_\mathrm{A} \sim \beta_\mathrm{B} \sim \beta_\mathrm{BA}$, as required for tomographic reconstruction [87].

The ideal continuous parity measurement gradually suppresses the unconditioned density matrix elements $\rho_{ij,kl} = \langle ij|\rho|kl\rangle$ connecting states with different parity (either $i \neq k$ or $j \neq l$), and leaves all other coherences (off-diagonal terms) and all populations (diagonal terms) unchanged. The experimental tomography reveals the expected suppression of coherence between states of different parity (Fig. 8.15b, c). The temporal evolution of $|\rho_{11,10}|$, with near full suppression by $\tau_\mathrm{P} = 400$ ns, is quantitatively matched by a master-equation simulation of the two-qubit system. Tomography also unveils a non-ideality: albeit more gradually, our parity measurement partially suppresses the absolute coherence between equal-parity states, $|\rho_{01,10}|$ and $|\rho_{00,11}|$. The effect is also quantitatively captured by the model. Although intrinsic qubit decoherence contributes, the dominant mechanism is the different AC-Stark phase shift induced by intra-cavity photons on basis states of the same parity [86, 88, 89]. This phase shift has both deterministic and stochastic components, and the latter suppresses absolute coherence under ensemble averaging. We emphasize that this imperfection is technical rather than fundamental. It can be mitigated in the odd subspace by perfecting the matching of χ_B to χ_A, and in the even subspace by increasing $\chi_\mathrm{A,B}/\kappa$ ($\sim$1.3 in this experiment).

8.5.4 *Probabilistic Entanglement by Measurement and Postselection*

The ability to discern parity subspaces while preserving coherence within each opens the door to generating entanglement by parity measurement on $|\psi^0\rangle$. For every run of the sequence in Fig. 8.15, we discriminate V_int using the threshold V_th that maximizes the parity measurement fidelity F_p (Fig. 8.16a). Assigning $M_\mathrm{P} = +1$ (-1) to V_int below (above) V_th, we bisect the tomographic measurements into two groups, and obtain the density matrix for each. We quantify the entanglement achieved in each case using concurrence $\mathcal{C}$ as the metric [90], which ranges from 0 % for an unentangled state to 100 % for a Bell state. As τ_P grows (Fig. 8.16b), the optimal balance between increasing F_p at the cost of measurement-induced dephasing and intrinsic decoherence is reached at $\sim$300 ns (Fig. 8.16c). Postselection on $M_\mathrm{P} = \pm 1$ achieves $\mathcal{C}_{|M_\mathrm{P}=-1} = 45 \pm 3$ % and $\mathcal{C}_{|M_\mathrm{P}=+1} = 17 \pm 3$ %, with each case occurring with probability $p_\mathrm{success} \sim 50$ %. The higher performance for $M_\mathrm{P} = -1$ results from lower measurement-induced dephasing in the odd subspace, consistent with Fig. 8.15.

The entanglement achieved by this probabilistic protocol can be increased with more stringent postselection. Setting a higher threshold $V_\mathrm{th-}$ achieves $\mathcal{C}_{|M_\mathrm{P}=-1} = 77 \pm 2$ % but keeps $p_\mathrm{success} \sim 20$ % of runs. Analogously, using $V_\mathrm{th+}$ achieves

$\mathcal{C}_{|M_P=+1} = 29 \pm 4\,\%$ with similar p_{success} (Fig. 8.16d, e). However, increasing $\mathcal{C}$ at the expense of reduced p_{success} is not evidently beneficial for QIP. For the many tasks calling for maximally-entangled qubit pairs (ebits), one may use an optimized distillation protocol [90] to prepare one ebit from $N = 1/E_{\mathcal{N}}(\rho)$ pairs in a partially-entangled state ρ, where $E_{\mathcal{N}}$ is the logarithmic negativity [90]. The efficiency $\mathcal{E}$ of ebit generation would be $\mathcal{E} = p_{\text{success}} E_{\mathcal{N}}(\rho)$. For postselection on $M_P = -1$, we calculate $\mathcal{E} = 0.31$ ebits/run using V_{th} and $\mathcal{E} = 0.20$ ebits/run using $V_{\text{th}-}$. Evidently, increasing entanglement at the expense of reducing p_{success} is counterproductive in this context.

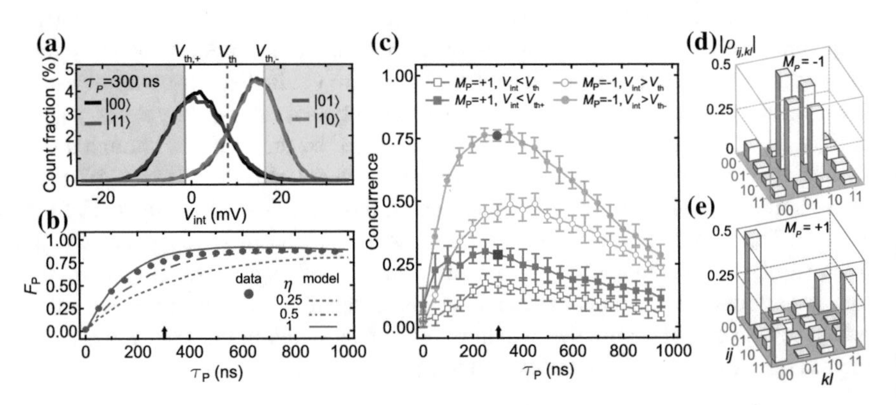

Fig. 8.16 Probabilistic entanglement generation by postselected parity measurement. **a** Histograms of V_{int} ($\tau_P = 300$ ns) for the four computational states. The results are digitized into $M_P = 1(-1)$ for V_P below (above) a chosen threshold. **b** Parity readout fidelity F_p as a function of τ_P. We define $F_p = 1 - \epsilon_e - \epsilon_o$, with $\epsilon_e = p(M_P = -1|\text{even})$ the readout error probability for a prepared even state, and similarly for ϵ_o. Data are corrected for residual qubit excitations (1–2 %). Error bars are smaller than the dot size. Model curves are obtained from 5 000 quantum trajectories for each initial state and τ_P, with quantum efficiencies $\eta = 0.25$, 0.5, and 1 for the readout amplification chain. No single value of η matches the dependence of F_p on τ_P. We attribute this discrepancy to low-frequency fluctuations in the parametric amplifier bias point, not included in the model. **c** Concurrence $\mathcal{C}$ of the two-qubit entangled state obtained by postselection on $M_P = -1$ (*orange*) and on $M_P = +1$ (*green squares*). *Empty symbols* correspond to the threshold V_{th} that maximizes F_p, binning $p_{\text{success}} \sim 50\,\%$ of the data into each case. Solid symbols correspond to a threshold $V_{\text{th}-}(V_{\text{th}+})$ for postselection on $M_P = -1(+1)$, at which $\epsilon_o(\epsilon_e) = 0.01$. Concurrence is optimized at $\tau_P \sim 300$ ns, where $p_{\text{success}} \sim 20\,\%$ in each case. We employ maximum-likelihood estimation [87] (MLE) to ensure physical density matrices, but concurrence values obtained with and without MLE differ by less than 3 % over the full data set. **d**, **e** State tomography conditioned on $V_P > V_{\text{th}-}$ (**d**) and $V_P < V_{\text{th}+}$ (**e**), with $\tau_P = 300$ ns, corresponding to the *dark symbols* in (**c**). Figure reproduced with permission of Nature Publishing Group from Ref. [3]

8.5.5 Deterministic Entanglement by Measurement and Feedback

Motivated by the above observation, we finally demonstrate the use of feedback control to transform entanglement by parity measurement from probabilistic to deterministic, i.e., $p_{\text{success}} = 100\,\%$. While initial proposals in cQED focused on analog feedback schemes [91], here we adopt a digital strategy. Specifically, we use our homebuilt programmable controller (Sect. 8.3 to apply a π pulse on Q_A conditional on measuring $M_\text{P} = +1$ (using V_{th}, Fig. 8.17). In addition to switching the two-qubit

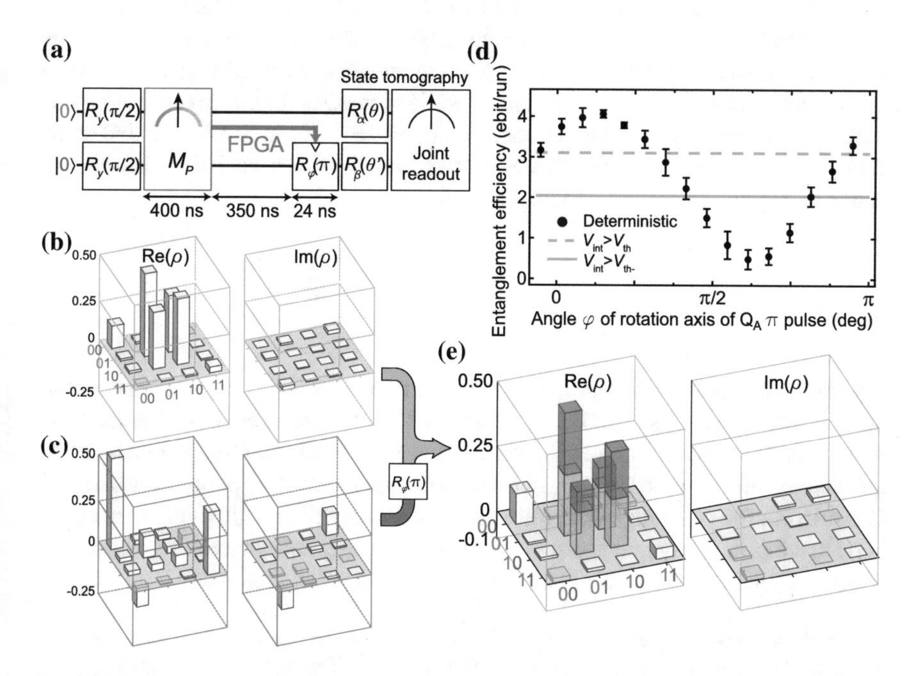

Fig. 8.17 Deterministic entanglement generation using feedback. **a** We close a digital feedback loop by triggering (via the FPGA) a π pulse on Q_A conditional on parity measurement result $M_\text{P} = +1$. This π pulse switches the two-qubit parity from even to odd, and allows the deterministic targeting of $|\Phi^+\rangle = (|01\rangle + |10\rangle)/\sqrt{2}$. **b**, **c** Parity measurement results $M_\text{P} = -1$ and $M_\text{P} = +1$ each occur with $\sim$50 % probability. The deterministic AC Stark phase acquired between $|01\rangle$ and $|10\rangle$ during parity measurement (due to residual mismatch between χ_A and χ_B) is compensated by a global phase rotation in the tomography pulses. A different AC Stark phase is acquired between $|00\rangle$ and $|11\rangle$, resulting in the state shown in (**c**), with the maximal overlap with even Bell state $[|00\rangle + \exp(-i\varphi_\text{e})|11\rangle]/\sqrt{2}$ at $\varphi_\text{e} = 0.73\pi$. **d** Generation rate of entanglement using feedback, as a function of the phase φ of the π pulse. The deterministic entanglement generation efficiency outperforms the efficiencies obtained with postselection (Fig. 8.16). Error bars are the standard deviation of 7 repetitions of the experiment at each φ. **e** Full state tomography for deterministic entanglement $[\varphi = (\pi - \varphi_\text{e})/2]$, achieving fidelity $\langle\Phi^+|\rho|\Phi^+\rangle = 66\,\%$ to the targeted $|\Phi^+\rangle$, and concurrence $C = 34\,\%$. Colored bars highlight the contribution from cases $M_\text{P} = -1$ (*orange*) and $M_\text{P} = +1$ (*green*). Figure reproduced with permission of Nature Publishing Group from Ref. [3]

parity, this pulse lets us choose which odd-parity Bell state to target by selecting the phase φ of the conditional pulse. To optimize deterministic entanglement, we need to maximize overlap to the same odd-parity Bell state for $M_\mathrm{P} = -1$ (Fig. 8.17b) as for $M_\mathrm{P} = +1$ (Fig. 8.17c). For the targeted state $|\Phi^+\rangle$, this requires cancelling the deterministic AC Stark phase $\varphi_\mathrm{e} = 0.73\pi$ accrued between $|00\rangle$ and $|11\rangle$ when $M_\mathrm{P} = +1$. This is accomplished by choosing $\varphi = (\pi - \varphi_\mathrm{e})/2$, which clearly maximizes the entanglement obtained when no postselection on M_P is applied (Fig. 8.17c, d). The highest deterministic $C = 34\,\%$ achieved is lower than for our best probabilistic scheme, but the boost to $p_\mathrm{success} = 100\,\%$ achieves a higher $\mathcal{E} = 0.41$ ebits/run.

A parallel development realized the probabilistic entanglement by measurement between two qubits in separate 3D cavities [92], establishing the first quantum connection between remote superconducting qubits. In another two-qubit, single-cavity system, feedback has been recently applied to enhance the fidelity of the generated entanglement [93]. Following the first realizations in 3D cQED, parity measurements have been implemented using an ancillary qubit [62, 94] in 2D. Compared to the cavity-based scheme, the use of an ancilla evades measurement-induced dephasing and is better suited to scaling to larger circuits.

8.6 Conclusion

We have presented the first implementation of digital feedback control in superconducting circuits, and its evolution to faster, simpler, and more configurable feedback loops. In particular, we showed the use of digital feedback for fast and deterministic qubit reset and for deterministic generation of entanglement by parity measurement. Considering the vast range of applications for feedback in quantum computing, we hope that this development is just the start of an exciting new phase of measurement-assisted digital control in solid-state quantum information processing.

Acknowledgments We thank all the collaborators who have contributed to the experiments here presented: J.G. van Leeuwen, C.C. Bultink, M. Dukalski, C.A. Watson, G. de Lange, H.-S. Ku, M.J. Tiggelman, K.W. Lehnert, Ya. M. Blanter, and R.N. Schouten. We acknowledge L. Tornberg and G. Johansson for useful discussions. Funding for this research was provided by the Dutch Organization for Fundamental Research on Matter (FOM), the Netherlands Organization for Scientific Research (NWO, VIDI scheme), and the EU FP7 projects SOLID and SCALEQIT.

References

1. D. Ristè, J.G. van Leeuwen, H.-S. Ku, K.W. Lehnert, L. DiCarlo, Initialization by measurement of a superconducting quantum bit circuit. Phys. Rev. Lett. **109**, 050507 (2012)
2. D. Ristè, C.C. Bultink, K.W. Lehnert, L. DiCarlo, Feedback control of a solid-state qubit using high-fidelity projective measurement. Phys. Rev. Lett. **109**, 240502 (2012)

3. D. Ristè, M. Dukalski, C.A. Watson, G. de Lange, M.J. Tiggelman, Y.M. Blanter, K.W. Lehnert, R.N. Schouten, L. DiCarlo, Deterministic entanglement of superconducting qubits by parity measurement and feedback. Nature **502**, 350 (2013)
4. G.G. Gillett et al., Experimental feedback control of quantum systems using weak measurements. Phys. Rev. Lett. **104**, 080503 (2010)
5. C. Sayrin et al., Real-time quantum feedback prepares and stabilizes photon number states. Nature **477**, 73 (2011)
6. P. Bushev et al., Feedback cooling of a single trapped ion. Phys. Rev. Lett. **96**, 043003 (2006)
7. M. Koch, C. Sames, A. Kubanek, M. Apel, M. Balbach, A. Ourjoumtsev, P.W.H. Pinkse, G. Rempe, Feedback cooling of a single neutral atom. Phys. Rev. Lett. **105**, 173003 (2010)
8. S. Brakhane, W. Alt, T. Kampschulte, M. Martinez-Dorantes, R. Reimann, S. Yoon, A. Widera, D. Meschede, Bayesian feedback control of a two-atom spin-state in an atom-cavity system. Phys. Rev. Lett. **109**, 173601 (2012)
9. G. de Lange, D. Ristè, M.J. Tiggelman, C. Eichler, L. Tornberg, G. Johansson, A. Wallraff, R.N. Schouten, L. DiCarlo, Reversing quantum trajectories with analog feedback. Phys. Rev. Lett. **112**, 080501 (2014)
10. M.S. Blok, C. Bonato, M.L. Markham, D.J. Twitchen, V.V. Dobrovitski, R. Hanson, Manipulating a qubit through the backaction of sequential partial measurements and real-time feedback. Nat. Phys. **10**, 189 (2014)
11. J.P. Groen, D. Ristè, L. Tornberg, J. Cramer, P.C. de Groot, T. Picot, G. Johansson, L. DiCarlo, Partial-measurement backaction and nonclassical weak values in a superconducting circuit. Phys. Rev. Lett. **111**, 090506 (2013)
12. D.P. DiVincenzo, The physical implementation of quantum computation. Fortschr. Phys. **48**, 771 (2000)
13. C. Monroe, D. Meekhof, B. King, S. Jefferts, W. Itano, D. Wineland, P. Gould, Resolved-sideband Raman cooling of a bound atom to the 3D zero-point energy. Phys. Rev. Lett. **75**, 4011 (1995)
14. M. Atatüre, J. Dreiser, A. Badolato, A. Högele, K. Karrai, A. Imamoglu, Quantum-dot spin-state preparation with near-unity fidelity. Science **312**, 551 (2006)
15. S.O. Valenzuela, W.D. Oliver, D.M. Berns, K.K. Berggren, L.S. Levitov, T.P. Orlando, Microwave-induced cooling of a superconducting qubit. Science **314**, 1589 (2006)
16. V.E. Manucharyan, J. Koch, L.I. Glazman, M.H. Devoret, Fluxonium: single cooper-pair circuit free of charge offsets. Science **326**, 113 (2009)
17. M.D. Reed, B.R. Johnson, A.A. Houck, L. DiCarlo, J.M. Chow, D.I. Schuster, L. Frunzio, R.J. Schoelkopf, Fast reset and suppressing spontaneous emission of a superconducting qubit. Appl. Phys. Lett. **96**, 203110 (2010)
18. M. Mariantoni et al., Implementing the quantum von Neumann architecture with superconducting circuits. Science **334**, 61 (2011)
19. L. Robledo, L. Childress, H. Bernien, B. Hensen, P.F.A. Alkemade, R. Hanson, High-fidelity projective read-out of a solid-state spin quantum register. Nature **477**, 574 (2011)
20. P. Schindler, J.T. Barreiro, T. Monz, V. Nebendahl, D. Nigg, M. Chwalla, M. Hennrich, R. Blatt, Experimental repetitive quantum error correction. Science **332**, 1059 (2011)
21. D. Ristè, C.C. Bultink, M.J. Tiggelman, R.N. Schouten, K.W. Lehnert, L. DiCarlo, Millisecond charge-parity fluctuations and induced decoherence in a superconducting transmon qubit. Nat. Commun. **4**, 1913 (2013)
22. L. Sun et al., Tracking photon jumps with repeated quantum non-demolition parity measurements. Nature **511**, 444 (2014)
23. R. Ruskov, A.N. Korotkov, Entanglement of solid-state qubits by measurement. Phys. Rev. B **67**, 241305 (2003)
24. M.A. Nielsen, I.L. Chuang, *Quantum Computation and Quantum Information* (Cambridge University Press, Cambridge, 2000)
25. H.-J. Briegel, W. Dür, J.I. Cirac, P. Zoller, Quantum repeaters: the role of imperfect local operations in quantum communication. Phys. Rev. Lett. **81**, 5932 (1998)

26. C.H. Bennett, D.P. DiVincenzo, J.A. Smolin, W.K. Wootters, Mixed-state entanglement and quantum error correction. Phys. Rev. A **54**, 3824 (1996)
27. D. Mermin, *Quantum Computer Science: An Introduction*, 1st edn. (Cambridge University Press, Cambridge, 2007)
28. A.G. Fowler, M. Mariantoni, J.M. Martinis, A.N. Cleland, Surface codes: towards practical large-scale quantum computation. Phys. Rev. A **86**, 032324 (2012)
29. D.S. Wang, A.G. Fowler, L.C.L. Hollenberg, Surface code quantum computing with error rates over 1%. Phys. Rev. A **83**, 020302 (2011)
30. J. Kelly et al., State preservation by repetitive error detection in a superconducting quantum circuit. Nature **519**, 66 (2015)
31. D. Ristè, S. Poletto, M.-Z. Huang, A. Bruno, V. Vesterinen, O.-P. Saira, L. DiCarlo, Detecting bit-flip errors in a logical qubit using stabilizer measurements. Nat. Commun. **6**, 6983 (2015)
32. A.G. Fowler, Time-optimal quantum computation (2012). arXiv:1210.4626
33. H.J. Briegel, D.E. Browne, W. Dür, R. Raussendorf, M. Van den Nest, Measurement-based quantum computation. Nat. Phys. **5**, 19 (2009)
34. M. Riebe, T. Monz, K. Kim, A. Villar, P. Schindler, M. Chwalla, M. Hennrich, R. Blatt, Deterministic entanglement swapping with an ion-trap quantum computer. Nat. Phys. **4**, 839 (2008)
35. A. Furusawa, J.L. Sørensen, S.L. Braunstein, C.A. Fuchs, H.J. Kimble, E.S. Polzik, Unconditional quantum teleportation. Science **282**, 706 (1998)
36. M.D. Barrett et al., Deterministic quantum teleportation of atomic qubits. Nature **429**, 737 (2004)
37. M. Riebe et al., Deterministic quantum teleportation with atoms. Nature **429**, 734 (2004)
38. J.F. Sherson, H. Krauter, R.K. Olsson, B. Julsgaard, K. Hammerer, I. Cirac, E.S. Polzik, Quantum teleportation between light and matter. Nature **443**, 557 (2006)
39. H. Krauter, D. Salart, C.A. Muschik, J.M. Petersen, H. Shen, T. Fernholz, E.S. Polzik, Deterministic quantum teleportation between distant atomic objects. Nat. Phys. **9**, 400 (2013)
40. M.S. Tame, R. Prevedel, M. Paternostro, P. Böhi, M.S. Kim, A. Zeilinger, Experimental realization of Deutsch's algorithm in a one-way quantum computer. Phys. Rev. Lett. **98**, 140501 (2007)
41. R. Prevedel, P. Walther, F. Tiefenbacher, P. Böhi, R. Kaltenbaek, T. Jennewein, A. Zeilinger, High-speed linear optics quantum computing using active feed-forward. Nature **445**, 65 (2007)
42. K. Chen, C.-M. Li, Q. Zhang, Y.-A. Chen, A. Goebel, S. Chen, A. Mair, J.-W. Pan, Experimental realization of one-way quantum computing with two-photon four-qubit cluster states. Phys. Rev. Lett. **99**, 120503 (2007)
43. G. Vallone, E. Pomarico, F. De Martini, P. Mataloni, Active one-way quantum computation with two-photon four-qubit cluster states. Phys. Rev. Lett. **100**, 160502 (2008)
44. R. Ukai, N. Iwata, Y. Shimokawa, S.C. Armstrong, A. Politi, J.-I. Yoshikawa, P. van Loock, A. Furusawa, Demonstration of unconditional one-way quantum computations for continuous variables. Phys. Rev. Lett. **106**, 240504 (2011)
45. B.A. Bell, D.A. Herrera-Martí, M.S. Tame, D. Markham, W.J. Wadsworth, J.G. Rarity, Experimental demonstration of a graph state quantum error-correction code. Nat. Commun. **5**, 3658 (2014)
46. C. Vitelli, N. Spagnolo, L. Aparo, F. Sciarrino, E. Santamato, L. Marrucci, Joining the quantum state of two photons into one. Nat. Photonics **7**, 521 (2013)
47. R. Vijay, C. Macklin, D.H. Slichter, K.W. Murch, R. Naik, A.N. Korotkov, I. Siddiqi, Stabilizing Rabi oscillations in a superconducting qubit using quantum feedback. Nature **490**, 77 (2012)
48. P. Campagne-Ibarcq, E. Flurin, N. Roch, D. Darson, P. Morfin, M. Mirrahimi, M.H. Devoret, F. Mallet, B. Huard, Persistent control of a superconducting qubit by stroboscopic measurement feedback. Phys. Rev. X **3**, 021008 (2013)
49. L. Steffen et al., Deterministic quantum teleportation with feed-forward in a solid state system. Nature **500**, 319 (2013)
50. W. Pfaff et al., Unconditional quantum teleportation between distant solid-state quantum bits. Science **345**, 532 (2014)

51. A. Blais, R.-S. Huang, A. Wallraff, S.M. Girvin, R.J. Schoelkopf, Cavity quantum electrodynamics for superconducting electrical circuits: an architecture for quantum computation. Phys. Rev. A **69**, 062320 (2004)
52. A. Wallraff, D.I. Schuster, A. Blais, L. Frunzio, R.-S. Huang, J. Majer, S. Kumar, S.M. Girvin, R.J. Schoelkopf, Strong coupling of a single photon to a superconducting qubit using circuit quantum electrodynamics. Nature **431**, 162 (2004)
53. H. Paik et al., Observation of high coherence in Josephson junction qubits measured in a three-dimensional circuit QED architecture. Phys. Rev. Lett. **107**, 240501 (2011)
54. M.A. Castellanos-Beltran, K.D. Irwin, G.C. Hilton, L.R. Vale, K.W. Lehnert, Amplification and squeezing of quantum noise with a tunable Josephson metamaterial. Nat. Phys. **4**, 929 (2008)
55. R. Vijay, M.H. Devoret, I. Siddiqi, Invited review article: the Josephson bifurcation amplifier. Rev. Sci. Instrum. **80**, 111101 (2009)
56. D.I. Schuster et al., Resolving photon number states in a superconducting circuit. Nature **445**, 515 (2007)
57. J. Majer et al., Coupling superconducting qubits via a cavity bus. Nature **449**, 443 (2007)
58. A.A. Houck et al., Controlling the spontaneous emission of a superconducting transmon qubit. Phys. Rev. Lett. **101**, 080502 (2008)
59. A. Lupaşcu, S. Saito, T. Picot, P.C. de Groot, C.J.P.M. Harmans, J.E. Mooij, Quantum nondemolition measurement of a superconducting two-level system. Nat. Phys. **3**, 119 (2007)
60. N. Boulant et al., Quantum nondemolition readout using a Josephson bifurcation amplifier. Phys. Rev. B **76**, 014525 (2007)
61. J.E. Johnson, C. Macklin, D.H. Slichter, R. Vijay, E.B. Weingarten, J. Clarke, I. Siddiqi, Heralded state preparation in a superconducting qubit. Phys. Rev. Lett. **109**, 050506 (2012)
62. J.M. Chow et al., Implementing a strand of a scalable fault-tolerant quantum computing fabric. Nat. Commun. **5**, 4015 (2014)
63. E. Jeffrey et al., Fast accurate state measurement with superconducting qubits. Phys. Rev. Lett. **112**, 190504 (2014)
64. Y. Lin, J.P. Gaebler, F. Reiter, T.R. Tan, R. Bowler, A.S. Sørensen, D. Leibfried, D.J. Wineland, Dissipative production of a maximally entangled steady state of two quantum bits. Nature **504**, 415 (2013)
65. K. O'Brien, C. Macklin, I. Siddiqi, X. Zhang, Resonant phase matching of Josephson junction traveling wave parametric amplifiers. Phys. Rev. Lett. **113**, 157001 (2014)
66. J.Y. Mutus et al., Strong environmental coupling in a Josephson parametric amplifier. Appl. Phys. Lett. **104**, 263513 (2014)
67. C. Eichler, Y. Salathe, J. Mlynek, S. Schmidt, A. Wallraff, Quantum-limited amplification and entanglement in coupled nonlinear resonators. Phys. Rev. Lett. **113**, 110502 (2014)
68. D. Hover, S. Zhu, T. Thorbeck, G.J. Ribeill, D. Sank, J. Kelly, R. Barends, J.M. Martinis, R. McDermott, High fidelity qubit readout with the superconducting low-inductance undulatory galvanometer microwave amplifier. Appl. Phys. Lett. **104**, 152601 (2014)
69. V. Schmitt, X. Zhou, K. Juliusson, B. Royer, A. Blais, P. Bertet, D. Vion, D. Esteve, Multiplexed readout of transmon qubits with Josephson bifurcation amplifiers. Phys. Rev. A **90**, 062333 (2014)
70. E. Garrido, L. Riesebos, J. Somers, S. Visser, Feedback system for three-qubit bit-flip code. MSc project, Delft University of Technology (2014)
71. D.T. McClure, H. Paik, L.S. Bishop, M. Steffen, J.M. Chow, J.M. Gambetta, Rapid driven reset of a qubit readout resonator. Phys. Rev. Applied **5**, 011001 (2016)
72. N. Ofek et al., Demonstrating real-time feedback that enhances the performance of measurement sequence with cat states in a cavity. Bull. Am. Phys. Soc. (2015). http://meetings.aps.org/link/BAPS.2015.MAR.Y39.12
73. A.D. Córcoles, J.M. Chow, J.M. Gambetta, C. Rigetti, J.R. Rozen, G.A. Keefe, M. Beth Rothwell, M.B. Ketchen, M. Steffen, Protecting superconducting qubits from radiation. Appl. Phys. Lett. **99**, 181906 (2011)

74. R.T. Thew, K. Nemoto, A.G. White, W.J. Munro, Qudit quantum-state tomography. Phys. Rev. A **66**, 012303 (2002)
75. R. Bianchetti, S. Filipp, M. Baur, J.M. Fink, M. Göppl, P.J. Leek, L. Steffen, A. Blais, A. Wallraff, Dynamics of dispersive single-qubit readout in circuit quantum electrodynamics. Phys. Rev. A **80**, 043840 (2009)
76. C. Rigetti et al., Superconducting qubit in a waveguide cavity with a coherence time approaching 0.1 ms. Phys. Rev. B **86**, 100506 (2012)
77. B. Trauzettel, A.N. Jordan, C.W.J. Beenakker, M. Büttiker, Parity meter for charge qubits: an efficient quantum entangler. Phys. Rev. B **73**, 235331 (2006)
78. R. Ionicioiu, Entangling spins by measuring charge: a parity-gate toolbox. Phys. Rev. A **75**, 032339 (2007)
79. N.S. Williams, A.N. Jordan, Entanglement genesis under continuous parity measurement. Phys. Rev. A **78**, 062322 (2008)
80. G. Haack, H. Förster, M. Büttiker, Parity detection and entanglement with a Mach-Zehnder interferometer. Phys. Rev. B **82**, 155303 (2010)
81. C.W.J. Beenakker, D.P. DiVincenzo, C. Emary, M. Kindermann, Charge detection enables free-electron quantum computation. Phys. Rev. Lett. **93**, 020501 (2004)
82. H.-A. Engel, D. Loss, Fermionic Bell-state analyzer for spin qubits. Science **309**, 586 (2005)
83. C. Ahn, A.C. Doherty, A.J. Landahl, Continuous quantum error correction via quantum feedback control. Phys. Rev. A **65**, 042301 (2002)
84. W. Pfaff, T.H. Taminiau, L. Robledo, H. Bernien, M. Markham, D.J. Twitchen, R. Hanson, Demonstration of entanglement-by-measurement of solid-state qubits. Nat. Phys. **9**, 29 (2013)
85. C.L. Hutchison, J.M. Gambetta, A. Blais, F.K. Wilhelm, Quantum trajectory equation for multiple qubits in circuit QED: generating entanglement by measurement. Can. J. Phys. **87**, 225 (2009)
86. K. Lalumière, J.M. Gambetta, A. Blais, Tunable joint measurements in the dispersive regime of cavity QED. Phys. Rev. A **81**, 040301 (2010)
87. S. Filipp et al., Two-qubit state tomography using a joint dispersive readout. Phys. Rev. Lett. **102**, 200402 (2009)
88. L. Tornberg, G. Johansson, High-fidelity feedback-assisted parity measurement in circuit QED. Phys. Rev. A **82**, 012329 (2010)
89. K.W. Murch, S.J. Weber, C. Macklin, I. Siddiqi, Observing single quantum trajectories of a superconducting quantum bit. Nature **502**, 211 (2013)
90. R. Horodecki, P. Horodecki, M. Horodecki, K. Horodecki, Quantum entanglement. Rev. Mod. Phys. **81**, 865 (2009)
91. M. Sarovar, H.-S. Goan, T.P. Spiller, G.J. Milburn, High-fidelity measurement and quantum feedback control in circuit QED. Phys. Rev. A **72**, 062327 (2005)
92. N. Roch et al., Observation of measurement-induced entanglement and quantum trajectories of remote superconducting qubits. Phys. Rev. Lett. **112**, 170501 (2014)
93. Y. Liu, S. Shankar, N. Ofek, M. Hatridge, A. Narla, K. Sliwa, L. Frunzio, R.J. Schoelkopf, M.H. Devoret, Comparing and combining measurement-based and driven-dissipative entanglement stabilization (2015). arxiv:1509.00860
94. O.-P. Saira, J.P. Groen, J. Cramer, M. Meretska, G. de Lange, L. DiCarlo, Entanglement genesis by Ancilla-based parity measurement in 2D circuit QED. Phys. Rev. Lett. **112**, 070502 (2014)

Chapter 9
Quantum Acoustics with Surface Acoustic Waves

Thomas Aref, Per Delsing, Maria K. Ekström, Anton Frisk Kockum, Martin V. Gustafsson, Göran Johansson, Peter J. Leek, Einar Magnusson and Riccardo Manenti

Abstract It has recently been demonstrated that surface acoustic waves (SAWs) can interact with superconducting qubits at the quantum level. SAW resonators in the GHz frequency range have also been found to have low loss at temperatures compatible with superconducting quantum circuits. These advances open up new possibilities to use the phonon degree of freedom to carry quantum information. In this chapter, we give a description of the basic SAW components needed to develop quantum circuits, where propagating or localized SAW-phonons are used both to study basic

---

T. Aref · P. Delsing (✉) · M.K. Ekström · A.F. Kockum · M.V. Gustafsson · G. Johansson
Microtechnology and Nanoscience, Chalmers University of Technology,
Kemivägen 9, 41296 Göteborg, Sweden
e-mail: per.delsing@chalmers.se

T. Aref
e-mail: thomas.aref@chalmers.se

M.K. Ekström
e-mail: ekstromm@chalmers.se

G. Johansson
e-mail: goran.l.johansson@chalmers.se

A.F. Kockum
Present Address: Center for Emergent Matter Science, RIKEN,
2-1 Hirosawa, Wako, Saitama 351-0198, Japan
e-mail: anton.frisk.kockum@gmail.com

M.V. Gustafsson
Department of Chemistry, Columbia University, NWC Building
550 West 120th Street, New York, NY 10027, USA
e-mail: mg3465@columbia.edu

P.J. Leek · E. Magnusson · R. Manenti
Clarendon Laboratory, Department of Physics, University of Oxford,
Oxford OX1 3PU, UK
e-mail: peter.leek@physics.ox.ac.uk

E. Magnusson
e-mail: Einar.Magnusson@physics.ox.ac.uk

R. Manenti
e-mail: Riccardo.Manenti@physics.ox.ac.uk

© Springer International Publishing Switzerland 2016

R.H. Hadfield and G. Johansson (eds.), *Superconducting Devices in Quantum Optics*, Quantum Science and Technology,
DOI 10.1007/978-3-319-24091-6_9

physics and to manipulate quantum information. Using phonons instead of photons offers new possibilities which make these quantum acoustic circuits very interesting. We discuss general considerations for SAW experiments at the quantum level and describe experiments both with SAW resonators and with interaction between SAWs and a qubit. We also discuss several potential future developments.

9.1 Introduction

Quantum optics studies the interaction between light and matter. Systems of electromagnetic waves and atoms can be described by quantum electrodynamics (QED) in great detail and with amazingly accurate agreement between experiment and theory. A large number of experiments have been carried out within the framework of cavity QED, where the electromagnetic field is captured in a 3D cavity and allowed to interact with individual atoms. Cavity QED has been developed both for microwaves interacting with Rydberg atoms [1] and for optical radiation interacting with ordinary atoms [2]. In this chapter, we will discuss the acoustic analogue of quantum optics, which we might call "quantum acoustics", where acoustic waves are treated at the quantum level and allowed to interact with artificial atoms in the form of superconducting qubits.

The on-chip version of quantum optics is known as circuit QED and has been described in Chaps. 6–8. In circuit QED, superconducting cavities are coupled to artificial atoms in the form of superconducting electrical circuits that include Josephson junctions [3, 4]. The nonlinearity of the Josephson junctions is used to create a nonequidistant energy spectrum for the artificial atoms. By isolating the two lowest levels of this energy spectrum, the artificial atoms can also be used as qubits. The most commonly used circuit is the transmon [5], which is described in Chap. 6. Often the junction is replaced by a superconducting quantum interference device (SQUID), consisting of two Josephson junctions in parallel. This allows the level splitting of the artificial atom to be tuned in situ by a magnetic field so that the atom transition frequency can be tuned relative to the cavity resonance frequency. However, as described in Chap. 6, a cavity is not necessary to study quantum optics. The interaction between a transmission line and the artificial atom can be made quite large even without a cavity. Placing one or more atoms in an open transmission line provides a convenient test bed for scattering between microwaves and artificial atoms in one dimension. This subdiscipline of circuit QED is referred to as waveguide QED [6–9].

In a number of experiments, systems exploiting the mechanical degree of freedom have been investigated, and in several cases they have reached the quantum limit [10–13]. Typically, systems containing micro-mechanical resonators in the form of beams or drums are cooled to temperatures low enough that the thermal excitations of the mechanical modes are frozen out. This can be done in two different ways: either the frequency is made so high that it suffices with ordinary cooling in a dilution refrigerator, or alternatively, for mechanical resonators with lower frequencies, active cooling mechanisms such as sideband cooling can be employed.

With the development of radar in the mid-20th century, a need arose for advanced processing of radio frequency (RF) and microwave signals. An important class of components created to fill this need is based on surface acoustic waves (SAWs), mechanical ripples that propagate across the face of a solid. When SAWs are used for signal processing, the surface of a microchip is used as the medium of propagation. An electrical RF signal is converted into an acoustic wave, processed acoustically, and then converted back to the electrical domain. The substrate is almost universally piezoelectric, since this provides an efficient way to do the electro-acoustic transduction. The conversion is achieved using periodic metallic structures called interdigital transducers (IDTs), which will be discussed later.

Since the speed of sound in solids is around five orders of magnitude lower than the speed of light, SAWs allow functions like delay lines and convolvers to be implemented in small packages [14–16]. The acoustic wavelength is correspondingly short, and thus reflective elements and gratings can readily be fabricated on the surface of propagation by lithography. These features enable interference-based functionality, such as narrow-band filtering, with performance and economy that is unmatched by all-electrical devices. Since the heyday of radar development, SAW-based components have found their way into almost all wireless communication technology, and more recently also into the field of quantum information processing.

The most well-explored function of SAWs in quantum technology has hitherto been to provide a propagating potential landscape in semiconductors, which is used to transport carriers of charge and spin [17–19]. Here we are concerned with an altogether different kind of system. Rather than using SAWs to transport particles, we focus on quantum information encoded directly in the mechanical degree of freedom of SAWs. This use of SAWs extends the prospects of mechanical quantum processing to propagating phonons, which can potentially be used to transport quantum information in the same way as itinerant photons.

Most solid-state quantum devices, such as superconducting qubits, are designed to operate at frequencies in the microwave range ($\sim$5 GHz). These frequencies are high enough that standard cryogenic equipment can bring the thermal mode population to negligible levels, yet low enough that circuits much smaller than the electrical wavelength can be fabricated. When cooled to low temperatures, SAW devices show good performance in this frequency range, where suspended mechanical resonators tend to suffer from high losses. Indeed, recent experiments report Q-values above 10^5 at millikelvin temperatures for SAWs confined in acoustic cavities [20]. Another recent experiment has shown that it is possible to couple SAW waves to artificial atoms in a way very similar to waveguide QED [21]. Combining these results opens up the possibility to study what corresponds to cavity QED in the acoustic domain.

The advantageous features of sound compared to light are partly the same in quantum applications as in classical ones. The low speed of sound offers long delay times, which in the case of quantum processing can allow electrical feedback signals to be applied during the time a quantum spends in free propagation. The short wavelength allows the acoustic coupling to a qubit to be tailored, distributed, and enhanced compared with electrical coupling.

In this chapter we will discuss the exciting possibilities in this new area of research. First, in Sect. 9.2, we describe the SAW devices and how they are fabricated. Then, Sect. 9.3 provides a theoretical background. In Sect. 9.3.1, we describe the classical theory needed to evaluate and design the SAW devices, in Sect. 9.3.2 we present a semiclassical model for the coupling between a qubit and SAWs, and in Sect. 9.3.3 we present the quantum theory. In the following section, we describe the SAW resonators and their characterization. The SAW-qubit experiment is discussed in Sect. 9.5 and in the last section we give an outlook for interesting future experiments.

9.2 Surface Acoustic Waves, Materials and Fabrication

There are several different types of SAWs, but here we will use the term to denote Rayleigh waves [14, 15, 22]. These propagate elastically on the surface of a solid, extending only about one wavelength into the material. At and above radio frequencies, the wavelength is short enough for the surface of a microchip to be used as the medium of propagation.

The most important component in ordinary SAW devices is the IDT, which converts an electrical signal to an acoustic signal and vice versa. In its simplest form, the IDT consists of two thin film electrodes on a piezoelectric substrate. The electrodes are made in the form of interdigitated fingers so that an applied AC voltage produces an oscillating strain wave on the surface of the piezoelectric material. This wave is periodic in both space and time and radiates as a SAW away from each finger pair. The periodicity of the fingers p defines the acoustic resonance of the IDT, with the frequency given by $\omega_{IDT} = 2\pi f_{IDT} = 2\pi v_0/p$, where v_0 is the SAW propagation speed. When the IDT is driven electrically at $\omega = \omega_{IDT}$, the SAWs from all fingers add constructively which results in the emission of strong acoustic beams across the substrate surface, in the two directions perpendicular to the IDT fingers. The number of finger pairs N_p determines the bandwidth of the IDT, which is approximately given by f_{IDT}/N_p.

9.2.1 *Materials for Quantum SAW Devices*

The choice of substrate strongly influences the properties of the SAW device. The material must be piezoelectric to couple the mechanical SAW to the electrical excitation of the IDT. Because SAW devices have many commercial applications, a wide search for suitable piezoelectric crystals has been performed. Theoretical and experimental material data are available for many substrates, albeit until recently not at the millikelvin temperatures required for quantum experiments. Piezoelectric materials used for conventional SAW devices [15] include bulk piezoelectric substrates such as gallium arsenide (GaAs), quartz, lithium niobate ($LiNbO_3$), and lithium tantalate ($LiTaO_3$), and piezoelectric films such as zinc oxide (ZnO) and aluminum nitride

(AlN) deposited onto nonpiezoelectric substrates. For any given cut of a piezoelectric crystal, there are only a few directions where SAWs will propagate in a straight line without curving (an effect known as beam steering). Thus it is common to specify both the cut and the direction of a substrate, which also affect piezoelectric properties. For example, YZ lithium niobate is a Y-cut with propagation in the Z direction. Additional effects that need to be considered are diffraction, bulk wave coupling, other surface wave modes, ease of handling, etc.

Several material properties affect the suitability for quantum experiments and must be fine-tuned with various trade-offs. The two primary factors that play into the design of IDTs and similar structures are the piezoelectric coupling coefficient, K^2, and the effective dielectric constant, C_S. C_S is given as the capacitance of an IDT, per finger pair and unit length of finger overlap, W. Other parameters of importance are propagation speed v_0, attenuation rate α, and coefficients for diffraction, γ, and thermal expansion, δ. Table 9.1 summarizes our current knowledge of these parameters for a selection of substrates.

The material properties relevant for SAW devices may be substantially altered from known textbook values in the high vacuum and low temperature environment used for quantum devices. For example, the attenuation coefficient for SAW propagation loss is given at room temperature in air by [15]

$$\alpha = \alpha_{\text{air}}(P)\frac{f}{10^9} + \alpha_{\text{vis}}(T)\left(\frac{f}{10^9}\right)^2 \left[\frac{\text{dB}}{\mu\text{s}}\right], \tag{9.1}$$

Table 9.1 Material properties for selected substrates

Cut/Orient.	Temp.	$LiNbO_3$	GaAs	Quartz	ZnO
		Y-Z	{110}-<100>	ST-X	(0001)
K^2 (%)	300 K	4.8 [15]	0.07 [14]	0.14 [15]	1.14 [23]
C_S (pF/cm)	300 K	4.6 [14]	1.2 [14]	0.56 [15]	0.98 [24]
Diffraction γ	300 K	−1.08 [25]	−0.537* [25]	0.378 [25]	–
δ (ppm/K)	300 K	4.1 [26]	5.4 [27]	13.7 [28]	2.6 or 4.5 [29]
$\Delta l/l$ (%)	4 K	0.08 [30]	0.10 [30]	0.26 [30]	0.05 or 0.09 [30]
v_0 (m/s)	300 K	3488 [14]	2864 [14]	3158 [14]	–
v_e (m/s)	300 K	–	2883 ± 1	3138 ± 1	–
v_e (m/s)	10 mK	–	2914 ± 0.8	3135 ± 1.5	2678 ± 0.5 [20]
α_{vis} (dB/μsGHz^2)	300 K	0.88 [25]	0.9 [31]	2.6 [15]	–
α_{vis} (dB/μsGHz^2)	10 mK	–	<0.1	<0.01	<0.01 [20]

The effective velocity and attenuation rates are extracted from measurements of 1-port SAW resonators with aluminum electrodes of thickness 100 nm (GaAs and quartz) or 30 nm (ZnO) thick aluminum electrodes. GaAs data is for devices on the {110} plane and SAW propagation in the <100> direction. ZnO data is for devices on the (0001) plane. Diffraction can be quantified by the derivative of the power flow angle γ and is minimized when this parameter approaches −1 [15]
δ is the thermal expansion coefficient
*There seems to be some disagreement between different articles [25, 32, 33]

where the first term is due to air loading, the second due to viscous damping in the substrate, and f is the frequency. α_{air} can be neglected in high vacuum, and although reliable values are not yet available at low temperature for α_{vis}, an upper limit can be placed on it from known resonator quality factors.

When an IDT device is cooled to low temperature, the frequency of the device changes for two reasons. First, there is a length contraction that alters the distance between the IDT fingers, and also the length of cavities. Second, the speed of sound changes. The fractional length contraction $\Delta l/l$ when cooling down from room temperature to liquid helium temperatures can be estimated as $\Delta l/l = (l_{RT} - l_{4K})/(l_{RT}) \approx -190\,\delta$ if the Debye temperature is in the range 250–450 K [30]. Several of the materials have somewhat higher Debye temperatures, which should lower the values in Table 9.1 slightly.

In Table 9.1, two different propagation velocities are reported: the free velocity v_0 on a bare piezoelectric substrate and the effective velocity v_e. The effective velocity is a result of perturbation by, for instance, the metal strips in the resonator and relates to the free velocity as $v_e = v_0 + \Delta v_e + \Delta v_m$. This assumes that the electrical loading Δv_e and the mechanical loading Δv_m are independent.

Although comprehensive information is available on suitable materials for SAW devices at room temperature, new materials can become viable at low temperature. One example of this is ZnO, which although commonly used as a thin film transducer on nonpiezoelectric substrates such as sapphire, diamond or SiO_2/Si [15, 34], is not viable as a bulk crystal substrate at room temperature due to a substantial electrical conductivity, which damps the SAWs. This problem disappears at low temperature [20], making a very low loss SAW substrate, as discussed in Sect. 9.4.

9.2.2 Fabrication of SAW Devices

Regardless of the desired substrate material, SAW devices can be fabricated using standard lithographic techniques. The metal forming the IDT is typically aluminum since it is both very light (which prevents mass loading effects) and an excellent superconductor at low temperatures. If top contact is to be made to the aluminum, a thin layer of palladium can be deposited on top to prevent formation of aluminum oxide.

A condition for operating in the quantum regime is $\hbar\omega \gg k_B T$, where T is typically 10–50 mK for a dilution refrigerator. This implies that f_{IDT} must be on the order of GHz, and considering that SAW velocities are typically around 3000 m/s the required wavelength is below one micron. An IDT emits SAWs efficiently when the finger distance p is half a wavelength and therefore lithography on the submicron scale is needed. These dimensions are difficult to reach with photolithography, so electron-beam lithography is typically used. In principle, etching could be used to fabricate the IDT but care would need to be taken to ensure that the surface where the SAW propagates is not damaged by the etching process. A lift-off process avoids the danger of possible surface damage, but contamination from remnants of resist is a concern.

9.3 Theory

9.3.1 Classical IDT Model

An IDT can be considered a three-port electrical device with two acoustic channels (rightward and leftward going waves are represented by + and − superscripts, respectively), as shown in Fig. 9.1. Thus it can be described by a scattering matrix equation:

$$\begin{pmatrix} \phi^-_{out} \\ \phi^+_{out} \\ V^- \end{pmatrix} = \begin{pmatrix} S_{11} & S_{12} & S_{13} \\ S_{21} & S_{22} & S_{23} \\ S_{31} & S_{32} & S_{33} \end{pmatrix} \begin{pmatrix} \phi^+_{in} \\ \phi^-_{in} \\ V^+ \end{pmatrix}. \tag{9.2}$$

Here, $\phi^{\pm}_{in/out}$ represents the complex amplitude of an incoming (outgoing) SAWs on the left (right) side of the IDT and $V^{\pm}$ represents the complex amplitude of the incoming (outgoing) voltage wave on the IDT electrodes. Assuming reciprocity (a SAW travelling through the device left to right is given by time reversing a SAW traveling in the opposite direction) gives $S_{12} = S_{21}$, $S_{13} = S_{31}$, and $S_{23} = S_{32}$ (receiving a SAW is the time reversal of emitting a SAW). Assuming symmetry (the IDT looks the same to a SAW regardless of whether the SAW travels to the right or to the left) we have $S_{13} = S_{23}$, $S_{11} = S_{22}$, and $S_{31} = S_{32}$. In some cases, one might also be able to assume power conservation, in which case S would be unitary as well.

Using just symmetry and reciprocity leaves four independent terms in S: S_{11}, S_{12}, S_{13}, and S_{33}. S_{33} is the electrical reflection coefficient for a drive tone arriving at the IDT from the electrical transmission line S_{11} is the reflection of the SAW off the IDT; it is a combination of the pure mechanical reflection and outgoing SAW regenerated by the voltage induced by the incoming SAW. S_{12} is the transmission of SAW through an IDT, which again has both a mechanical component and a voltage regeneration component. S_{13} represents the electro-acoustic transduction and is thus proportional to the transmitter response function μ, which will be discussed below.

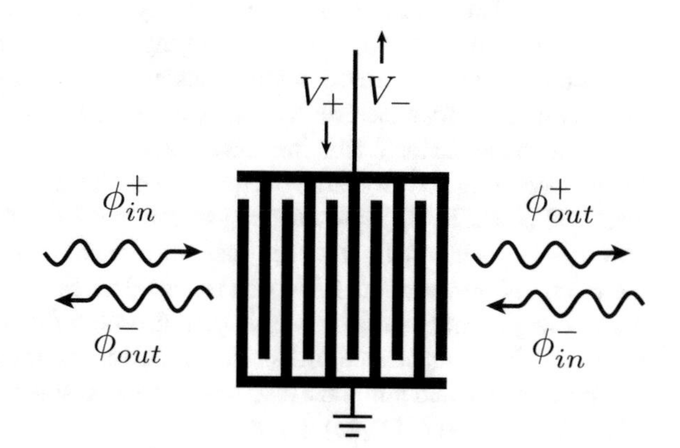

Fig. 9.1 An illustration of the three-port model for an IDT, featuring one electrical port and two acoustic ones. The IDT shown here has a single-finger structure

The simple single-finger form of the IDT, shown in Fig. 9.1, suffers from internal mechanical reflections which complicates the response and makes the scattering matrix only possible to estimate numerically, e.g., using the techniques known as the Reflective Array Method (RAM) and Coupling Of Modes (COM) [15]. Fortunately, one can eliminate these internal reflections using the double-finger structure, where each finger in the single-finger structure is replaced by two. The spacing between these two fingers can then be chosen such that reflection from the first finger interferes destructively with the reflection from the second finger. We will thus proceed with ideal single-finger and double-finger structures, assuming no mechanical reflection and no loss. This is approximately true for a superconducting double-finger structure but not always a good approximation for the single-finger structure.

Following Datta [14], the IDT is assumed to have a transmitter function $\mu(f)$ such that the emitted surface potential wave with amplitude ϕ_{em} is given by $\phi_{em} = \mu V_t$ when a voltage of amplitude $V_t = V^+ - V^-$ with frequency f is applied. Likewise, the IDT has a receiver response function $g(f)$ such that an incoming SAW induces a current $I = g\phi_{in}$. We define a characteristic impedance Z_0 such that a SAW of voltage amplitude ϕ carries power P_{SAW} where $Z_0 = 1/Y_0 = \left|\phi^2\right|/2P_{SAW}$. It can be shown using reciprocity that $g = 2\mu Y_0$.

The L and C parameters of the equivalent transmission line model for SAWs are given by the usual expressions $C = 1/(Z_0 v_0)$ and $L = Z_0/v_0$, where v_0 is the free surface wave velocity. Since $v_0 = 1/\sqrt{LC}$, small changes in velocity are related to small changes in capacitance: $|\Delta v_0/v_0| = \Delta C/(2C)$. In a conducting metal film on top of the SAW substrate, the surface potential ϕ induces a surface charge density q_s resulting in $\Delta C = -q_s W/\phi$ for a film of width W (the effective length of the IDT finger) and a corresponding Δv_0. The resulting connection between applied voltage and SAW is called the piezoelectric coupling constant K^2, defined such that $2\Delta v_0/v_0 = K^2$. K^2 is a material property and has been calculated and measured for a variety of SAW substrates, see Table 9.1. By considering the discontinuity caused by the surface charge density and using Maxwell's laws, it can be shown that $Y_0 = \omega_{IDT} W C_S/K^2$.

Without piezoelectric effects, the IDT is just a geometric capacitor C_t with admittance $i\omega C_t$. The presence of piezoelectricity adds a complex admittance element $Y_a(\omega) = G_a(\omega) + iB_a(\omega)$ [14–16], capturing the electro-acoustic properties. Including a current source representing the transduction from SAW to electricity, we have the circuit model for a receiver IDT, shown in Fig. 9.2. We can also consider a voltage source with the characteristic impedance $Z_C = 1/Y_C$ and get the equivalent circuit for a transmitting IDT. When a voltage V_t is applied to an emitting IDT, dissipation of electrical power in G_a represents conversion into SAW power $P = \frac{1}{2}|V_t|^2 G_a$. This power is divided equally between the two directions of propagation. By equating the electrical power to the SAW power, one derives $G_a = 2|\mu|^2 Y_0$. Because μ and g are always purely imaginary, this can also be written as $G_a = -\mu g$. Matching G_a to the 50 Ω impedance line is important for the transmitting/receiving IDT to maximize signal and minimize electrical reflection since $S_{33} = (Y_C - Y)/(Y_C + Y)$ where $Y = G_a(\omega) + iB_a(\omega) + i\omega C_t$.

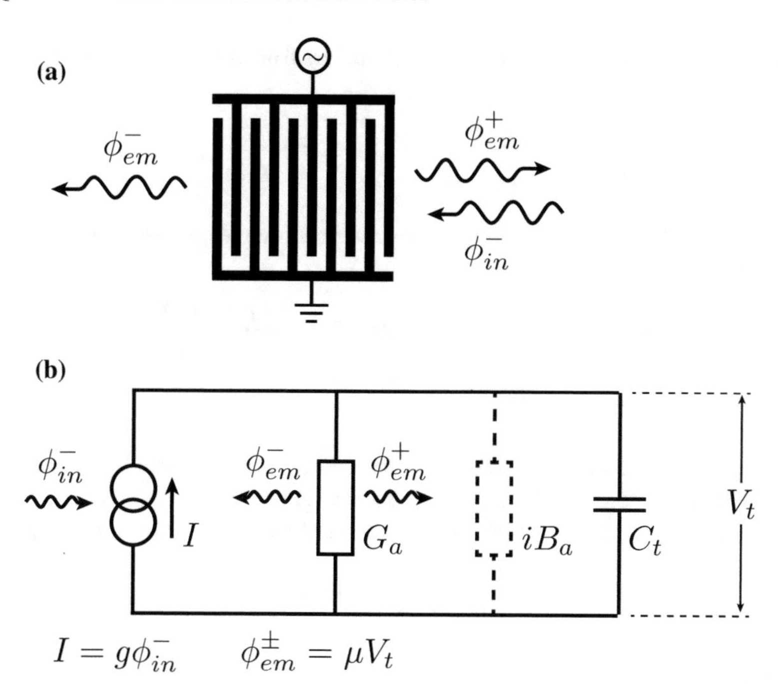

Fig. 9.2 Classical model for the IDT. **a** Layout of the IDT, with incoming and outgoing SAW components shown. **b** Equivalent circuit, see text

The imaginary component of the acoustic admittance B_a is the Hilbert transform of G_a due to causality. Thus specifying μ and the capacitance C_t yields the entire electro-acoustic behavior of the IDT. It can be shown that μ depends on the Fourier transform of the surface charge density, though this is either a complicated algebraic formula for regular (evenly spaced) transducers or must be evaluated numerically for nonregular transducers. Likewise the capacitance relates to the surface charge induced by an applied voltage, and is again a complicated algebraic formula for regular transducers, for more complicated structures it needs to be evaluated numerically. In the particular case of single-finger and double-finger electrodes with metallization ratio of 50 %, the results are

$$\mu_0^{sf} = 1.6i\Delta v_0/v_0 \approx 0.8iK^2, \quad \mu_0^{df} = 1.2i\Delta v_0/v_0 \approx (0.8/\sqrt{2})iK^2, \tag{9.3}$$

$$C_t^{sf} = N_p C_S W, \qquad C_t^{df} = \sqrt{2}N_p C_S W. \tag{9.4}$$

Here, $\mu_0^{sf/df}$ is the response of one finger pair when all other fingers are grounded, and C_t is the capacitance of an IDT with N_p finger periods. Using superposition, this allows separating out the response of a single finger (called an element factor) from the effect of superimposing several fingers (the array factor) for regular transducers.

The array factor is the sum of the phase factors from all the different fingers

$$A(\omega)^{sf} = \sum_{n=0}^{N_p-1} \exp\left(-i2\pi n \frac{\omega}{\omega_{IDT}}\right), \tag{9.5}$$

which can be shown by a geometric series and small-angle approximation argument to yield

$$A(\omega)^{sf} = N_p \frac{\sin(X)}{X}, \quad A(\omega)^{df} = \sqrt{2} N_p \frac{\sin(X)}{X}, \tag{9.6}$$

$$X = N_p \pi \frac{\omega - \omega_{IDT}}{\omega_{IDT}}. \tag{9.7}$$

In the approximation that the element factor for the double-finger case is smaller by roughly a factor of $\sqrt{2}$ while the array factor is greater by a factor of $\sqrt{2}$, the two cancel and we get the same form for the total response function of double-finger and single-finger IDTs:

$$\mu = 0.8 i K^2 N_p \frac{\sin(X)}{X}. \tag{9.8}$$

Thus, we get that

$$G_a(\omega) = G_{a0} \left[\frac{\sin(X)}{X}\right]^2, \tag{9.9}$$

$$B_a(\omega) = G_{a0} \left[\frac{\sin(2X) - 2X}{2X^2}\right], \tag{9.10}$$

where $G_{a0} \approx 1.3 K^2 N_p^2 \omega_{IDT} W C_S$. On acoustic resonance, $\omega = \omega_{IDT}$, G_a is at its maximum and the imaginary element B_a is zero. The sinc function dependence of μ on frequency implies a bandwidth of approximately $0.9 f_{IDT}/N_p$.

9.3.2 Semiclassical Theory for SAW-Qubit Interaction

In the experiment coupling a transmon qubit to SAWs [21], we take advantage of the similarities between the interdigitated structure of a classical IDT and that of the transmon. The transmon consists of a SQUID connecting two superconducting islands that form a large geometric capacitance C_{tr} in an interdigitated pattern. The SQUID acts as a nonlinear inductance L_J (the Josephson inductance) and forms an electrical resonance circuit together with C_{tr}. The nonlinearity of L_J gives the transmon an anharmonic energy spectrum. L_J can be tuned by an external magnetic flux and in that way the frequency of the transmon can be adjusted. In this section, we consider a semiclassical model, valid when the incoming SAW power is low enough that the qubit is never excited beyond the $|1\rangle$ state. In that case, the SQUID can be

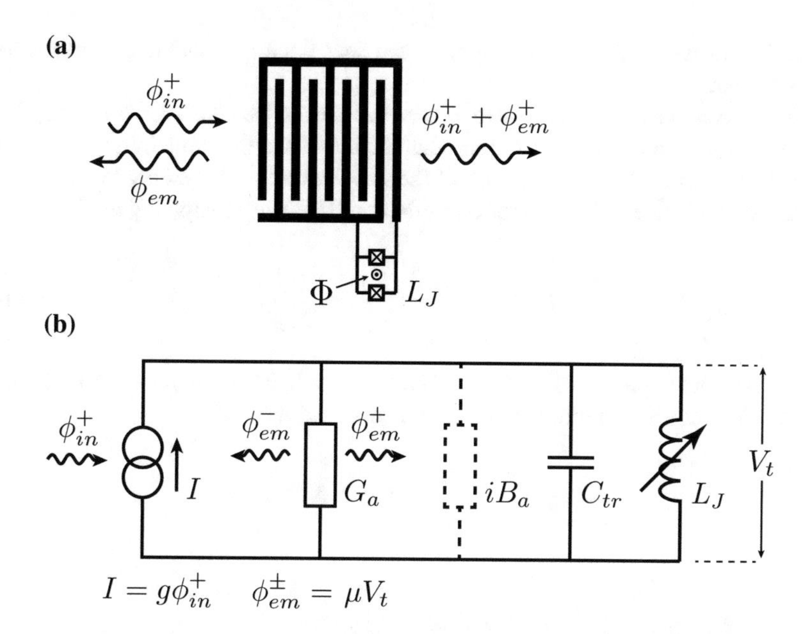

Fig. 9.3 Semiclassical model for the acoustically coupled qubit. **a** Layout of the qubit, with incoming and outgoing SAW components shown. The capacitively coupled gate is not included in this model. **b** Equivalent circuit, see text

approximated as a linear inductance. We can then use the circuit model shown in Fig. 9.3.

Lithography fixes the acoustic resonance frequency ω_{IDT}, but the electrical resonance frequency of the transmon can be tuned by adjusting L_J with a magnetic field. The two resonances coincide when

$$L_J = \frac{1}{\omega_{IDT}^2 C_{tr}}. \tag{9.11}$$

At this point, the impedance of the LC circuit approaches infinity, and we are left with only the acoustic element G_a. As a result, the current generated by an incoming SAW beam with amplitude ϕ_{in}^{+} in the rightward direction produces a voltage over G_a,

$$V_t = I/G_a = g\phi_{in}^{+}/G_a, \tag{9.12}$$

which in turn gives rise to re-radiation of SAW in the rightward and leftward directions with amplitudes

$$\phi_{em}^{\pm} = \mu V_t = (\mu g/G_a)\phi_{in}^{+} = -\phi_{in}^{+}. \tag{9.13}$$

Hence, the net transmission of SAW in the rightward direction is $\phi_{out}^{+} = \phi_{in}^{+} + \phi_{em}^{+} = 0$, and the emission in the leftward direction is $\phi_{out}^{-} = -\phi_{in}^{+}$. This explains the acoustic reflection S_{11} we observe experimentally (see Sect. 9.5 for details) in the limit of low SAW power.

The semiclassical model can also be used to estimate the relaxation rate (coupling) Γ_{ac} of the qubit to phonons, assuming that the qubit is at the same frequency, ω_{10} as its IDT structure. The damping rate of the RLC circuit tells us how fast electrical energy stored in the LC resonator converts into SAWs by dissipating in G_a, giving

$$\Gamma_{ac} = \omega_{10} \frac{G_a}{2} \sqrt{\frac{L_J}{C_{tr}}} = \frac{G_a}{2C_{tr}}. \tag{9.14}$$

Thus, using our expressions for G_{a0} and C_{tr} from Sect. 9.3.1, we get a simple expression for the acoustic coupling between the qubit and the SAWs:

$$\Gamma_{ac} = \frac{1.3\, K^2 N_p \omega_{10}}{2\sqrt{2}} \approx 0.5\, K^2 N_p \omega_{IDT}. \tag{9.15}$$

9.3.3 Quantum Theory for Giant Atoms

To go beyond the semiclassical model and understand the behavior of the transmon coupled to SAWs in regimes where stronger drive strengths are used and more levels of the artificial atom come into play, a fully quantum description is needed. From a quantum optics perspective, one of the main reasons that the transmon coupled to SAWs is a very interesting system is that it forms a "giant artificial atom". Natural atoms used in traditional quantum optics typically have a radius $r \approx 10^{-10}$ m and interact with light at optical wavelengths $\lambda \approx 10^{-7} - 10^{-6}$m [2, 35]. Sometimes the atoms are excited to high Rydberg states ($r \approx 10^{-8} - 10^{-7}$m), but they then interact with microwave radiation ($\lambda \approx 10^{-3} - 10^{-1}$m) [36, 37]. Microwaves also interact with superconducting qubits, but even these structures are typically measured in hundreds of micrometers (although some recent designs approach wavelength sizes [38]). Consequently, theoretical investigations of atom-light interaction usually rely on the dipole approximation that the atom can be considered point-like when compared to the wavelength of the light. This is clearly not the case for the transmon coupled to SAWs, since here each IDT finger is a connection point and the separation between fingers is always on the order of wavelengths.

Conceptually, the SAW-transmon setup is equivalent to a model where an atom couples to a 1D continuum of bosonic modes at a number of discrete points, which can be spaced wavelengths apart. Such a setup should also be possible to realize with a variation of the transmon design, the "xmon" [39], coupled to a meandering superconducting transmission line for microwave photons. In Ref. [40], we investigated the physics of this model, a sketch of which is shown in Fig. 9.4. Here, we summarize the main results from that paper.

The Hamiltonian for our model is

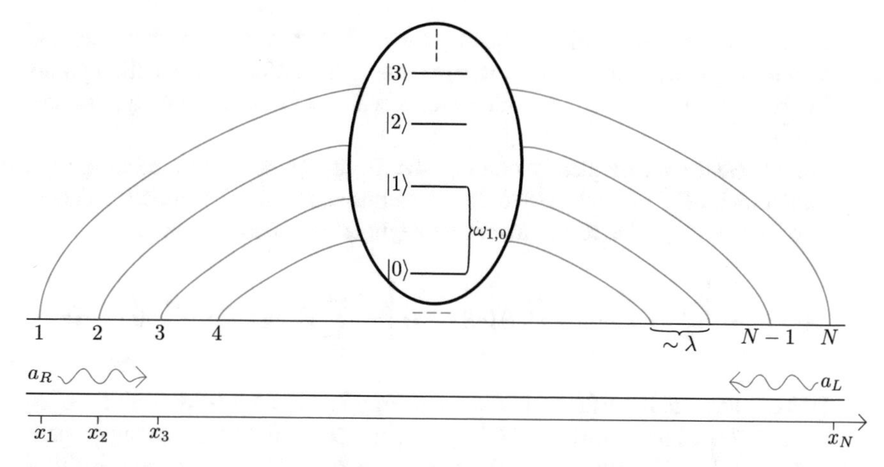

Fig. 9.4 The quantum model for a transmon coupled to SAWs: a giant artificial multi-level atom, connected at N points to left- and right-moving excitations in a 1D transmission line

$$H = H_a + H_{tl} + H_{int}, \tag{9.16}$$

where we define the multi-level-atom Hamiltonian

$$H_a = \sum_m \hbar\omega_m |m\rangle \langle m| , \tag{9.17}$$

the transmission line Hamiltonian

$$H_{tl} = \sum_j \hbar\omega_j \left(a_{Rj}^\dagger a_{Rj} + a_{Lj}^\dagger a_{Lj} \right), \tag{9.18}$$

and the interaction Hamiltonian

$$H_{int} = \sum_{j,k,m} \hbar g_{jkm} \left(\sigma_-^m + \sigma_+^m\right) \left(a_{Rj} e^{-i\frac{\omega_j x_k}{v_0}} + a_{Lj} e^{i\frac{\omega_j x_k}{v_0}} + \text{H.c.} \right), \tag{9.19}$$

where $\sigma_-^m = |m\rangle\langle m+1|$, $\sigma_+^m = |m+1\rangle\langle m|$, and H.c. denotes Hermitian conjugate. The atom has energy levels $m = 0, 1, 2, \ldots$ with energies $\hbar\omega_m$. It is connected to right- and left-moving bosonic modes Rj and Lj of the transmission line with some coupling strength g_{jkm} at N points with coordinates x_k. We assume that the time it takes for a transmission line excitation to travel across all the atom connection points is negligible compared to the timescale of atom relaxation. Thus, only the phase shifts between connection points need to be included in the calculations, not the time delays. In addition, we assume that the coupling strengths g_{jkm} are small compared to the relevant ω_m and ω_j and that they can be factorized as $g_{jkm} = g_j g_k g_m$, which is the case for the transmon [5]. In general, the mode coupling strength g_j can be considered constant over a wide frequency range. The factors g_k are dimensionless and only describe the relative coupling strengths of the different connection points. Finally, for the transmon [5] and other atoms with small anharmonicity, we have $g_m = \sqrt{m+1}$.

Using standard techniques, including the Born, Markov, and rotating-wave approximations [41, 42], we derive the master equation for the effective density matrix of the atom, ρ. The result, assuming negligible temperature, is

$$\dot{\rho}(t) = -i \left[\sum_m (\omega_m + \Delta_m) \left| m \right\rangle \left\langle m \right| , \rho(t) \right] + \sum_m \Gamma_{m+1,m} \mathcal{D} \left[\sigma_-^m \right] \rho(t), \quad (9.20)$$

where $\hbar\Delta_m$ are small shifts of the atom energy levels (see below) and we use the notation $\mathcal{D}[O]\rho = O\rho O^\dagger - \frac{1}{2} O^\dagger O \rho - \frac{1}{2} \rho O^\dagger O$ for the Lindblad superoperators [43] that describe relaxation. The relaxation rates $\Gamma_{m+1,m}$ for the transitions $|m+1\rangle \to |m\rangle$ are given by

$$\Gamma_{m+1,m} = 4\pi g_m^2 J(\omega_{m+1,m}) \left| A(\omega_{m+1,m}) \right|^2, \quad (9.21)$$

where $A(\omega_j) = g_j \sum_k g_k e^{i\omega_j x_k / v_0}$ contains the array factor from the classical SAW theory above, $J(\omega)$ is the density of states for the bosonic modes, and $\omega_{r,s} = \omega_r - \omega_s$. $A(\omega)$ enters squared, just like $G_a \propto |\mu|^2$; this gives a frequency-dependent coupling set by interference between the coupling points. Thus, we can design our artificial atom such that it relaxes fast at certain transition frequencies, but remains protected from decay at others.

The shift of the atom energy levels by $\hbar\Delta_m$ in Eq. (9.20) is an example of a Lamb shift [44, 45], which is a renormalization of the atom energy levels caused by the interaction with vacuum fluctuations of the 1D continuum. The shifts are given by

$$\Delta_m = 2\mathcal{P} \int_0^\infty d\omega \frac{J(\omega)}{\omega} |A(\omega)|^2 \left(\frac{g_m^2 \omega_{m+1,m}}{\omega + \omega_{m+1,m}} - \frac{g_{m-1}^2 \omega_{m,m-1}}{\omega - \omega_{m,m-1}} \right), \quad (9.22)$$

where $\mathcal{P}$ denotes the principal value.

Again, this is essentially equivalent to the imaginary acoustic admittance iB_a, which shifts the LC resonance frequency in the semiclassical calculation above if the atom is not on electrical resonance with the IDT structure.

In conclusion, the giant artificial atom differs from an ordinary, "small" atom in that both its relaxation rates and its Lamb shifts become frequency-dependent. The intuition for this frequency-dependence carries over from the classical SAW theory.

9.4 SAW Resonators for Quantum Devices

Seeing that a qubit can be coupled to SAWs, it is natural to consider the prospects for trapping SAW phonons in resonant structures, for example for implementation of a SAW phonon version of circuit QED. High-quality SAW resonators have long been used in conventional SAW devices [15], for example to implement high-quality oscillators, having been first introduced in the 1970s [46]. A resonator is made by creating high-reflectivity mirrors for a SAW, which can be achieved using a shorted or open grating of many electrodes, with period $p = \lambda/2$ (see Fig. 9.5). Such gratings operate in much the same way as a Bragg grating in optics, with constructive interference occurring between the multiple reflections from the electrodes when the device is on resonance. A signal can be coupled into and out of the device with an IDT, and the frequency response measured to obtain information about the resonator modes and their quality factors.

A schematic of a one-port SAW resonator and its frequency response are shown in Fig. 9.5. The frequency response close to resonance can be modelled as an RLC resonator to obtain the following formula for the reflection coefficient:

$$r_R(f) = \frac{(Q_e - Q_i) + 2i Q_i Q_e \Delta f / f_0}{(Q_e + Q_i) + 2i Q_i Q_e \Delta f / f_0}, \tag{9.23}$$

where f_0 is the resonant frequency for the SAW cavity, $\Delta f = f - f_0$ and Q_i, Q_e are the internal and external quality factors, respectively. Note that here the actual capacitance of the IDT is neglected.

9.4.1 *Resonator Quality Factors at Low Temperature*

The external quality factor Q_e of a SAW resonator is determined by the IDT geometry and external circuit parameters, whereas the internal quality factor Q_i has contributions from multiple sources, including diffraction, conversion to bulk phonons, finite grating reflectivity and resistivity. In the following, we describe preliminary measurements of SAW resonator quality factors on ST-cut quartz measured at low temperature, characterizing two of these contributions specifically—the Q_i due to finite grating reflectivity, and Q_e due to the IDT. In all cases, parameters are extracted from fits to Eq. (9.23), after appropriate calibration of imperfections from the measurement circuit.

Figure 9.6 shows measurements of Q_i for a set of devices with $p = 3\,\mu\text{m}$, the cavity width $W = 750\,\mu\text{m}$, the cavity length $L_c \approx 460\,\mu\text{m}$ ($f_0 \approx 0.522\,\text{GHz}$). The number of grating electrodes N_g is varied. The contribution to Q_i due to the grating reflectivity, Q_g, can be derived by summing the multiple reflections arising from the grating electrodes, which in the limit of large N_g is given by [15]

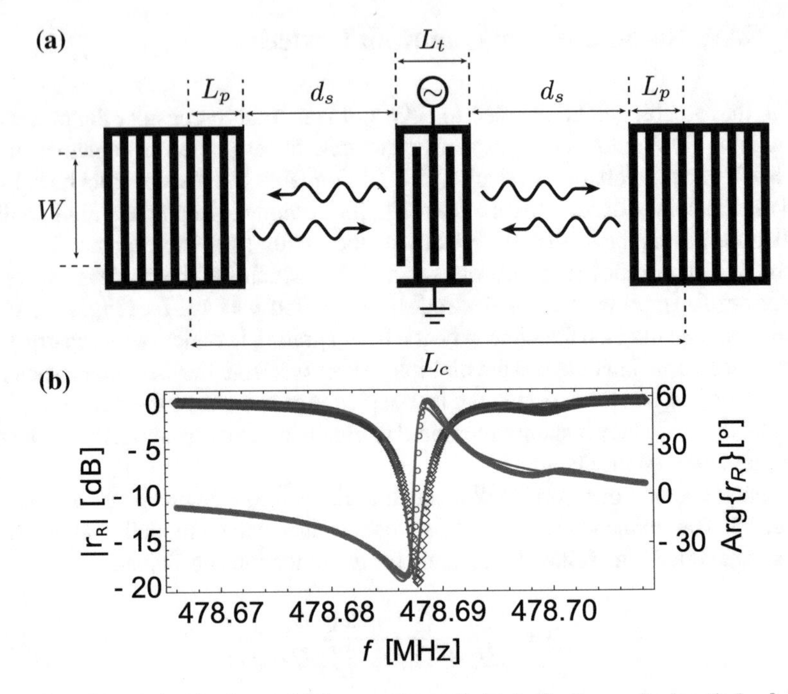

Fig. 9.5 **a** Schematic of a 1-port SAW resonator, with total effective cavity length L_c. **b** Measurement of the reflection coefficient r_R of a 1-port SAW resonator on GaAs at a wavelength of $p = 3\,\mu\text{m}$. *Blue* and *green circles* are the magnitude and phase of r_R respectively, and the *solid line* is a fit to Eq. (9.23)

$$Q_g(L_c, N_g) = \frac{\pi L_c}{\lambda_0 \left(1 - \tanh\left(|r_s|\, N_g\right)\right)}, \tag{9.24}$$

with L_c the cavity length, λ_0 the cavity wavelength, and r_s the reflectivity of a single grating electrode. The data fit extremely well to this equation, indicating that the

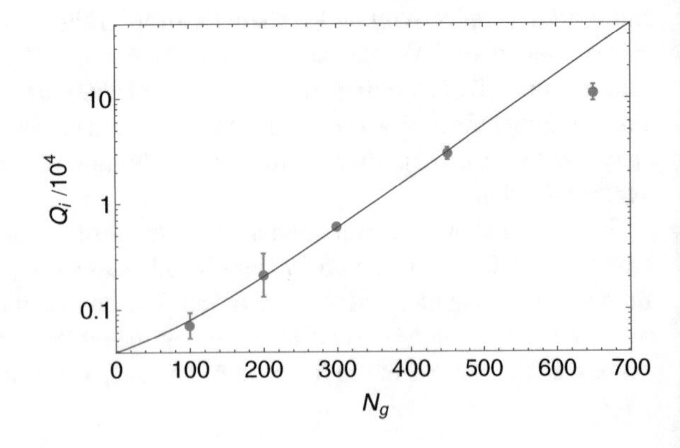

Fig. 9.6 Measurements of internal quality factor Q_i of a set of SAW resonators on ST-cut quartz with $p = 3\,\mu\text{m}$, as a function of number of electrodes in the grating reflectors N_g, measured at 10 mK. The *blue line* is a fit to Eq. (9.24)

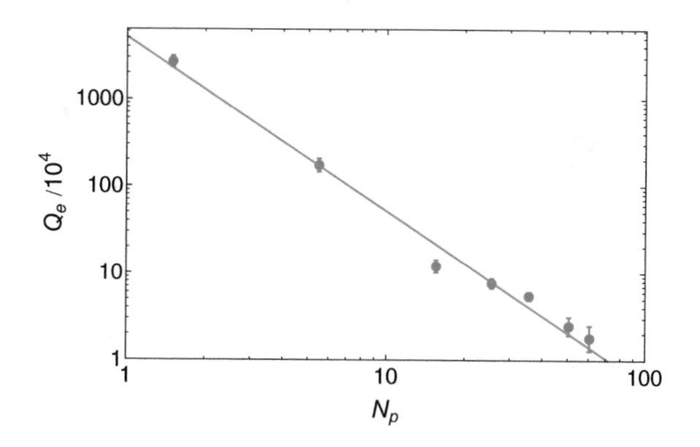

Fig. 9.7 Measurements of external quality factor Q_e of SAW resonators on ST-cut quartz at $p = 3\,\mu\text{m}$, as a function of number of finger pairs in the IDT N_p, measured at 10 mK. The *green line* is a fit to Eq. (9.26)

gratings indeed behave according to this simple model, and we can conclude that any other contributions to the dissipation are at the $Q_i > 10^5$ level. Further experiments at larger N_g will enable determination of contributions from other sources such as diffraction, which is expected to follow [47]

$$Q_d(W) = \frac{5\pi}{|1+\gamma|}\left(\frac{W}{\lambda_0}\right)^2, \tag{9.25}$$

where the diffraction parameter $\gamma = 0.378$ for ST-X quartz [25].

Figure 9.7 shows measurements of Q_e for a set of devices with $p = 3\,\mu\text{m}$, $W = 160\,\mu\text{m}$, and $L_c \approx 200\,\mu\text{m}$ in which the number of electrodes in the IDT, N_p, is varied. A larger IDT naturally couples more strongly to the resonator, giving lower Q_e. A full expression for the expected dependence is given by [15]

$$Q_e(L_c, N_p) = \frac{1}{5.74\, v_0 Z_C C_S W K^2}\frac{L_c}{N_p^2}, \tag{9.26}$$

where Z_C is the characteristic impedance of the electrical port coupled to the IDT. The data fit extremely well to this equation, which shows that a wide range of external quality factors, from 10^4 to above 10^7, can be engineered. Combining this information with the observed internal quality factors of up to 10^5 shows that strongly under- or over-coupled resonators can easily be fabricated.

9.4.2 *ZnO for High Q SAW Devices at Low Temperature*

Bulk ZnO has intrinsic carriers at elevated temperatures, and is thus only feasible as a material for SAW devices at low temperatures. In [20], results are reported on a delay line and a one-port resonator fabricated on a 0.5 mm thick high quality ZnO substrate, measured in a dilution refrigerator down to 10 mK. We summarize the important results in Fig. 9.8, in which we show the temperature dependence of the transmission of the delay line, and of the quality factor of the resonator. Remarkably,

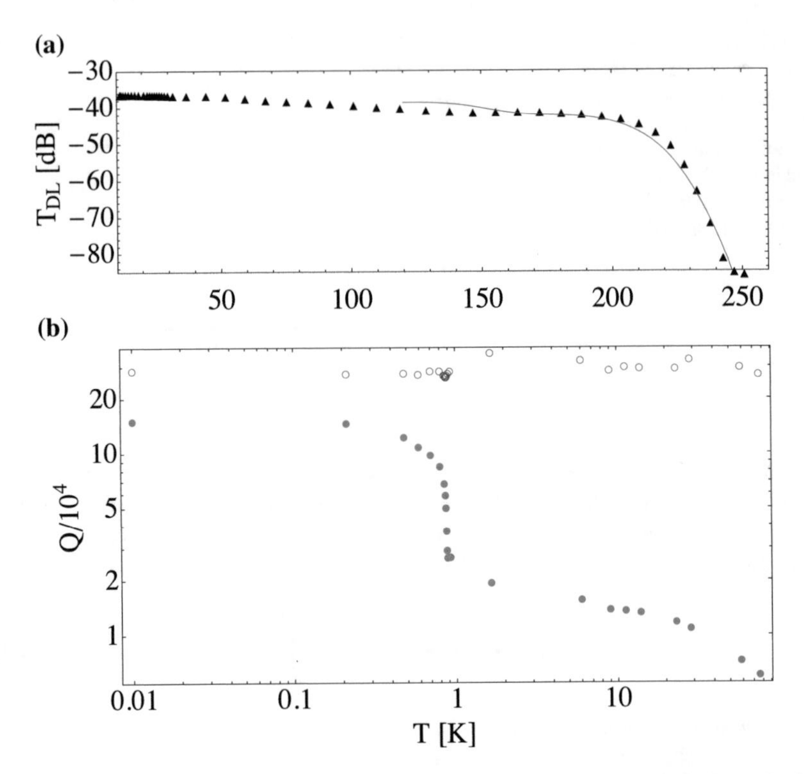

Fig. 9.8 **a** Transmission amplitude T_{DL} of the fundamental mode ($f_{DL} = 446$ MHz) of a ZnO delay line as a function of temperature T. The *solid line* is a fit of transmission attenuating linearly with the measured conductivity. **b** Temperature dependence of internal quality factor Q_i (*filled circles*) and external quality factor Q_e (*empty circles*) of a one-port SAW resonator on ZnO, at $f_R \simeq 1.7$ GHz. Figures taken from [20]. Copyright 2015, AIP Publishing LLC

not only does the substrate become viable for these SAW devices at low temperature, but it also proves to have very low loss, with the resonator at 1.7 GHz reaching an internal quality factor of $Q_i \simeq 1.5 \times 10^5$ at 10 mK.

The delay line device has a parallel IDT design with $p = 2\,\mu$m and $p = 3\,\mu$m transducers, and a mirrored output IDT at 2 mm separation. As Fig. 9.8a shows, the transmittance through the delay line (at the fundamental frequency f_{DL}) is quenched around 200–240 K due to the onset of conductance. The line is a fit to the data of $T_{DL} = a \cdot e^{-b/\rho_e(T)}$, with parameters $a = -38.3$ dB and $b = 2.65 \times 10^8$ Ωm ($\rho_e(T)$ is the resistivity of the substrate). The agreement with the data demonstrates that the attenuation is indeed inversely proportional to the measured resistivity ρ_e, showing that this is the dominant source of loss in the high temperature range. The resonator has a single IDT with 21 fingers, and gratings of 1750 fingers on either side, with $p = 0.8\,\mu$m, with a resonance seen at $f_R = 1.677784$ GHz at 10 mK. As can be seen in Fig. 9.8b, the internal quality factor drops by almost an order of magnitude in the range $0.01 < T < 1$ K, indicating a strong contribution from the superconductivity of the Ti/Al bilayer electrodes.

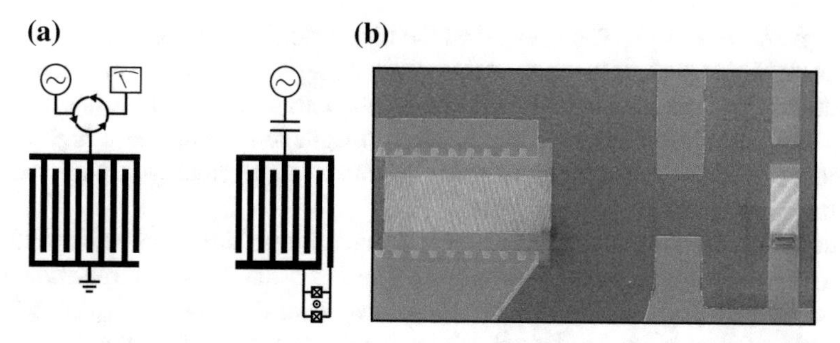

Fig. 9.9 **a** Simplified layout for the SAW-qubit sample. A surface acoustic wave is generated by the IDT and launched towards the transmon qubit, with its capacitance shaped into an IDT. This setup allows both to test SAW reflection on the qubit, and to listen to phonons emitted by the relaxing qubit. **b** False color picture of the sample. The *bluish parts* are the IDT (to the *left*) and the qubit with its gate (to the *right*). The *yellow parts* are the coplanar waveguide and surrounding ground plane

9.5 SAW-Qubit Interaction in Experiment

Having seen the experimental results for SAW resonators and the theory for IDTs and qubits, we now turn to an overview of the results presented in [21], where the coupling between SAWs and a superconducting qubit was demonstrated. We highlight some of the technical considerations that apply to the design of hybrid quantum-acoustical devices.

The sample used in [21] (see Fig. 9.9) consisted of a polished GaAs substrate. A single IDT on one end of the chip is used to convert electrical microwaves into SAWs and vice versa. The IDT is aligned with the [011] direction of the crystal. 100 μm away from the IDT, the transmon qubit is deposited, with the shunt capacitance fashioned into a finger structure as described above, aligned with the IDT. When the IDT is excited electrically, it emits a coherent SAW beam in the direction of the qubit, and the phonons that are reflected or emitted by the qubit can be detected by the IDT. One of the qubit electrodes is grounded, and the other one couples to an electrical gate through a small capacitance.

Although both the IDT and the qubit have interdigitated structures, they are subject to different constraints, and this presents a challenge in the choice of materials and layout. To achieve optimal electro-acoustic conversion in the IDT, its impedance should be matched as well as possible to that of the electrical transmission line it is connected to, which is typically 50 Ω. This is generally easiest to achieve on a strongly piezoelectric substrate material such as $LiNbO_3$. In that case only a few finger periods are needed to bring the real part of the acoustic impedance, G_0, close to 50 Ω. This gives the IDT a large acoustic bandwidth and reduces the influence of internal mechanical reflections, as well as of the shunt capacitance C_{IDT}.

The acoustic impedance of the qubit, on the other hand, does not need to be matched to any electrical transmission line, since its coupling to the electrical gate is designed to be weak. The strength of the piezoelectric coupling constant and the number of finger periods determines the acoustic coupling rate Γ_{ac}, which represents the rate at which the qubit can absorb and emit phonons (see Eq. (9.15)). It is desirable that this rate exceeds any coupling to electromagnetic modes, and a high rate of phonon processing also relaxes the requirements on signal fidelity through the IDT and the amplifier chain.

An essential feature of a qubit, however, is that the transitions between its different energy levels can be separately addressed. The separation between transition energies, $a = \omega_{21} - \omega_{10}$ is known as the *anharmonicity* of the qubit. If Γ_{ac} approaches $|a|$, a signal used to excite the qubit from the ground state $|0\rangle$ to the first excited state $|1\rangle$ is also capable of exciting the $|2\rangle$ state. This means that the qubit ceases to work as a two-level system. The transmon design is a good candidate for coupling to SAWs since its large shunt capacitance can be shaped into an IDT-like structure. However, this design also comes with inherently low anharmonicity [5], which puts an upper bound on Γ_{ac}. A straightforward way to reduce Γ_{ac} is to lower the number of finger periods, N_{tr}, but the finger structure of the qubit needs to have at least a few finger pairs in order to couple preferentially to the desired Rayleigh modes. On a strongly piezoelectric material, it is not necessarily possible to achieve a good balance between coupling strength and anharmonicity.

In the sample discussed here, the trade-off between IDT and qubit performance was managed by using GaAs, which is only weakly piezoelectric ($K^2 \approx 7 \times 10^{-4}$). With a moderate number of finger periods in the qubit, $N_{tr} = 20$, spaced for operation around $\omega_{10}/2\pi = 4.8$ GHz, this gives a coupling of $\Gamma_{ac}/2\pi = 30$ MHz according to Eq. (9.15). The width of the SAW beam (length of the fingers) is $W = 25\,\mu$m and the fingers are pairwise alternating, in a design that minimizes internal mechanical reflections [14, 15, 48] (c.f. Sect. 9.3.1). With this design, the acoustic bandwidth of the qubit substantially exceeds Γ_{ac} and C_{tr} is sufficiently high for the qubit to operate well in the transmon regime.

To match the IDT to 50 Ω with the same kind of ideal (nonreflective) finger structure would, however, not be possible on this substrate. Several thousand fingers would be required to reach optimal impedance matching, which makes fabrication infeasible and introduces problems due to the high shunt capacitance. To achieve the required impedance matching, the IDT instead consists of a single-finger structure where internal mechanical reflections are prominent. These reflections confine SAWs within the finger structure, achieving a stronger coupling (lower impedance) with a manageable number of fingers. The optimal number of fingers in this configuration was found experimentally to be $N_p = 125$. This value depends on the mechanical reflection coefficient of each IDT finger, which in turn depends on the material and thickness of the metallization. The resonant operation of the IDT, along with the relatively high value of N_p compared with N_{tr}, makes the bandwidth of the IDT very slim, ~1 MHz. This is the band in which phonons can be launched toward the qubit and phonons emanating from the qubit detected.

The capacitive gate suffers no such bandwidth limitation, and can be used to address transitions in the qubit also outside the acoustic frequency band of the IDT.

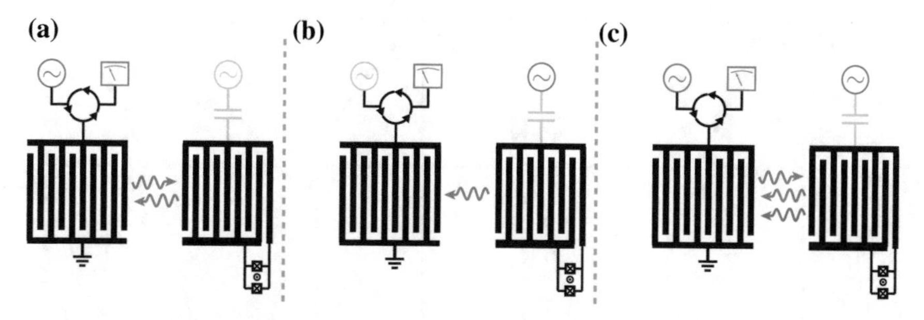

Fig. 9.10 Three different experiments. **a** Acoustic reflection. In the first experiment an acoustic wave is launched towards the qubit, and the acoustic reflection is measured. **b** Listening. The qubit is excited through the gate (by a continuous RF signal or by an RF pulse), and the emission of phonons is detected by the IDT. **c** Two-tone spectroscopy. The acoustic reflection is measured while irradiating the qubit with microwaves through the gate

The coupling to the gate is engineered to be sufficiently weak that the excited qubit preferentially relaxes by emitting SAW phonons.

In the article by Gustafsson et al., three different experiments were presented (see Fig. 9.10). The experimental data demonstrate the following key features:

(1) On electrical and acoustical resonance, the reflection of SAW power from the qubit is nonlinear in the excitation power. For low powers, $P_{SAW} \ll \hbar\omega_{01}\Gamma_{ac}$, the qubit reflects the incoming SAW perfectly. As the power increases and the $|1\rangle$ state of the qubit becomes more populated, the reflection coefficient decreases. For $P_{SAW} \gg \hbar\omega_{01}\Gamma_{ac}$, the reflection coefficient tends to zero.
(2) The electrical resonance of the qubit can be tuned by applying a magnetic field through its SQUID loop, periodically bringing the electrical qubit resonance frequency ω_{10} in and out of the acoustic band of the IDT. This can be seen in Fig. 9.11.
(3) When the qubit is excited through the gate at coinciding electrical and acoustic resonance frequencies, it relaxes by emitting SAW phonons, which can be detected by the IDT. The transmission from the gate to the acoustic channel is nonlinear in the same way as the acoustic reflection coefficient.
(4) Since the electrical gate has a high bandwidth, it can be used to excite the qubit with short pulses at ω_{10}. The emission from the qubit arrives at the IDT after a delay of ~40 ns compared with the applied electrical pulse. This corresponds to the acoustic propagation time between the qubit and the IDT. In Fig. 9.12 we show how a 25 ns pulse is bouncing back and forth between the qubit and the IDT. The first peak is due to electrical cross-talk between the qubit gate and the IDT. The subsequent peak, which arrives 40 ns after the excitation pulse is applied, is the acoustic signal emitted by the qubit. The SAW is then partially reflected by the IDT and returns to the qubit, where it is reflected again. This echo signal arrives 80 ns after the pulse is applied. Two echoes spaced by 80 ns can be observed.

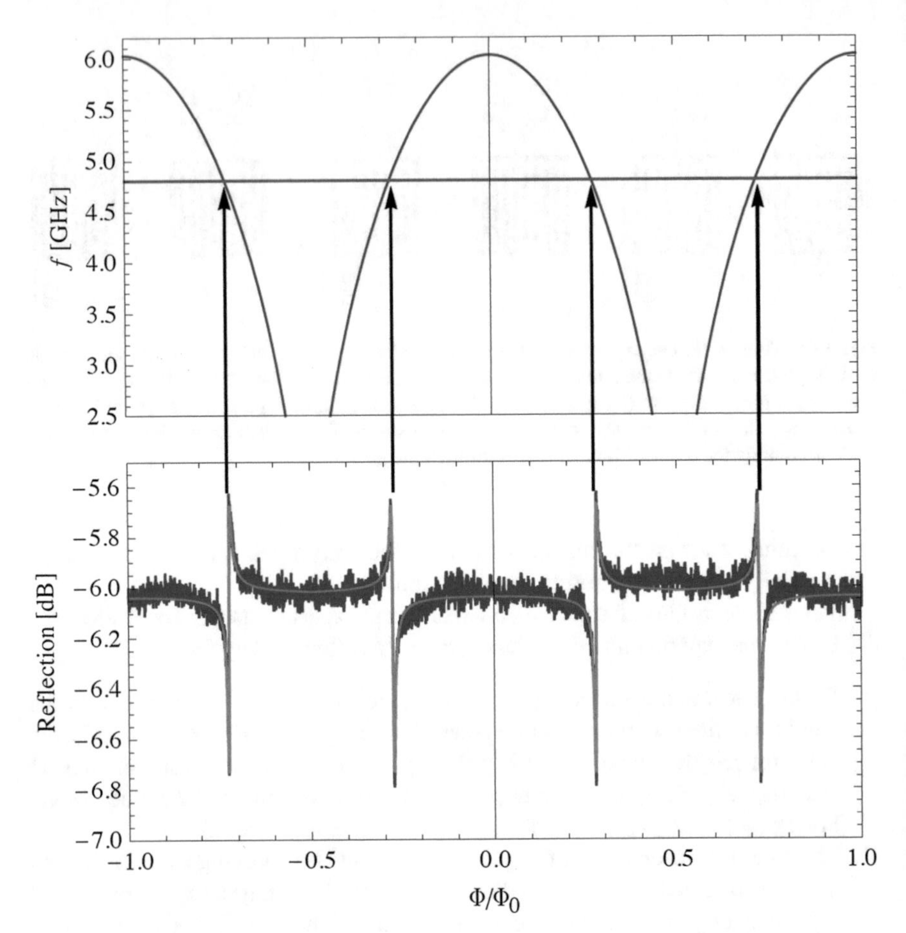

Fig. 9.11 **a** Qubit frequency as a function of external magnetic flux. The *blue line* is the calculated qubit frequency and the *red line* is the IDT frequency at which we can listen to the SAW. **b** Measured reflection coefficient from the IDT. In the flat regions the qubit is far detuned from the IDT frequency and the signal is just reflected from the IDT. When the qubit comes into resonance with the IDT frequency, there is an additional signal due to acoustic reflection at the qubit. The phase of the acoustic signal varies with the detuning and interferes with the signal which is directly reflected by the IDT. The *blue trace* is the measured data and the *red trace* is a fit to the theory

(5) When the qubit is probed with a weak acoustic tone, its reflection coefficient depends in a complex way on the frequency and power of an electrical signal applied to the gate, as well as the electrical detuning of the qubit with respect to the acoustic center frequency. The features observed include an enhanced acoustic reflection at the $|1\rangle \rightarrow |2\rangle$ transition frequency when the $|0\rangle \rightarrow |1\rangle$ transition is addressed electrically, and Rabi splitting of the various energy levels when the electrical signal is strong.

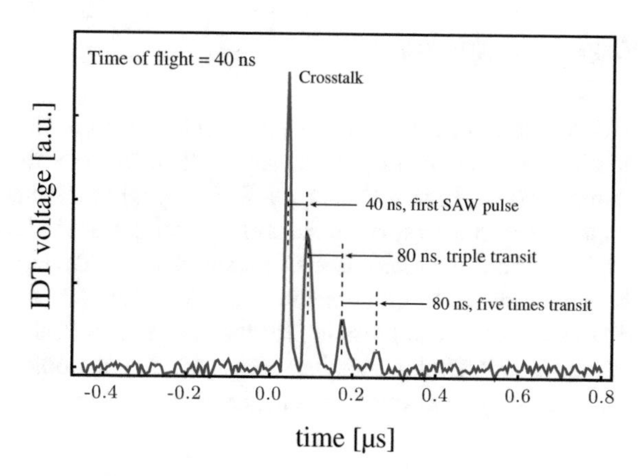

Fig. 9.12 Recorded IDT signal for short RF pulses applied through the gate. A 25 ns pulse is sent to the qubit gate at $t = 0$. Due to a capacitive coupling between the qubit gate and the IDT, there is an immediate cross-talk response. The acoustic signal arrives after 40 ns, which agrees well with the time of flight. Subsequent reflections of the acoustic signal give rise to additional echo signals spaced by 80 ns

All experiments show good agreement with the theory of Sect. 9.3. The acoustic coupling rate also agrees with the semiclassical estimate given by Eq. (9.15). Points 1 to 3 show that the qubit works as a two-level system, where the $|0\rangle \rightarrow |1\rangle$ transition can be addressed separately from all transitions to higher energy states. Point 4 proves directly that the qubit primarily relaxes by emitting SAW phonons. Point 5 underscores the nonclassical nature of the qubit, and demonstrates that the system is well described by the quantum theory.

9.6 Future Directions

As we have seen in the previous sections, quantum SAW devices have many similarities to circuit QED devices. It is important to realize that there are also differences between the two. Although one may argue that photons will always be more coherent than phonons and therefore ask why one would be interested in SAW phonons, not all the differences are detrimental for experiments. On the contrary, below, we show that there are several interesting research directions to pursue and that the SAW phonons indeed offer new and interesting physics. The much lower propagation speed of SAWs compared to electromagnetic waves plays a crucial role here.

9.6.1 In-Flight Manipulation

The speed of SAWs is of the order of 3000 m/s in most of the interesting piezoelectric materials, which is five orders of magnitude slower than for light in vacuum. This means, for instance, that the time it takes for a SAW signal to traverse a 3 cm long chip which is approximately 10 μs, while the corresponding traversal time for a light signal is just 300 ps. There is ample time to manipulate a SAW signal "in flight" but it would be very difficult to do something similar with photons, especially if one wants to do measurements and provide feedback signals in real time. Since the typical tuning time for a SQUID is less than 10 ns [49, 50], it would be possible to tune a transmon or a cavity in or out of resonance with the SAW wave many times during a 10 μs traversal.

9.6.2 Coupling to Optical Photons

Circuit QED systems need to be cold, both to maintain superconductivity and to avoid thermal excitations. On the other hand, it is clear that photons propagating either in free space or in optical fibers is the preferred choice for sending quantum information over large distances. The frequencies at which circuit QED devices work is five orders of magnitude lower than optical frequencies. This results in very similar wavelengths for SAW waves and optical signals. In the paper by Gustafsson et al. [21], the wavelength used was 600 nm. This wavelength could easily be modified to match with photons that could travel in optical fibers. Of course, the details of how such a conversion device would look like remain to be worked out. We note that many of the piezoelectric materials are transparent at optical frequencies and are routinely used in optical applications. It is also interesting that some of these materials have strong nonlinearities, especially $LiNbO_3$, which is used in optical modulators and other applications [51].

9.6.3 Ultrastrong Coupling Between SAWs and Artificial Atoms

In the SAW-qubit experiment, which is described above and in greater detail in Ref. [21], the coupling between the transmon and the SAW was 38 MHz. In spite of the fact that GaAs is a very weakly piezoelectric material, this is similar to the coupling which has been shown between a transmon and open transmission lines in waveguide QED. Using other materials such as $LiNbO_3$ would allow much stronger coupling. Since the value of K^2 is 70 times larger for $LiNbO_3$ than for GaAs (see Table 9.1), it should be possible to reach couplings exceeding 1 GHz, entering the ultrastrong coupling regime, or even the deep ultrastrong regime [52, 53]. In that situation, it is necessary to use a qubit which has a larger anharmonicity than the transmon, since the maximum anharmonicity of the transmon is about 10 % of the qubit frequency. One possibility is to use a single-Cooper-pair box (SCB) [54, 55], which has much larger anharmonicity. The fact that the SCB has a shorter coherence time is less of a problem since the important timescale is the inverse coupling rate, which is made very small in the ultrastrong coupling regime. Another possibility would be to use a capacitively shunted flux qubit [56].

9.6.4 Large Atoms

For experiments in quantum optics using natural atoms and laser light at visible wavelengths, the wavelength exceeds the size of the atom by several orders of magnitude. Rydberg atoms and superconducting qubits are much larger than regular atoms, but nonetheless substantially smaller than the wavelength of the radiation they interact with. In quantum acoustics with SAW waves, this small-atom approximation no longer holds. The transmon used in a SAW experiment can be of similar size to those used for circuit QED experiments, but the wavelength of the SAWs is orders of magnitude lower. Such devices thus operate in an unexplored regime, where the frequency dependence of the coupling strength and the Lamb shift are modified. These modifications are discussed in Sect. 9.3.3 and in more detail in Ref. [40]. Further work could also explore the combination of large atoms with additional large atoms or other systems, and also the regime where the traveling time across the large atom is no longer negligible compared to the relaxation time of the atom.

9.6.5 SAW Resonators

Initial experiments indicate that SAW resonators can reach quality factors of order 10^5 at GHz frequencies, close to what is seen in superconducting transmission line resonators used in circuit QED [57]. Experiments with a transmon in an open SAW transmission line [21] indicate that coupling strengths can also be similar to those found in circuit QED, and higher still if strongly piezoelectric materials are used. These early results bode well for realizing strong coupling SAW-based circuit QED. Beyond such a basic realization, there are several experiments that can highlight differences between SAW resonators and conventional circuit QED. For example, combining the in-flight manipulation discussed above with a SAW resonator could allow generation of exotic cavity phonon states, while the multimodal nature of long SAW cavities (similar to optical Fabry-Perot cavities) could enable new regimes of quantum optics to be reached. Advancing further the understanding of internal quality factors may push SAW resonators into a regime in which they may be useful as on-chip quantum memories, much smaller than their electromagnetic counterparts. Finally, it is worth noting that high Q SAW resonators could also be employed to implement a new form of microwave cavity optomechanics [58], using circuit designs that realize a modulation of the qubit frequency by the SAW amplitude. Such an implementation may be of strong interest due to the combined high frequency and quality factor of the SAW compared to other mechanical systems used in the field.

9.6.6 Analogues of Quantum Optics

Finally, we note that there are a number of interesting quantum optics experiments which could be repeated in the quantum acoustics domain. For instance, a single phonon generator should be possible to make in a similar way to single photon generators in circuit QED. Engineering the couplings of a three level atom should allow the creation of population inversion and thus "SAW-lasing".

9.7 Conclusions

In this chapter, we have discussed new possibilities of performing quantum physics experiments using surface acoustic waves. We have showed that a superconducting transmon qubit can couple strongly to propagating SAWs and that the SAWs can be confined in high Q resonators. Taken together, these results indicate that experiments conceived for quantum optics with photons can now be performed in quantum acoustics using SAW phonons. Furthermore, the slow propagation velocity and short wavelength of the phonons promises access to new regimes which are difficult to reach with traditional all-electrical circuits, such as giant atoms and improved feedback setups.

Acknowledgments This work was supported by the Swedish Research Council, the European Research Council, the Knut and Alice Wallenberg Foundation, the UK Engineering and Physical Sciences Research Council. We also acknowledge support from the People Programme (Marie Curie Actions) and the FET-project SCALEQIT of the European Unions Seventh Framework Programme.

References

1. S. Haroche, J.M. Raimond, *Exploring the Quantum* (Oxford University Press, Oxford, 2006)
2. R. Miller, T.E. Northup, K.M. Birnbaum, A. Boca, A.D. Boozer, H.J. Kimble, J. Phys. B: At. Mol. Opt. Phys. **38**, S551 (2005)
3. A. Wallraff, D.I. Schuster, A. Blais, L. Frunzio, R.S. Huang, J. Majer, S. Kumar, S.M. Girvin, R.J. Schoelkopf, Nature **431**(7005), 162 (2004)
4. R. Schoelkopf, S. Girvin, Nature **451**(7179), 664 (2008)
5. J. Koch, T.M. Yu, J. Gambetta, A.A. Houck, D.I. Schuster, J. Majer, A. Blais, M.H. Devoret, S.M. Girvin, R.J. Schoelkopf, Phys. Rev. A **76**(4), 042319 (2007)
6. H. Zheng, D.J. Gauthier, H.U. Baranger, Phys. Rev. A **82**, 063816 (2010)

7. H. Zheng, D.J. Gauthier, H.U. Baranger, Phys. Rev. Lett. **111**, 090502 (2013)
8. D. Valente, Y. Li, J.P. Poizat, J.M. Gerard, L.C. Kwek, M.F. Santos, A. Auffeves, New J. Phys. **14**, 083029 (2012)
9. I.C. Hoi, C.M. Wilson, G. Johansson, J. Lindkvist, B. Peropadre, T. Palomaki, P. Delsing, New J. Phys. **15**, 025011 (2013)
10. J.D. Teufel, T. Donner, M.A. Castellanos-Beltran, J.W. Harlow, K.W. Lehnert, Nat. Nanotechnol. **4**(12), 820 (2009)
11. M.D. LaHaye, J. Suh, P.M. Echternach, K.C. Schwab, M.L. Roukes, Nature **459**(7249), 960 (2009)
12. A.D. O'Connell, M. Hofheinz, M. Ansmann, R.C. Bialczak, M. Lenander, E. Lucero, M. Neeley, D. Sank, H. Wang, M. Weides, J. Wenner, J.M. Martinis, A.N. Cleland, Nature **464**(7289), 697 (2010)
13. J.M. Pirkkalainen, S.U. Cho, J. Li, G.S. Paraoanu, P.J. Hakonen, M.A. Sillanpää, Nature **494**(7436), 211 (2013)
14. S. Datta, *Surface Acoustic Wave Devices* (Prentice-Hall, Englewood Cliffs, 1986)
15. D. Morgan, *Surface Acoustic Wave Filters*, 2nd edn. (Academic Press, Waltham, 2007)
16. C. Campbell, *Surface Acoustic Wave Devices for Mobile and Wireless Communications* (Academic Press, New York, 1998)
17. C. Barnes, J. Shilton, A. Robinson, Phys. Rev. B **62**(12), 8410 (2000)
18. S. Hermelin, S. Takada, M. Yamamoto, S. Tarucha, A.D. Wieck, L. Saminadayar, C. Bäuerle, T. Meunier, Nature **477**(7365), 435 (2011)
19. R.P.G. McNeil, M. Kataoka, C.J.B. Ford, C.H.W. Barnes, D. Anderson, G.A.C. Jones, I. Farrer, D.A. Ritchie, Nature **477**(7365), 439 (2011)
20. E.B. Magnusson, B.H. Williams, R. Manenti, M.S. Nam, A. Nersisyan, M.J. Peterer, A. Ardavan, P.J. Leek, Appl. Phys. Lett. **106**, 063509 (2015)
21. M.V. Gustafsson, T. Aref, A.F. Kockum, M.K. Ekström, G. Johansson, P. Delsing, Science **346**(6206), 207 (2014)
22. J.W. Strutt, Lord Rayleigh, Proc. Lond. Math. Soc. **17**, 4 (1885)
23. J. Pedros, L. Garcia-Gancedo, C. Ford, C. Barnes, J. Griffiths, G. Jones, A. Flewitt, J. Appl. Phys. **110**, 103501 (2011)
24. O. Madelung, U. Rössler, M. Schulz (eds.), *II–VI and I–VII compounds; semimagnetic compounds*, Landolt-Börnstein—Group III Condensed Matter (Springer, Berlin, 1999)
25. A.J. Slobodnik, in *Acoustic Surface Waves*, ed. by A.A. Oliner (Springer, Heidelberg, 1978), p. 225
26. J.S. Browder, S.S. Ballard, Appl. Opt. **16**, 3214 (1977)
27. T.F. Smith, G. White, J. Phys. C: Solid State Phys. **8**, 2031 (1975)
28. K. Hashimoto, *Surface Acoustic Wave Devices in Telecommunications: Modelling and Simulation* (Springer, Heidelberg, 2000)
29. B. Yates, R.F. Cooper, M.M. Kreitman, Phys. Rev. B **4**, 1314 (1971)
30. G.K. White, P.J. Meeson, *Experimental Techniques in Low-Temperature Physics*, 4th edn. (Clarendon Press, Oxford, 2002)
31. W.D. Hunt, R.L. Miller, B.J. Hunsinger, J. Appl. Phys. **60**, 3532 (1986)
32. W.D. Hunt, Y. Kim, F.M. Fliegel, J. Appl. Phys. **69**(4), 1936 (1991)
33. J.M.M. de Lima, F. Alsina, W. Seidel, P.V. Santos, J. Appl. Phys. **94**, 7848 (2003)
34. A. Weber, G. Weiss, S. Hunklinger, in *IEEE 1991 Ultrasonics Symposium* (IEEE, 1991), pp. 363–366
35. D. Leibfried, R. Blatt, C. Monroe, D. Wineland, Rev. Mod. Phys. **75**, 281 (2003)
36. S. Haroche, Rev. Mod. Phys. **85**, 1083 (2013)
37. H. Walther, B.T.H. Varcoe, B.G. Englert, T. Becker, Rep. Prog. Phys. **69**, 1325 (2006)
38. G. Kirchmair, B. Vlastakis, Z. Leghtas, S. Nigg, H. Paik, E. Ginossar, M. Mirrahimi, L. Frunzio, S. Girvin, R. Schoelkopf, Nature **495**, 205 (2013)
39. R. Barends, J. Kelly, A. Megrant, D. Sank, E. Jeffrey, Y. Chen, Y. Yin, B. Chiaro, J. Mutus, C. Neill, P. O'Malley, P. Roushan, J. Wenner, T.C. White, A.N. Cleland, J.M. Martinis, Phys. Rev. Lett. **111**, 080502 (2013)

40. A.F. Kockum, P. Delsing, G. Johansson, Phys. Rev. A **90**, 013837 (2014)
41. H.J. Carmichael, *Statistical Methods in Quantum Optics 1* (Springer, Berlin, 1999)
42. C.W. Gardiner, P. Zoller, *Quantum Noise*, 3rd edn. (Springer, Berlin, 2004)
43. G. Lindblad, Commun. Math. Phys. **48**, 119 (1976)
44. W.E. Lamb, R.C. Retherford, Phys. Rev. **72**, 241 (1947)
45. H.A. Bethe, Phys. Rev. **72**, 339 (1947)
46. E. Ash, in *G-MTT 1970 International Microwave Symposium*, vol. 70 (IEEE, 1970), pp. 385–386
47. D.L.T. Bell Jr., R.C.M. Li, Proc. IEEE **64**(5), 711 (1976)
48. T. Bristol, W. Jones, P. Snow, W. Smith, in *1972 Ultrasonics Symposium* (IEEE, 1972), pp. 343–345
49. M. Sandberg, C.M. Wilson, F. Persson, T. Bauch, G. Johansson, V. Shumeiko, T. Duty, P. Delsing, Appl. Phys. Lett. **92**, 203501 (2008)
50. M. Pierre, I.M. Svensson, S.R. Sathyamoorthy, G. Johansson, P. Delsing, Appl. Phys. Lett. **104**(23), 232604 (2014)
51. R.W. Boyd, *Nonlinear Optics*, 3rd edn. (Academic Press, Orlando, 2008)
52. T. Niemczyk, F. Deppe, H. Huebl, E.P. Menzel, F. Hocke, M.J. Schwarz, J.J. Garcia-Ripoll, D. Zueco, T. Hummer, E. Solano, A. Marx, R. Gross, Nat. Phys. **6**, 772 (2010)
53. D. Ballester, G. Romero, J.J. Garcia-Ripoll, F. Deppe, E. Solano, Phys. Rev. X **2**(2), 021007 (2012)
54. M. Büttiker, Phys. Rev. B (Condensed Matter) **36**(7), 3548 (1987)
55. V. Bouchiat, D. Vion, P. Joyez, D. Esteve, M.H. Devoret, Phys. Scr. T **76**, 165 (1998)
56. F. Yan, S. Gustavsson, A. Kamal, J. Birenbaum, A.P. Sears, D. Hover, T.J.Gudmundsen, J.L. Yoder, T.P. Orlando, J. Clarke, A.J. Kerman, W.D. Oliver The Flux Qubit Revisited (2015). arXiv:1508.06299
57. M. Göppl, A. Fragner, M. Baur, R. Bianchetti, S. Filipp, J.M. Fink, P.J. Leek, G. Puebla, L. Steffen, A. Wallraff, J. Appl. Phys. **104**(11), 113904 (2008)
58. M. Aspelmeyer, T.J. Kippenberg, in *Cavity Optomechanics*, ed. by M. Aspelmeyer, T.J. Kippenberg, F. Marquardt (Springer, Berlin, 2014)

Index

A
Acoustic coupling (to superconducting qubits), 219, 226–228
Acoustic reflection (of SAWs), 231–233
AC-Stark phase shift (in superconducting qubits), 209
Active area (in SNSPD practical detection efficiency), 4, 7
Active area (of SNAPs), 18, 20
Active area (of TES), 40
Active area (performance trade-offs in SNSPDs), 13
Adiabatic demagnetization refrigerator (ADR), 40
Admittance (of IDT), 224
Afterpulsing (in SNAP), 17
Afterpulsing (in SNSPDs), 5
Amorphous superconductors (for SNSPDs and SNAPs), 8, 12, 13, 25
Analog feedback (for superconducting qubits), 164, 179
Anharmonicity (in superconducting qubits), 140, 143, 144, 230, 236, 241
Anti-bunching (of photons), 159
Arm-trigger regime (in SNAPs), 17
Array
  of SQUIDs for TES readout, 36
Array factor (for IDT), 225
Arrays (of optical fibers), 95
Arrays (of single photon sources), 64
Arrays (of SNSPDs), 24, 25
Arrays (of superconducting qubits), 183
Artificial atom, 140, 145, 151, 153, 160, 168, 218, 228–230, 240
Attenuation (in optical fiber), 99, 100, 116
Attenuation (in optical waveguide circuits), 92
Attenuation (of SAW), 221, 234
Avalanche photodiode (semiconductor), 32, 41, 75, 149, 166
Avalanche regime (of SNAP), 15–17, 21

B
Backaction (in superconducting qubit measurements), 164–167, 169, 170, 172–174, 177, 179, 183
Bandwidth (in microwave photon detection), 151
Bandwidth (of coupling to superconducting qubit), 158, 174, 182, 194, 197, 236
Bandwidth (of IDT), 220, 226, 235, 237
Bandwidth (of optical stack), 37, 38
Bandwidth (of SQUID readout for TES), 42
Bandwidth (SNSPD readout), 24, 68, 100
Bandwidth, spectral (of SNSPD), 89, 116
Bell measurement, 130
Bell test, 46, 49–51
Bell's inequality, 49, 50, 146
Black-body radiation/photons, 5, 36, 43, 101
Bloch sphere, 149, 168, 173, 175, 197
Born's rule, 164, 165, 230
Bragg grating, 127, 128, 231

© Springer International Publishing Switzerland 2016
R.H. Hadfield and G. Johansson (eds.), *Superconducting Devices in Quantum Optics*, Quantum Science and Technology,
DOI 10.1007/978-3-319-24091-6

Bragg mirror/reflector, 64
Bunching (of photons), 47

C
Cascade recombination (in quantum dots), 63
Cascade switching (in superconducting nanowires), 4, 14, 20
Cavity QED, 146, 158, 168, 218, 219
Circuit model (of IDT), 224, 227
Circuit model (of SNAP), 15, 18
Circuit model (of SNSPD array), 24
Circuit model (of SNSPD), 10, 11
Circuit QED (cQED), 140, 146, 158, 167–169, 172, 174, 192, 193, 197, 198, 201, 204, 206, 207, 211, 212, 218, 231, 239–242
Coherence, 128, 129, 131, 139–141, 146, 168, 177, 191, 201, 204, 208, 241
Coherent state superposition (CSS), 47, 48
Collapse (of quantum superposition), 164, 208
Collapse (of wavefunction), 165, 177, 188
Constrictions, 8, 9, 12, 13
Controlled-NOT gate (CNOT), 63
Conversion efficiency (in wavelength conversion), 128
Cooper pair, 88, 89
Cooper-pair box (CPB), 141–143, 147, 241
Coupling (in cQED), 140, 147, 153, 157, 169, 172, 202, 219
Coupling efficiency (optical), 5, 6, 51, 55, 56, 69, 70, 72, 78, 120
Coupling of modes (COM) method, 224
Coupling, electron-phonon, 34, 41, 53
Cross-Kerr, 142, 148, 149, 153–157
Crosstalk (between optical fibers), 109
Crosstalk (between SNSPD elements), 74, 75
Crosstalk (SAW), 239

D
Dark count rate (DCR), 4, 5, 7, 17, 32, 62, 71, 72, 86, 88, 89, 96, 99, 101, 108, 120
Dark counts, 5, 23, 36, 50, 54, 79, 117, 120, 129, 151
Dark fiber, 115, 116
Dead time (for feedback in superconducting circuits), 183
Dead time (of SNSPD), 10, 13, 18, 74, 89, 122
Dead time (of TES), 41
Decoherence, 62, 168, 174, 179–181, 206, 209
Decoy-state QKD protocol, 110, 112, 120
Detection efficiency (DE), 4, 6, 8, 9, 12, 17, 19, 22, 26, 32, 33, 36, 40, 45, 46, 50, 52–56, 69, 71, 77, 86, 89, 95, 108, 111, 129, 151, 153
Detection loophole (for Bell test), 46
Detector Packaging, 40
Detector tomography, 44
Deterministic entanglement, 189, 205, 207, 211, 212
Deterministic reset (for superconducting qubits), 201
Device detection efficiency (DDE), 5, 13, 16, 17
Differential phase-shift quantum key distribution (DPS-QKD), 110, 113
Digital feedback control, 188–190, 193–195, 197, 198, 200, 202, 204, 205, 207, 208, 210, 212
Dilution refrigerator, 88, 182, 198, 201, 218, 222, 233

E
Electron beam lithography (EBL), 9, 65, 67, 68, 88, 92, 98, 222
Electron-phonon coupling, 53
Electrothermal (ET) model, 10, 24, 78
Electrothermal feedback, 34
End-fire coupling, 68
Entangled photon pair source, 121, 123–125, 128, 129, 131
Entanglement, entangled states, 49, 50, 56, 126, 130, 166, 167, 170, 189, 190, 205–207, 209–212
Evanescent coupling, 51, 52, 93, 97

F
Fabry-Perot, 69, 71, 78, 168, 241
Fair-sampling loophole (in Bell test), 49, 50
Feedback, Feedback protocol, 163, 164, 179–183, 188–192, 198–207, 211, 212, 219, 240, 242
Feedforward (for qubit control), 189
Field-programmable gate array (FPGA), 113, 114, 199, 200, 205, 207, 211
Figure of merit (single-photon detectors), 75, 86, 87
Fock state, 44, 145, 157–159

G
GaAs/AlGaAs waveguide, 64, 69, 70
Gallium arsenide (GaAs) waveguide circuit, 67, 71, 73, 76, 95
Grating coupler (optical), 99

H
Heralded photon-subtraction, 54
Heralded single-photon source, 121, 122, 125, 126, 130
Heralded state detection, 46–48, 54
Homodyne detection, 49, 54
Hong-Ou-Mandel (HOM) interference, 46, 47, 130
Hotspot (in superconducting nanowire), 9, 10, 12, 15, 74, 77

I
Indirect measurement (of qubit), 165, 166, 169
Instrument response function (IRF), 6, 21
Interdigital transducer (IDT), 219–226, 228, 230, 231, 233–239

J
Jaynes-Cummings Hamiltonian, 147, 158, 164, 168
Josephson coupling, 142
Josephson inductance, 226
Josephson junction, 36, 140, 141, 150, 171, 197, 218
Josephson parametric amplifier (JPA), 171, 192, 193, 197, 207
Josephson parametric converter (JPC), 174
Joule heating, 9, 11, 12, 34, 64

K
Kinetic inductance, 10, 11, 13, 72, 73, 76, 96
Kinetic inductance detector (KID), 23
Kramers-Kronig relation, 157

L
Lamb shift, 230, 241
Latching (of SNSPD), 10
Lattice parameters (NbN and substrates), 65
Lindblad decoherence, 168
Lindblad superoperators, 230
Linear detector, 145, 149, 153, 159
Loophole (of Bell test), 46, 49, 50

M
Magnesium diboride (MgB_2), 8
Magnetron sputtering, 65, 66
Markovian approximation, 154, 230
Master equation, 155, 168, 174–177, 201, 205, 209, 230
Measurement-based quantum computing, 190
Metastable state, 149–151
Microcalorimeter, 33, 34
Microring resonator, 91, 97, 98
Microwave photon generation, 140, 158
Microwave photon source, 159
Microwave quantum photonics, 140–142, 144, 145, 147–151, 153–160
Microwave single-photon detection, 140, 151, 160
Mid-infrared (mid-IR) single-photon detection, 5, 7, 12, 37
Molybdenum germanium (MoGe) superconducting nanowire, SNSPD, 8, 9, 13
Molybdenum silicide (MoSi) superconducting nanowire, SNSPD, 7–9, 13
Multi-photon correlation function, 22
Multi-time correlation function, 149, 159

N
Niobium nitride (NbN) refractive index, 70
Niobium nitride (NbN) superconducting nanowire, SNSPD, 7–9, 11–13, 16, 18, 65, 67, 69, 70, 72–76, 79, 88, 93–96, 116, 122
Niobium nitride (NbN) thin film, 65–68, 71, 72, 88
Niobium titanium nitride (NbTiN) superconducting nanowire, SNSPD, 7, 12, 88, 95, 96, 99, 122
Noise equivalent power (NEP), 87, 88, 96, 99
Non-linear detector, 43
Non-locality, quantum, 46, 49
Number resolution, 51

O
On-chip detection efficiency (OCDE), 69, 71, 77, 89, 95, 96, 101
On-chip detector calibration, 55
One-time pad (OTP), 108, 110, 114, 115
Optical coupling, 13, 40, 50, 55, 64, 68–70, 87, 91, 108
Optical stack, 9, 37, 38

Optical time domain reflectometry (OTDR), 99–101, 109, 118

P

Parametric amplification, 146
Parity measurement, 183, 188, 189, 205–212
Performance metrics (for SNSPDs), 4, 6, 13, 26, 86, 101
Performance trade-offs (for SNSPDs), 13
Periodically-poled $KTiOPO_4$ (PPKTP), 121–126
Periodically-poled lithium niobate (PPLN), 127, 128
Phase qubit, 141, 151, 158
Phase-sensitive amplifier, 170, 172, 173
Photon statistics, 46, 47, 49, 64, 78, 124, 126, 159
Photon subtraction, 48, 51, 54, 56
Photon-number resolution, 4, 23, 26, 36, 43, 44, 53, 56
Photon-number-resolving (PNR) detector, 4, 21, 36, 43, 56, 71, 74, 75, 77–79
Photonic crystal, 62, 91
Photonic qubit, 63
Piezoelectric materials, 220
Pointer system, 166, 169
Positive-operator-valued measure (POVM), 44, 45, 164–167
Postselection, 194, 196, 202, 203, 209–212
Projective measurement (of superconducting qubit), 165–167, 176, 179, 188, 190–192, 195, 197

Q

Quantum acoustics, 218, 242
Quantum bit error rate (QBER), 111, 112, 114
Quantum computation, quantum computing, 47, 61, 62, 86, 101, 102, 146, 188–190, 212
Quantum dot (QD), 62, 64, 70, 139
Quantum efficiency (QE), 65, 71–73, 78, 87, 89, 96, 97, 151, 167
Quantum Electrodynamics, QED (in superconducting circuits), 140, 146, 149, *see also* Circuit QED (cQED)
Quantum feedback, 163, 179, 188
Quantum information (QI) network, 107, 108, 121
Quantum interface, 107, 108, 126, 127, 129–131
Quantum key distribution (QKD), 7, 62, 74, 87, 108–111, 113–116, 120
Quantum non-demolition (QND) measurement, 148, 150, 153, 158, 160, 165, 166, 176
Quantum non-locality, 46, 49
Quantum photonic circuit, 64, 85, 86, 102
Quantum photonic integrated circuit (QPIC), 61, 65, 70–72
Quantum repeater, 126, 128, 130, 131, 189
Quantum simulation, 61, 62
Quantum state tomography, 124, 129, 175, 182, 183, 201, 208–211
Quantum trajectory, 167, 174–177
Qubit, 62, 140–143, 146–148, 151, 152, 158, 159, 164–170, 172–183, 188–195, 197, 198, 200–212, 218, 219, 226–228, 231, 235–242

R

Rabi oscillations, 150, 163, 164, 177, 179–182, 190, 195, 196
Rabi splitting, 238
Ramsey cavity, 149, 150
Ramsey fringes, 150
Rayleigh waves, 220
Readout (for SNAP), 23
Readout (for SNSPD array), 4, 24, 25
Readout (for SNSPD), 6, 9–11, 14, 89, 91, 95, 102
Readout (for superconducting qubits), 148, 151, 167–169, 191–197, 199–203, 206–210
Readout (for TES), 35, 36, 42
Refractive index, 38, 39, 70, 90, 92, 94, 99, 100
Refrigeration (for SNSPD), 95, 108, 122, 129
Refrigeration (for superconducting qubits), 182, 198, 218
Refrigeration (for TES), 40
Reset (of superconducting qubit), 188, 189, 191, 199, 201–205, 212
Reset time (of SNAP), 17–19
Reset time (of SNSPD), 4, 10, 89, 94, 96
Rise time, 10, 20, 42, 78
Rotating-wave approximation, 230
Row-column readout (of SNSPD arrays), 24, 25
Rydberg atom, 149, 150, 158, 159, 218, 241
Rydberg states, 147, 149, 228

S
Sagnac interferometer, 123
Sagnac loop configuration, 50, 124
Schrödinger cat state, 47, 48
Second-order correlation function, 49, 62, 72, 74, 80
Signal-to-noise ratio (SNR) (SNSPDs and SNAPs), 14, 15, 22, 75
Silica waveguide, 51, 52, 74
Silicon nitride photonic circuits, 91, 92, 94–96
Single-Cooper-pair box, SCB, 241
Single-photon detection, 7, 12, 56, 62, 79, 87, 88, 101, 108, 140, 148, 151, 152, 155
Single-photon detector, 12, 15, 17, 32, 41, 51, 55, 56, 61, 62, 67, 69–71, 74, 77, 88, 99, 101, 108
Single-photon source, 61, 62, 65, 86, 108, 122, 125, 159
Spontaneous parametric down-conversion (SPDC), 47, 50, 62, 122
Sputtering, 39
Superconducting nanowire avalanche photodetector (SNAP), 4, 11, 14–19, 21, 22
Superconducting nanowire-single photon detector (SNSPD or SSPD), 4, 7–19, 22, 24, 25, 32, 41, 62, 64, 75, 76, 87–89, 95–101, 108–110, 112, 113, 115, 116, 118, 120–126, 129–131
Superconducting quantum interference device (SQUID), 35, 36, 42, 53, 140, 141, 226, 237, 240
Surface Acoustic Wave (SAW), 218–223, 226–230, 232, 233, 235–242
Surface Acoustic Wave (SAW) device fabrication, 220, 222
Surface Acoustic Wave (SAW) device materials, 220, 221, 233
Surface Acoustic Wave (SAW) resonator, 221, 225, 231–233, 241
Surface Acoustic Wave (SAW)-lasing, 242
Surface Acoustic Wave (SAW)-qubit interaction, 226, 228, 235, 240
System detection efficiency (SDE), 4, 19, 33, 50, 69, 89, 108

T
Tantalum nitride (TaN) superconducting nanowire, SNSPD, 7, 88
Thermal expansion coefficients, 39, 221
Time-correlated single-photon counting (TCSPC), 6, 89, 100
Time symmetry, 178
Timing jitter, 4, 6, 7, 13, 20, 26, 39, 41, 42, 51, 74, 87, 89, 96, 98, 108, 122, 129
Tokyo QKD network, 109, 110, 114, 116, 131
Transition edge sensor (TES), 31–39, 41, 42, 44–46, 48–57
Transmon qubit, 141–145, 147, 148, 150, 153–159, 181, 191, 192, 194, 197, 201, 202, 204–207, 218, 226–230, 235, 236, 240–242
Tungsten (for TES), 33, 34, 37–39, 41, 42, 53
Tungsten silicide (WSi) superconducting nanowire, SNSPD, 7–10, 12, 13, 20, 25, 65, 88

U
Ultrastrong coupling, 240, 241
Unitary evolution, 164, 176, 177, 179
Unstable regime (in SNAP operation), 17, 18
UV laser writing, 52

V
Vacuum Rabi splitting, 142, 147, 148, 159

W
Waveguide autocorrelator, 70, 72–74
Waveguide circuit, 32, 95
Waveguide-integrated SNSPD, 68, 71, 90, 96
Waveguide-integrated TES, 55
Waveguide photon-number-resolving detector (WPNRD), 75–79
Waveguide PPLN, 128
Waveguide QED, 218, 219, 240
Wavelength conversion, 108, 126–128, 130, 131
Weak measurement (of superconducting qubit), 163, 164, 167, 172–174, 176, 177, 179, 180, 183, 188
Wigner function, 48

Printed in the United States
By Bookmasters